Biotechnological Approaches for the Integrated Management of CROP DISEASES

(Professor L.V. GANGAWANE Festschrift Volume)

— *Editors* —

Prof. C.D. Mayee, New Delhi

Prof. C. Manoharachary, Hyderabad

Prof. K.V.B.R. Tilak, New Delhi

Prof. D.S. Mukadam, Aurangabad

Dr. Jayashree Deshpande, Aurangabad

2024

Daya Publishing House®

A Division of

Astral International Pvt. Ltd.

New Delhi – 110 002

First Published, 2004

Reprinted, 2024

ISBN 978-81-7035-352-2 (HB)

Disclaimer:

Every possible effort has been made to ensure that the information contained in this book is accurate at the time of going to press, and the publisher and author cannot accept responsibility for any errors or omissions, however caused. No responsibility for loss or damage occasioned to any person acting, or refraining from action, as a result of the material in this publication can be accepted by the editor, the publisher or the author. The Publisher is not associated with any product or vendor mentioned in the book. The contents of this work are intended to further general scientific research, understanding and discussion only. Readers should consult with a specialist where appropriate.

Every effort has been made to trace the owners of copyright material used in this book, if any. The author and the publisher will be grateful for any omission brought to their notice for acknowledgement in the future editions of the book.

Published by : **Daya Publishing House®**
A Division of
Astral International Pvt. Ltd.
– ISO 9001: 2015 Certified Company –
4736/23, Ansari Road, Darya Ganj
New Delhi-110 002
Ph. 011-43549197, 23278134
E-mail: info@astralint.com
Website: www.astralint.com

Digitally Printed at : Replika Press Pvt. Ltd.

Foreword

It has been my good fortune to know Professor L. V. Gangawane for the last 30 years. I consider it good fortune because it is so uncommon these days to find teachers and researchers, who spend their lifetime with complete dedication and keep on inspiring countless students by being the role-models, following the best of the traditional Indian values of a teacher. Professor Gangawane expanded the horizons of his activities to related disciplines, always keeping his research at the cutting edge of the science of crop health management. It is most befitting that a Commemoration Volume *"Biotechnological Approaches for the Integrated Management of Crop Diseases"* is being published on the occasion of Professor Gangawane's formal retirement. Many eminent authors have contributed articles to this volume and thus it should prove most useful to the teachers and researchers all over India and elsewhere.

Persons of the stature of Professor Gangawane never really retire because they continue to inspire generations of students.

I pay my highest regards to Professor L.V. Gangawane.

Y.L. NENE Ph.D., D.Sc (Hon.)
Chairman, Asian Agri-History Foundation
[Also formerly Deputy Director General, ICRISAT, Patancheru – 502 324 (A.P), India]
47, ICRISAT Colony-I, Tarbund,
Secunderabad – 500 009 (A.P.), India

सत्यमेव जयते

Dr. C.D. Mayee
Tel / Fax : 2338 3549

कृषि आयुक्त
भारत सरकार
कृषि मंत्रालय
(कृषि एवं सहकारिता विभाग)
कृषि भवन, नई दिल्ली-110 001
Agriculture Commissioner
Government of India
Ministry of Agriculture
(Department of Agriculture & Cooperation)
Krishi Bhawan, New Delhi-110 001

Preface

There are two major events in Research and Development that characterizes the beginning of 21st century; emergence of integrated pest management (IPM) as the viable alternative to use of pesticides as also an effective tool for reduction of damage due to insect pests, pathogens and weeds in crop plants, and extensive use of biotechnology in IPM including crop disease management. Over the last few years scientists have researched extensively in these two fields and brought about dramatic alternatives to disease management which are bio-safe, environmental friendly, ecologically compatible and economically sound. Excellent technological developments have taken place and a number of technologies are currently crossed the pilot tests for commercialization.

Dr. L.V. Gangawane, the Professor and Head, Department of Botany and Soil Microbiology and Pesticides Laboratories, Dr. Babasaheb Ambedkar Marathwada University, Aurangabad (Maharashtra) and to whom I know personally for the last 30 years is to retire soon from active service. It is, therefore, a matter of great opportunity to bring about a "festschrift volume" on such a focused subject of biotechnology in IDM as a tribute to Dr. Gangawane's services to the science. He has consistently contributed for the growth of disease management science and art. Not only the biotechnological approaches of improvements of beneficial strains of microorganisms but also developed tools of IDM for wide range of crops popularly grown in dryland agriculture. The unique feature of this volume is the compilation of articles written by internationally recognized plant pathologists and microbiologists. It is for the services of Dr. Gangawane that all of them have taken the trouble to write up the article with indepth study.

I am proud to be associated with the publication of this issue not only as contributor but also as editor. On behalf of the editorial board, I thank all the authors for their full cooperation and support in publication of this volume. I am sure that this publication will be a turning point in R&D of this subject. A total of twenty six theme reviews have been included in this publication. They will serve as reference point for young researchers to identify gaps in research. These articles will also provide an opportunity to formulate the IDM modules for wide scale testing and some of them can be useful for developing protocols of disease management. For students and teachers of plant pathology, I think this publication will give a basis of curriculum building in the area of biotechnological plants disease management.

18th May, 2004

(Dr. C.D. MAYEE)

Contents

A Short Curriculum Vitae of Professor L.V. Gangawane

Prof. Laxman Vithalrao Gangawane was born on 29th July 1944 in a village in Aurangabad District of Maharashtra. He was educated in Aurangabad itself. He obtained his M.Sc. and Ph.D. Degree from Dr. Babasaheb Ambedkar Marathwada University, Aurangabad. After M.Sc. in 1967 he joined as research fellow and demonstrator in Department of Botany and completed his Ph.D. on Rhizosphere mycoflora of groundnut in 1972. Later he joined as Lecturer and in 1977 he was Associate Professor of Botany in Government Institute of Science, Aurangabad and initiated the work on various aspects of Botany. Again 1979 he joined as Reader and was then selected as Professor. As soon as he joined Reader he established the Soil Microbiology and Pesticides Laboratory in the same Department and now this laboratory is world wide known.

After a training in Chemical management of plant pathogens at G.B.P. University of Agriculture and Technology, Pantnagar under Prof. Y.L. Nene he initiated the work in fungicides. He also had an opportunity to get acquainted with microbiology of industrial microorganisms in Hindustan Antibiotics Ltd., Pune with Dr. M.J. Thirumalachar. Later he completed post-graduate course in Fungicide Resistance in plant pathogens at University of Agriculture, Wageningen, Netherlands. He also worked with Dr. K.J. Brent at University of Bristol, U.K. on fungicide resistance. He was Visiting Fellow at the University of Swansea to work on phytoalexin production. He was visiting fellow and participated in conferences in Thailand, Singapore, Kaulalumpur, Al Ain University, U.A.E. He was also deputed to Egypt under Indo-Egypt cultural exchange programme and his talks on biofertilizer technology, fungicide resistance and integrated crop disease management were very well appreciated in different universities of Egypt.

He has guided 35 Ph.Ds. and published more than 215 research papers in scientific journals in India and abroad. He is also well known among the farming community as he was first to write a book in local language Marathi. He conducted many seminars and conferences. He is always interested in farmers and therefore he always keeps a session in Marathi for farmers. His book on *Jaiwik Khate Tantradnyan* in Marathi is well known throughout Maharashtra and many farmers have actually benefited. He also edited some books like Pesticides of crop plants in India, Soil borne pathogens in India (a compendium), Biofertilizer Technology Transfer, Chemical Management of Plant Pathogens in Western India and Vistas in Mycology and Plant Pathology. Moreover he participated in more than 90 conferences and many a times he was Chairman. In addition, he was invited as resource person to deliver lectures in this and other Universities.

Prof. Gangawane is associated with many Academic professional bodies. He is member of nearly 15 professional associations where he was Chairman of Association of Microbiologist in India (Aurangabad Unit) and Councillor of Plant Protection Association of India. He is Fellow of Botanical Society, Councillor and President (WZ) of Indian Phytopathological Society, member of Organizing Committee, International Conference on Integrated Management of Plant Pathogens for Sustainable Agriculture, IARI, New Delhi and founder President, Marathwada Botanical Society, Aurangabad. Besides he served on many academic committees of Dr. Babasaheb Ambedkar Marathwada University, Aurangabad and other Universities and Institutes in India.

Interestingly Prof. Gangawane is recipient of Prof. M.S. Pavgi Award, IPS, New Delhi, P.R. Verma, Indian Society of Mycology and Plant Pathology, Udaipur and Ideal Teacher Award, Govt. of Maharashtra, Mumbai. The importance of researches done by Prof. Gangawane can be described as follows:

1. Microorganisms are very important in the rhizosphere or root region of crop plants. They enhance the utilization of plant nutrients from the soil and also act in biological control agents for plant pathogens. The research work done on agronomic practices and rhizosphere microflora has shown manipulation of rhizosphere microorganisms to the sustainable crop development.
2. Viruses are known to cause various diseases to the plants and human beings. Hepatitis B is very important in the Indian society. It has been for the first time shown that Hepatitis B virus can be multiplied on tobacco plants. Now-a-days production of transgenic vaccine is prepared and administered in the various countries. It is really achievement based on the researches done by Prof. Gangawane. This work was published in USA.
3. In order to avoid chemical pollution of the soil, application of biofertilizers is suggested. The work on the development of biofertilizers from *Rhizobium, Azotobacter*, Blue green Algae and other nitrogen-fixing microorganisms is well appreciated. The book on Biofertilizer Technology Transfers by Dr. Gangawane was published by Central Government DST, New Delhi and provides guidelines to the researchers and academicians throughout the world. It is also proved that *Rhizobium* also acts as biocontrol agent, pesticide degradent and salt tolerant along with nitrogen fixation. This is the great achievement in the green revolution and has been appreciated by Dr. M.S. Swaminathan.
4. Management of crop diseases is important task for the development of agricultural crops. The work carried out by Dr. Gangawane is very important on this aspect. Many fungicides have been advocated for the management of crop diseases. However, it is observed that the fungicide resistance in plant pathogens, sometimes hampers the management of crop

diseases. This problem can overcome by various methods. Integrated management of crop diseases has been suggested by Dr. Gangawane and is widely accepted throughout the country.

5. Management of Municipal effluents and industrial effluents is an important aspect to maintain clear environment. But Marathwada is semi-arid-zone and farmers face the problem of irrigation. The work of Dr. Gangawane throws more light on this aspect. For example epidemiology of *Escherichia coli* the intestine pathogen can be controlled by various methods of sewage application. In addition how soil microbes are affected due to use of sewage is also worked out. Dr. Gangawane was once invited on TV during drought condition in Marathwada.

We all wishes him best life after superannuation and hope that this volume will serve best for the recent knowledge to the teachers, researchers, students and even to the common man.

Editors

Important Publications

1. Gangawane,L.V. and K.B. Deshpande (1973). Effect of groundnut rosett virus on the rhizosphere mycoflora of groundnut. *Ind. Phytopath.*, **26**: 325-329.
2. Gangawane, L.V. and K.B. Deshpande (1973). Studies on rhizosphere mycoflora of groundnut–III. A list of fungi isolated from rhizosphere, Rhizoplane and Soil of groundnut. *Sydowia*, **27**: 312-316.
3. Kelkar, S.S., S.R. Sengupta, L.V. Gangawane and K.B. Deshpande (1975). Lesion of Hepatitis B antigen containing sera in tobacco plants (*Nicotiana tabaccum Var. xanthi. Ind. J. Med. Res.*, **63**: 1198-1203.
4. Gangawane, L.V. and K.B. Deshpande (1975). Agronomic treatment and changes in rhizosphere mycoflora of groundnut–IV. Effect of crop rotation. *Ind. Nat. Sci. Acad.*, **41B**: 383-387.
5. Gangawane, L.V. and K.B. Deshpande (1976). Agronomic treatment and changes in rhizosphere mycoflora of groundnut–I. Effect of organic amendments. *Kavaka*, **4**: 21-24.
6. Kelkar, S.S. and L.V. Gangawane (1976). Effect of HB-Ag containing sera on tobacco plants. *Lancet.*, **2**: 519.
7. Gangawane, L.V. and K.B. Deshpande (1977). Agronomic treatment and change in rhizosphere mycoflora of groundnut–II. *Acta Agronomica Acad. Sci. Hung.*, **26**: 68-76.
8. Gangawane, L.V. and K.B. Deshpande (1977). Seasonal variation in the rhizosphere mycoflora of groundnut. *Ind. J. Bot. Soc.*, **56(4)**: 288-294.
9. Gangawane, L.V. and V.H. Talib (1977). A new medium for isolation of *Rhizobium. Ind. J. Microbial.*, **17(1)**: 33-34.

10. Gangawane, L.V. and V.V. Datar (1977). Efficacy of individual and composite rhizobial inoculants on the nodulation of some legume crops. *Fertilizer Technology*, **14 (1-2)**: 134-135.
11. Umalkar, G.V., D.B. Vishwe and L.V. Gangawane (1977). In vitro inhibition of *Alternaria brassiera* (Berk) Sacc. By some newly synthesized compounds. *Ind. Phytopath.*, **30(4)**: 563-564.
12. Gangawane, L.V. and K.B. Deshpande (1978). Antagonistic relationships between rhizosphere microfungi of groundnut. *Ind. J. Microbiology*, **18(4)**: 231-234.
13. Gangawane, L.V. and K.B. Deshpande (1978). Agronomic treatment and changes in rhizosphere mycoflora of groundnut–VII. Effect of foliar sprays. *Indian Phytopath.*, **31(4)**: 528.
14. Gangawane, L.V. and K.B. Deshpande (1978). Agronomic treatment and changes in rhizosphere mycoflora of groundnut–VI. Effect of partial sterilization of soil. *Rev. Du. Ecol. Soil Biol.*, **15(3)**: 311-321.
15. Gangawane, L.V. and V.V. Datar (1978). Ascorbic acid content of healthy and *Alternaria solani* infected tomato germplasm. *Ind. Phytopath.*, **31(2)**: 237-238.
16. Gangawane, L.V. (1978). Tolerance of Streptomycin and Aureofungin by N-fixing blue green algae. *Hindustan Antibiotic. Bull.*, **21(1)**: 36-37.
17. Gangawane, L.V. (1979). Tolerance of Thimet by N-fixing blue green algae. *Pesticides*, **13(3)**: 33-34.
18. Gangawane, L.V. and Leela Kulkarni (1979). Tolerance of certain pesticides by N-fixing blue green algae and their side effect on rice cultivars. *Pesticides*, **13(5)**: 37-38.
19. Gangawane, L.V. and R.S. Saler (1979). Tolerance of certain fungicides by nitrogen fixing blue green algae. *Curr. Sci.*, **48(7)**: 306-308.
20. Gangawane, L.V. and V.V. Datar (1979). Effect of dry algal manure on the nodulation and nitrogen fixation in groundnut. *Ind. J. Microbial.*, **19(2)**: 91-92.
21. Chaporkar, C.B., Kauser Khalil and L.V. Gangawane. Fungicidal control of Microfungi associated with blue green algae. *Ind. J. Plant Protection*, **7(3)**: 156-160.
22. Gangawane, L.V. and K.B. Deshpande (1980). Studies on rhizosphere mycoflora of groundnut–V. Root exudate and amino acid nutrition in relation to ecological behavior. In: *Advances in the Biology of Microorganisms* (Eds. Bilgrami and Vyas), pp. 63-69.
23. Gangawane, L.V. (1980). Tolerance of certain fungicides by N-fixing blue green algae. *Ind. J. Bot. Soc.*, **59(2)**: 157-160.
24. Saler, R.S. and L.V. Gangawane (1980). Tolerance of target and non-target microfungi of groundnut supermosphere to fungicides. *Pesticides*, **14(1)**: 22-26.
25. Saler, R.S. and L.V. Gangawane, (1980). Tolerance of brassicol by non-target microfungi of groundnut rhizosphere. *Ind. Phytopath.*, **33(1)**: 171.
26. Gangawane, L.V., R.S. Saler, C.B. Chaporkar and Kauser Khalil (1980). Tolerance of pesticide by blue green algae. *Pestology*, **6(11)**: 17-19.
27. Saler, R.S. and L.V. Gangawane (1981). Pesticide tolerance of microorganisms in the rhizosphere of groundnut. *Pesticide Annual*, pp.98-107.
28. Leela Kulkarni and L.V. Gangawane (1986). Rhizoplane mycoflora of groundnut as affected by sewage and sludge treatment to soil. *Ind. Phytopath.*, **38(4)**: 756-757.

29. Gangawane, L.V. (1981). Effect of Aureofungin and certain fungicides seed treatment on rhizosphere mycoflora of groundnut. *Pesticides,* **15(9)**:31-32.

30. Gangawane, L.V. (1981). Fungicide resistance in crop protection. *Pesticides,* **15(1)**: 12-16.

31. Gangawane, L.V. and Saler, S.S. (1981). Resistance to fungicides in *Aspergillus flavus. Netherlands J. Pl. Pathology,* **87**: 254.

32. Salve, P.B. and L.V. Gangawane (1982). Studies on *Rhizobium* from wild legumes–I. Isolates form *Indigofera* spp. and nitrogen fixation in groundnut. *Proc. Nat. Symp.* On Biol. Nitrogen fixation (DAE), BARC Publ. Bombay, pp. 556-566.

33. Kulkarni Leela and L.V. Gangawane (1982). Rhizosphere mycoflora of groundnut as influenced by household pesticides and antiseptics as sewage components. *Rev. Du. Ecol. Biol. Sol.,* **19(4)**: 525-533.

34. Salve, P.B. and L.V. Gangawane (1984). Effect of whey irrigation on nodulation of groundnut. In: *Progress in Leaf Protein Research* (Ed. N. Singh), pp. 391-397.

35. Chaporkar, C.B. and L.V. Gangawane (1984). Blue green algae of Marathwada, Maharashtra, *Phykos.,* **23(1&2)**: 55-58.

36. Gangawane, L.V. and B.R.C. Reddy (1985). Resistance of *Aspergillus flavus* to certain fungicides. *ISPP Newsletter U.K.,* **6**: 23.

37. Gangawane, L.V. (1985). Crop management and rhizosphere mycoflora: A review. *Ind. Bot. Reptr.,* **4(2)**: 160-168.

38. Chaporkar, C.B. and L.V. Gangawane (1985). Effect of plant growth substances on growth and N-fixation by *Nostoc hatei* Dixit. *Ind. Jour. Microbial.,* **24(3&4)**:264-265.

39. Gangawane, L.V. and B. Ramchandra Reddy (1986). Micronutrient reduce resistance to carbendazim in *Aspergillus flavus. ISPP Chemical Newsletter,* **7**: 19-20.

40. Gangawane, L.V. (1986). Phyllosphere microflora in relation to leaf protein extraction. *Ind. Bot. Reptr.,* **5(1)**:110-112.

41. Gangawane, L.V. (1987). Rational pesticide use. *Pesticides,* **21(12)**: 11-14.

42. Chaudhary, M. and L.V. Gangawane (1987). Antagonistic activity of *Actinomycetes* against *Phytophthora drechsleri* f. sp. *cajani* from sewage irrigated soils. *Science and Culture,* **53(2)**: 57-59.

43. Gangawane, L.V. and P.B. Salve (1987). Interaction between weed microflora *Rhizobium* and *Rhizoctonia bataticola* in groundnut. *Int. Tropical Ecology,* **28**: 193-199.

44. Gangawane, L.V. (1987). Biofertilizer Technology. An agrobased industry for farmers. In: *International seminar on technology transfer* 21st–23rd March, 1987. Aurangabad (Presidential Address), pp. 1-4.

45. Gangawane, L.V. (1988). Geocarposphere mycoflora and pod diseases of groundnut In: *Perspectives of Mycopathology,* (Eds. V.P. Agnihotri *et al.*), Malhotra Publ. House, New Delhi, pp. 149-163.

46. Gangawane, L.V. (1987). The Microbiology of sewage irrigated soils. *RAPA-FAO Bulletin,* **3**: 30-31.

47. Gangawane, L.V. and B.R.C. Reddy (1987). Development and management of carbendazim resistance in *Aspergillus flavus. Indian Phytopath.,* **40(2)**: 297.

48. Gangawane, L.V. and R.S. Saler (1988). A comparative study of tolerance of fungicides by *Rhizoctonia bataticola* and other microfungi in the rhizosphere of groundnut. *Pesticides,* **22(3)**: 27-37.

49. Gangawane, L.V. and B.R.C. Reddy (1987). Distribution and survival of toxicgenic strain of *Aspergillus flavus* resistant to fungicides in Marathwada. In: *Progress in Venom and Toxin Research,* (Eds. P. Gopalkrishnakone and C.K. Tan), National Univ. of Singapore Publications, pp. 658-668.

50. Gangawane, L.V. and S.A. Shaikh (1988). Management of resistance in *Pythium aphanidermarum* to aluminium ethyl phosphite. *Current Science,* **57(16)**: 905-906.

51. Gangawane, L.V. and S.A. Shaikh (1988). Development of resistance to aluminium ethyl phosphite in *Pythium aphanidermatum*. *Indian Phytopath.,* **41(4)**: 638-641.

52. Gangawane, L.V. (1988). Sewage application in agriculture. *Science Reporter,* **9**: 535-542.

53. Gangawane, L.V., Suman Waghmare and B.M. Kareppa (1988). Studies on metalaxyl resistance in *Phytophthora drechsleri* f. sp. *cajani*. *Ind. Phytopath.,* **41(2)**: 299.

54. Gangawane, L.V. and Shahana A. Shaikh (1988). Resistance of *Phytophthora nicotianae* to aluminium ethyl phosphite in India. *Ind. Phytopath.,* **41(2)**: 199-200.

55. Gangawane, L.V. (1989). Sewage farming and Microorganisms: A review. In: *Soil pollutant and soil organisms* (Ed. P.C. Mishra), Ashish Publishing House, New Delhi, pp. 87-96.

56. Gangawane, L.V. and S.T. Waghmare (1991). Synergistic effect of agrochemicals on Metalaxyl reesistance in *Phytophthora* blight of pigeonpea. *International Pl. Pathology Newsletter,* **17**: 39-40.

57. Gangawane, L.V. and P.B. Salve (1993). Salt tolerant rhizobia from wild legumes and nitrogen fixation in groundnut in semi arid tropics. In: *Towards the rational use of high salinity tolerant plants,* Vol.II, (Eds. H. Lieth and A. Al Masoom), Kluwer Academic Publishers, Netherlands, pp.53-58.

58. Gangawane, L.V. (1990). Fungicide resistance in plant pathogens in India. In: *Proc. 3rd International Conference on plant protection in tropics,* Vol. II, MAPPS PUBL. Malaysia, pp.177-121.

59. Gangawane, L.V. and S.A. Shaikh (1992). Synergistic effect of antibiotics on aluminum ethyl phosphite resistance in *Phytophthora nicotianae. International Phytophthora Newsletter,* **18**: 25-26.

60. Vidhale, N.N., L.V. Gangawane and V.B. Chavan (1992). Survival of *Escherichia coli* in sewage irrigated soil. *Ind. J. Microbial. Ecol.,* **2**: 83-84.

61. Chaudhary Manjiri and L.V. Gangawane (1992). Biological control of dry root rot of groundnut caused by *Rhizoctonia bataticola* using *Trichoderma harzianum* under condition of sewage and paper mill effluent irrigation. In: *New Trends in Biotechnology* (Eds. N.S. Subba Rao, C. Balagopalan and S.V. Ramkrishanan), Oxford and IBH Publ., New Delhi, pp. 213-218.

62. Arora, R.K., S.S. Kamble and L.V. Gangawane (1992). Resistance to metalaxyl in *Phytophthora infestans* in Nilgiri Hills of southern India. *Intrnat. Phytophthora Newletter UK,* **18**: 8-9.

63. Gangawane, L.V. and C.B. Chaporkar (1992). Interaction between blue green algae *Nostoc hatei* and soil microorganism. In: *Cyanobacterial Nitrogen Fixation* (Ed. B.D. Kaushik) IARI Publication, New Delhi, pp.431-436.

64. Salve, P.B. and L.V. Gangawane (1992). *Rhizobium* from wild legumes and Nitrogen fixation in groundnut. In: *Biofertilizer Technology Transfer,* Associated Publ. Co., New Delhi, pp. 95-100.
65. Gangawane, L.V., R.P. Francis and P.Vasudevan (1992). Biodegradation of malathion by *Rhizobium* from wild legumes. In: *Biofertilizer Technology Transfer,* Associated Publ. Co., New Delhi, pp. 339-342.
66. Gangawane, L.V., R.K. Arora and S.S. Kamble (1993). Effect of passage on the development of metalaxyl resistance in *Phytophthora infestans*. In: *Chemical Management of Plant Pathogens in Western India* (Eds. Gangawane L.V. *et al.*), IPS (WZ) Publ., Aurangabad, pp. 27-30.
67. Gangawane, L.V. (1993). Rational pesticides Use (Presidental address). In: *Chemical Management of Plant Pathogens in Western India* (Eds. Gangawane L.V. *et al.*), IPS (WZ) Publ., Aurangabad, pp. 8-13.
68. Gangawane, L.V. and B.M. Kareppa (1993). Survivability of carbendazim resistant strain of *Puccinia arachidis* in mixed population. In: *Proc. Vth Zonal Meeting of Western Zone*, IPS, New Delhi, PKV Akola (Eds. C.D. Mayee *et al.*), pp. 58-61.
69. Gangawane, L.V. (1995). Global and rational approach towards management of fungicides resistance in plant pathogen. *Ind. J. Mycol. & Pl. Pathol.*, **25(1-2)**: 49-50.
70. Gangawane, L.V. (1995). Management of fungicides resistance in soil-borne pathogens. *Ind. J. Myol. & Pl. Pathol.*, **25(1-2)**: 50-51.
71. Kareppa, B.M. and L.V. Gangawane (1995). Variation in the sensitivity of groundnut rust to tridemorph in central Maharashtra. *International Archis Newsletter*, ICRISAT, pp. 28-30.
72. Deshpande Jayashree and L.V. Gangawane (1995). On the possibility of airborne fungi in the biodeterioration of Buddhist paintings in Ajanta caves. In: *Proc. 3rd International Conference on Biodeterioration of Cultural Property*, (ICBCP–3 Committee Publ.), Bangkok, pp. 455-457.
73. Gangawane, L.V., S.S. Kamble and R.K. Arora (1995). Synergistic effect of other fungicides on metalaxyl resistance in *Phytophthora infestans* at Nilgiri hills. *Ind. J. Plant Protection*, **23(2)**: 159-162.
74. Gangawane, L.V. and S.A. Shaikh(1996). Production of phytoalexins by aluminum ethyl phosphite against sensitive and resistant strains of *Phythium aphanidermatum*. In: *Perspectives in Biological Science* (Eds. V. Rai, M.L. Naik and Manoharachari), Pt. R.S. University Publ., Raipur, pp. 287-289.
75. Gangawane, L.V. and N.N. Vidhale (1996). Sewage irrigation and *Escherichia coli* in the rhizosphere of crop plants. *J. Advances in Sci. and Tech.*, **2(1)**: 1-7.
76. Gangawane, L.V., N.S. Suryawanshi and Jayashree Deshpande (1998). Use of homeopathic medicines and ornamental cactus extracts in the management of streptomycin tolerant *Xanthomonas campestris pv. citri* citrus canker. In: *Plant Pathogenic Bacteria* (Ed. A. Mahadevan), CAS in Botany, Chennai Publ., pp. 460-463.
77. Suryawanshi, N.S., Leela Kulkarni and L.V. Gangawane (1998). Management of streptomycin resistant *Xanthomonas campestris pv. citri* causing citrus canker with medicinal plant extracts. In: *Plant Pathogenic Bacteria* (Ed. A. Mahadevan), CAS in Botany, Chennai Publ., pp. 531-536.
78. Deshpande Jayashree and L.V.Gangawane (1997). Microorganisms collected during solar eclipse in India. *Aerobiologia.*, **13**: 289-294.

79. Gangawane, L.V. and Jayashree Deshpande (1999). Herbicidal effects on soil fungi. In: *Modern Approaches and Innovations in Soil Management* (Eds. D.J. Bagyraj, *et al.*), Rastogi Publ., Meerut, pp. 185-188.

80. Khilare, V.C. and Gangawane L.V. (1999). Biochemical changes in thiophanate methyl resistant *Penicilliium digitatum* causing green mold of sweet orange. *Ind. J. Bot. Soc.*, **78**: 173-174.

81. Vidhale, N.N. and Gangawane L.V. (1999). Occurrence and serological grouping of enteropathogenic *Escherichia coli* from sewage. *Asian J. and Microbiol and Biotech. & Environ. Sci.*, **1(3-4)**: 149-152.

82. Gangawane, L.V. and Kamble, S.S. (2001). Use of other agrochemicals in the management of charcoal rot of potato caused by *Macrophomina phaseolina* resistant to carbendazim. In: *Frontiers in Fungal Biotechnology and Plant Pathogen Relations*, (Eds. C. Manoharachary *et al.*), Allied Publishers Ltd., New Delhi, pp. 58-62.

83. Kadam, K.S., Khilare, V.C. and Gangawane, L.V, (2001) Effect of passage on the development of Aureofungin resistance in *Alternaria alternata* causing fruit rot of grapes. In: *Frontiers in Fungal Biotechnology and Plant Pathogen Relations*, (Eds. C. Manoharachary *et al.*), Allied Publishers Ltd., New Delhi, pp. 260-263.

84. Khilare.V.C. and Gangawane L.V. (2000). Effect of carbon and nitrogen nutrition on the growth of thiophanate methyl resistant *Penicillium digitatum* causing green mold of sweet orange. *J. Phytol. Res.*, **13(1)**: 61-64.

85. Gangawane, L.V. (2002). Role of nitrogen fixing microorganisms in bioremediation of salt affected soils. In: *Frontiers in Microbial Biotechnology and Plant Pathology* (Eds. C. Manoharachary *et al.*), Scientific Publishers, Jodhpur, pp. 73-80.

86. Gangawane, L.V. (1997). Management of fungicide resistance in plant pathogens. *Ind. Phytopath.* **50(3)**: 305-315, (Prof. M.S. Pavgi award lecture delivered at 49th Annual Meeting of Indian Phytopathological Society, Jabalpur.

87. Solunke B.S., B.M. Kareppa and L.V., Gangawane (2001). Integrated management of Selerotium rot of potato using carbendazim and plant extract. *Ind. J. Plant Protection,* **29 (1-2)**: 142-143.

88. Khilare, V.C. and Gangawane, L.V. (2001). Role of oxidative enzymes in thiophanate methyl resistance in *Penicillium digitatum* the green mold of sweet orange. *Ind. Phytopathology,* **55(1)**: 73- 74.

89. Deokate, A.S., V.C. Khilare, and Gangawane, L.V. (2002). Management of anthracnose of grape by carbendazim in combination with agrochemical. *Ind. Phytopathology,* **55(1)**: 107-108.

90. Gangawane, L.V. and Francis, R.P. (1997). Rhizobium from wild legumes and pesticides degradation: the two in one. In: *Microbial Biotechnology*, (Eds. S.M. Reddy *et al.*), Scientific Publication, Jodhpur, India, pp. 59-63.

91. Deokate, A.S., V.C. Khilare, and Gangawane, L.V. (2002). Resistance to carbendazim in *Gleosporium ampelophagum* (Pass) Sacc. Causing anthracnose of grapevine in Maharashtra. *Ind. J. Plant. Prot.*, **30(1)**: 69-70.

92. Khilare, V.C., A.M. Chavan, and L.V. Gangawane (2002). Synergistic effects of antibiotics and insecticides on thiophanate methyl resistance in green mould of sweet orange. *J. Ind. Botanical Soc.*, **81(1-2)**: 177-178.

93. Gangawane, L.V. (2003). Rational approach towards the management of fungicide resistance in plant pathogens. *Ind. Phytopath.*, **56(3)**: 304.

List of Ph.D. Students

1. Phyllosphere mycoflora of wheat in Marathwada–L.D. Balkhande, 1982.
2. Investigation on microorganisms associated with sewage irrigated soils–Leela Kulkarni, 1982.
3. Investigations on tolerance of pesticidal tolerance in microorganisms in the rhizosphere of groundnut–R.S. Saler, 1982.
4. Effect of agrochemicals on nitrogen fixing blue green algae of Marathwada soils–C.B. Chaporkar, 1983.
5. Studies on nitrogen fixing blue green algae of sewage irrigated soils–Miss Kauser Khalil, 1983.
6. Studies on mycoflora associated with peanut pods of Marathwada–K.G. Jadhav, 1983.
7. Studies on Physiology of *Phoma herbarum* with special reference to cellulolytic enzymes–Mrs. Shanta R. Shinde, 1984.
8. Studies on rhizobia of non-cultivated legumes of Marathwada–P.B. Salve, 1985.
9. Studies on resistance of fungal pathogen to fungicides II–B.R.C. Reddy, 1985.
10. Studies on fungicide Alliete in relation to target and non-target microorganisms–Miss Shahana Shaikh, 1987.
11. Effect of sewage and industrial effluents on soil–borne pathogens and related microorganisms–Mrs. Anuradha G. Datar, 1987.
12. Studies on resistance of fungal pathogens to fungicides–IV–B.M. Kareppa, 1989.
13. Studies on resistance of fungal pathogens to fungicides–III–Miss S.T. Waghmare, 1990.
14. Studies on fungicide resistance in late blight of potato–S. S. Kamble, 1991.

15. Studies on *E. coli* in relation to sewage irrigated soil microorganisms–M.N. Vidhale, 1991.
16. Studies on *Rhizobium* from *Crotolaria spp.*–Miss Alka Dhumal, 1991.
17. Studies on resistance of fungal pathogens to certain fungicides–I–Miss Sahera Nasreen, 1992.
18. Studies on pesticide degradation by soil beneficial microorganisms–Romain Francis, 1993.
19. Studies on fungicide resistance in certain pathogens in Marathwada–I.S.V. Kamble, 1993.
20. Studies on fungicide resistance in *Sclerotium rolfsii* causing root rot of groundnut–B.A. Patil, 1994.
21. Studies on Nitrogen fixing microorganisms from saline soils in Marathwada–Sunita Nilekar, 1994.
22. Studies on pre and post harvest diseases of groundnut pods in Marathwada–B.R. Kamble, 1996.
23. Studies on fungicide resistance in Sclerotium rot of potato–B.S. Solunke, 1997.
24. Integrated management of *Alternaria* fruit rot of grapes in Maharashtra–K.S. Kadam, 1997.
25. Studies on fungicide resistance in certain pathogens causing post harvest diseases of sweet orange and their integrated management–V.C. Khilare, 1997.
26. Studies on fungicide resistance in certain pathogens causing grape diseases in Maharashtra–A. S. Deokate, 1998.
27. Studies on antibiotic resistance in *Xanthomonas campestris pv. Citri* causing citrus canker and its management–N.S. Suryawanshi, 1998.
28. Integrated management of post harvest diseases of banana–Shama C. Mahadik, 2002.
29. Studies on the effect of effluents from some pharmaceutical industries on soil microorganisms–Manojkumar S. Bhalkikar.
30. Studies on the integrated management of fruit disease of ber (*Zyzyphus spp.*) in Marathwada–Miss Purnima Sable.
31. Studies on mutation breeding of Rhizobium and other microbes for integrated management of certain soil–borne disease of groundnut–A.B. Ade, 2004.
32. Studies on pesticides degradation of free–living N-fixing microorganisms from Marathwada soils–T.A. Kadam, 2003.
33. Investigation on integrated management of certain diseases of onion in Maharashtra–M.R. Meshram, 2001.
34. Studies on white fly (*B. Tabaci)* transmitted geminivirus in India with special reference to virus vector relationship and screening of TLCV resistant sources of tomato–Astapure, 2004.
35. Studies on *Azotobacter* from sewage irrigated soils and Rhizosphere of crop plants–Mrs. Pushpalata Jadhav, 2004.

1

Biotechnological Approach for the Integrated Management of Crop Diseases

AMERIKA SINGH, O.P. SHARMA, O.M. BAMBAWALE & S.K. SINGH

National Centre for Integrated Pest Management (ICAR),
Pusa Campus, New Delhi – 110 012, India

Introduction

In India tremendous progress has been made in agriculture. In absence of check on population growth, which is increasing at the rate of 1.8 per cent per annum, it may cross 1.5 billions by 2030. As a consequence per capita availability of 0.13 ha land will further reduce to 0.1 hectare in 2020 A.D. The demand of growing population for food and fibre requires the country to produce higher yields from presently cultivated land. Currently an average of 2790 calories of food is available each day for every human on the planet, (23 per cent more than in 1961 and enough to feed everyone). On the basis of trends of consumption and income growth it is estimated that the total food grain requirements will be 240 million tonnes by 2006 A.D. The production targets and required yield to feed the ever increasing current population with presently available resources will require the growth rate of 2.35 and 2.2 per cent in rice and wheat, respectively and as high as 4.45 and 3.38 per cent for pulses and oil seeds, respectively. This requires national average yield to be increased by 30-50 per cent, which is not a cakewalk. India will require producing additional 5-6 million tonnes annually to keep pace with ever-increasing national food and nutritional requirements. Recurrent famines, starvation deaths changed and lack of resources to green revolution, which as of now has turned our begging bowl to agriculturally advanced cereal exporting country. Over the past four decades, worldwide food production has more than kept pace with the doubling of world population. Gains in food availability has been maximum in India, where the green revolution enabled a rise of 38 per cent between 1961 and 1998 to 2660 calories per person daily. In India the production has more than doubled since 1960, largely as a result of the introduction of new crop varieties and the intensification of agriculture–supported by increased applications of chemical fertilizers and pesticides. Plant diseases, caused by infectious fungi, bacteria, viruses, phytoplasmas and nematodes, result in economic losses. These problems include reduced yields, lower quality, decreased nutritional value, and, contamination with toxins. Control of plant diseases is essential to provide an adequate supply of food, feed, and fiber. Growers currently spend large sums to achieve partial control of pathogens that attack have remained crops and other plants, even then, crop losses significant because of diseases. Reducing such losses has long been a high priority objective for agriculture. Besides the obvious benefits for producers and processors, plant health protection is an important mechanism for increasing food supplies without increasing the land under cultivation.

The demand for certain food crops in 21st century warrants all out immediate remedial measures. Planner as well as policy makers are worried for tougher trade rules in the wake of implementation of World Trade Organization (WTO) agreement, and support the need and the aims of sustainable agriculture to produce sufficient food, in an environmentally and socially acceptable manner while maintaining the local natural resource base for future generations. But some scientists suggest that population and economic growth worldwide have raised the demand for food beyond the levels that can be supported by extensive and environmentally friendly farming. However, like others we are equally committed to Integrated Pest Management (IPM) as defined by the FAO, as an economically viable and socially acceptable approach to crop protection. It is the top priority mission of the ICAR and Government of India to provide safe and effective technologies to protect against unacceptable losses caused by weeds, diseases and pests, which according to an estimate works out to Rs. 90,000 crores per annum at the current price and even exceeds in the time of epidemics.

Plant biotechnology ushers in a new era for plant protection workers to maintain healthy crop to optimize crop yields with minimum pesticide use. The chief objective of biotechnology is to provide nutritional security to expanding population. Genetic engineering has the potential to provide cultivars with enhanced ability to withstand or resist attack by plant pathogens. New approaches to plant disease control are particularly important for pathogens that are difficult to control by existing conventional methods. In India the average percentage of crop losses caused by plant pathogens, insect pests, and weeds, has steadily increased to 18 per cent, while worldwide, plant diseases reduce crop productivity by 12 per cent. Worldwide, pesticide applications costing $26 billion dollars annually are applied to manage pest losses. Genetically engineered plants resistant to plant pathogens can prevent crop losses and reduce pesticide usage. This article provides a current perspective on areas of research and application of plant genetic engineering for resistance to plant pathogens.

Conventional Disease Management

Conventional strategies for the management of plant diseases include planting resistant crop varieties, cultural practices to create less favourable conditions for disease development, employing biological control, applying chemical fungicides, and using integrated disease management (combining two or more of the above approaches). For diseases where host plant resistance is unavailable, strategies integrate cultural, biological, and chemical control procedures. For example, different crop rotations or improved cultural practices can sometimes suppress development of pathogenic organisms while maintaining a high level of productivity. Biological control has great potential for disease management, but it is inconsistent or marginally effective. Additional research is needed to understand the interactions of microorganisms in the soil, plant parts, and on the mechanisms that biological control agents use to reduce or prevent the onset and development of diseases.

The ability to develop any of these strategies and its success depends first on correctly and quickly identifying the pathogen causing the disease, then learning how to interrupt its disease cycle. It is required to have rapid, reliable pathogen detection and identification procedures for disease diagnosis. The detailed knowledge about the ecology of pathogen and genetic, biochemical, and physiological processes that govern the host and pathogen, help in devising control methods. In addition, an understanding of the epidemiology of disease and spread of pathogens is important for deciding which control actions are most effective in a given situation.

The appropriate use of chemical fungicides will continue to be an important tool to manage plant diseases, especially where alternative controls are not effective. As with all living organisms, plant pathogens exhibit a remarkable ability to change and adapt. More virulent strains of pathogens continually evolve that overcome resistant crop varieties or are no longer controlled by strategies and chemicals that were once effective. Continuing research to develop new control methods is necessary to increase or even maintain current levels of crop production. Further complicating the situation, public awareness has grown in recent years regarding the use of chemicals to control diseases and their adverse consequences. As a result, pressure has increased to develop non-chemical approaches to plant disease control, which as of now is slow and less effective. Host plant resistance to plant diseases is a mandatory control strategy because of its effectiveness and acceptability with no additional expense to producers. To be most effective such resistance should be durable, (not readily overcome by mutations in pathogens that increase virulence). This requires a clear idea of the pathogens that are present, their ability to change, and the nature of the resistance processes themselves.

The overall goal of the National Agricultural Research System in India is to develop and improve ways to reduce crop losses caused by plant diseases. The program focuses on developing effective disease control strategies that are not environmentally harmful, and are compatible with sustainable crop production.

Biotechnology as a Tool for Diagnosis of Plant Diseases

The first step towards controlling a plant disease is to be able to detect it, and identify the disease-causing organism as early as possible so that appropriate treatment can be used (Donalds, 1992). However, many diseases are identifiable visually only when it is already too late to prevent major damage. Developing diagnostic kits for diseases has been a major breakthrough in limiting crop damage by diseases. Kits for diagnosing plant diseases such as bacterial canker of tomato and soybean root rot at an early stage of infection have been developed using biotechnology. Diseases can be diagnosed by identifying a unique feature of the organism, such as its genetic material (DNA) or a protein. These tests use antibodies that 'recognize' specific proteins from the disease-causing organism. Bananas and plantains are highly susceptible to viral, bacterial and fungal diseases. Attempts to breed new varieties with better yields need to start with disease-free plants, hence DNA-based diagnostics are being standardised to detect and distinguish between different viruses. Improved detection and identification procedures are also becoming more important as international trade of plant products increases and as trading partners seek to protect themselves from the introduction of unwanted diseases.

Molecular Diagnostics of Plant Diseases

The first and most important step in managing a plant disease is to correctly identify it. Although some diseases can be diagnosed quickly by visual examination, others require laboratory testing for diagnosis. These laboratory procedures may take days or even weeks to complete and are, in some cases, relatively insensitive, Delays are frustrating when a quick diagnosis is needed so that appropriate disease control measures may be taken to prevent plant injury, especially when high value cash crops are at stake. Fortunately, as the result of advances in biotechnology, new products and techniques are becoming available that will complement or replace time consuming laboratory procedures (McCartney, 2003).

Enzyme-Linked Immunosorbent Assay (ELISA) Diagnostic Kits

The Enzyme-Linked Immunosorbent Assay (ELISA), first introduced into plant virology about 23 years ago, is one of the most widely used serological tests for plant viruses detection, identification, quantification and comparison owing to its simplicity, sensitivity and adaptability (Bar-Joseph and Garnsey, 1981; Clark, 1981). A number of detection kits have been developed for use at the site where a disease is suspected. These kits, which in most cases do not require laboratory equipment, are especially useful to growers and take five minutes to perform. The diagnostic kits are based on a method that uses proteins (called antibodies) to detect disease causing organisms of plants (plant pathogens). The technique used is called ELISA (enzyme-linked immunosorbent assay). This assay is based on the ability of an antibody to recognize and bind to a specific antigen, a substance associated with a plant pathogen. The antibodies used in the diagnostic kits are produced by injecting a rabbit with an antigen (highly purified proteins) associated with particular plant disease. The rabbits react to the antigen and produce specific antibodies. The antibodies produced recognize and react only with the proteins (antigen) associated with the causal agent of specific plant disease. Colour changes on the unit's surface indicate a positive (disease present) reaction. Examples of diseases ELISA kits can detect include bacterial canker of tomato, soybean root rot, viral diseases of potatoes etc.

Polymerase Chain Reaction (PCR)

Polymerase Chain Reaction (PCR) which is now a key tool in molecular biology, provides a rapid and powerful technique for the *in vitro* amplification of DNA sequence (Mullis *et al.*, 1986). PCR is now routinely used for gene cloning, analysis and their manipulation. It has great potential for raising the sensitivity of various assays that use nucleic acid probes. PCR is used to produce enormous numbers of copies of a specified nucleic acid sequence. This technique enables the detection of very small amount of a pathogen in a sample by amplifying the pathogen sequences to a detectable level. PCR is especially useful in plant quarantine.

Development of New Fungicides

To develop a new fungicide, on an average more than twenty thousand chemicals are biologically screened (Mel-nikov, 1986). To prove anti-fungal activity, *in vitro* assay, such as inhibition of mycelial growth and spore germination and *in vivo* efficacy test in greenhouse and agronomic fields are adopted. Although *in vitro* screening has been used as an effective method, the results of *in vitro* tests were not always consistent and did not correlate with field tests. In pharmaceutical field, however, target-site specific strategies for biological screening have been widely adopted for new drug development. In spite of limitation in our understanding on the basic biochemistry of pathogenesis by plant pathogenic fungi, application of technical innovations in pharmaceutical field has not been applicable for development of more efficient screening methods in agrochemical area. Recently, however, much research is being conducted on the target discovery, target selection, and screening technologies in the field of agricultural biotechnology. These include genomics, chemistry and high screening to maximize the potential of the agricultural industry.

Molecular Approach

Scientists from all over the world are investigating the biochemical nature of, and the signals involved in, a plant's reaction to pathogen invasion and disease development. Plant resistance genes and the genes involved in resistance reactions are being identified and engineered into crop plants to protect them against plant diseases. Understanding of pathogenesis at biochemical and molecular biological level is prerequisite to the development of novel strategies to control plant diseases. The rice blast fungus *Magnaporthe grisea* has emerged as a model system for understanding fungal-plant interactions. The mechanism of plant infection by *M. grisea* is better understood than any other cereal disease. Reciprocal genetic and molecular studies of rice have allowed a holistic view of the interaction to emerge, encompassing infection- related development, disease symptom production, host metabolic responses and host resistance. Most important research aims considered during the last several years include:

1. Determination of the molecular basis of rice blast disease;
2. Determination of resistance of rice cultivars to perceive and respond to the rice blast fungus, and
3. Understanding of the genetic mechanisms leading to generation of newly virulent forms of the fungus.

Now it has not remained a secret science and has become routine to move/transfer genes from one organism to another. It is now possible to introduce genes conferring disease resistance into crop plants. It can be accomplished by either by direct methods or vector-mediated methods. Gene gun and *Agrobacterium*-mediated method are the best examples of the direct method and vector-mediated method respectively.

Transgenic Plant Disease Management

Controlling of plant diseases has been dependent on the breeding of resistant cultivars and application of chemical fungicides. However, the frequent appearance of new races and fungicide resistant strains has reduced the effectiveness of resistant cultivars in the field (Brough *et al.*, 1995). Recent environmental regulations, especially WTO has also restricted the use of chemical fungicides. Therefore, it is required to develop alternative and novel strategies to control plant diseases. Disease resistance genes could be sourced from plant pathogens themselves, as was possible with coat protein-mediated plant viral resistance and with toxin-inactivating protein-mediated, bacterial resistance. Host plants can also contribute an enormous number of disease resistance genes such as those encoding pathogenesis-related (PR) proteins, which have been used against fungal diseases. Plants can be protected from diseases with transgenes (genes that are engineered into plants) that are derived from the pathogens themselves, a concept referred to as pathogen-derived resistance. For example, plant viral transgenes can protect plants from infection by the virus from which the transgene was derived. One of the most successful examples, as of date, of the use of transgenic resistance against plant diseases is that of management of papaya ring spot virus (PRSV) in Hawaii. Traditional breeding in bringing about resistance was of no avail, as cross ability barrier was a constraint. Under these circumstances, coat-protein-mediated resistance using coat protein gene sourced from a Hawaiian strain of PRSV was attempted resulting in resistant transgenic line. Recently, a gene silencing mechanism has been put to productive use in containing rice yellow mottle virus. An open reading frame of the virus itself is expressed in rice in order to stop the viral spread in an effective manner. Similar attempts also have been made in containing multiple viral infections (tomato spotted wilt virus and turnip mosaic virus) in plants. *Xa21*, a wide-spectrum bacterial blight resistance gene isolated from an African rice, *Oryza longistaminata* was back crossed into a cultivated variety at International Rice Research Institute, Philippines. The resistance gene was cloned using molecular techniques. Wild fire disease of tobacco caused by *Psuedomonas syringae* pv. *tabaci* is a serious disease. A phytotoxin secreted by the pathogen drastically modifies the amino acid metabolism of the plant with the eventual accumulation of ammonia in tobacco leaves, which causes extensive blighting. Interestingly, the pathogen that synthesises the phytotoxin remains unaffected by the toxin. This formed the basis for a search of the candidate gene from the pathogen itself. A toxin-inactivating gene, '*ttr*' was successfully isolated from the pathogen and cloned into tobacco cultivars, which showed excellent wildfire resistance. PR protein genes appear to be a very potential source for candidate genes for fungal resistance. These proteins may play a direct role in defence by attacking and degrading pathogen cell wall components. Typical candidate genes are that encoding chitinases and ß-1,3 glucanases. Increasing expression of individual and multiple PR-proteins in various crops have demonstrated some success in enhancing disease resistance in particular pathogens (*e.g.*, *Rhizoctonia solani*, the sheath blight pathogen). A recent research shows a chitinase gene from an anti-fungal biocontrol fungus species (*Trichoderma viride*) confers transgenic resistance against the rice sheath blight pathogen. A rice PR-5 protein gene in wheat delays onset of symptoms caused by the wheat scab pathogen.

Biocontrol of Plant Pathogens

The environmental concern over pesticide use has been instrumental in a large upsurge of biological disease control. Among the various antagonists used for the management of plant diseases, *Trichoderma* and *Pseudomonas* play a vital role. Among the various species of *Trichoderma*, *T. viride*, *T. harzianum*, *T. virens* and *T. hamatum* are used against the management of various soil-borne pathogens causing diseases of crop plants. It has many advantages as a bio-control agent owing to its high rhizosphere competence, ability to synthesize polysaccharide-degrading enzymes, amenability for mass multiplication, broad spectrum of action against various pathogens and above all its environmental friendliness (Baker, 1980).

Fluorescent pseudomonads suppress the pathogens either directly through the secondary metabolites or indirectly by inducing plant-mediated defense reactions. The crucial factor in the success of biological control by fluorescent pseudomonads is their ability to colonize the rhizosphere and their persistence throughout the growing season. Fluorescent pseudomonads are root colonizers because they occur in the natural habitat of rhizosphere and thus when they are reintroduced to roots through seed they colonize root surface profusely. Fluorescent pseudomonads suppress the pathogens by antibiosis through the production of various antibiotic substances such as 2,4-diacetyl phloroglucinol, phenazine-1-carboxylic acid, oomycin A, oxychlororaphine, pyoluteorin, pyrrolnitrin and pyocyanine. Siderophores are extracellular, low molecular weight substances which selectively complex iron with high affinity. Fluorescent pseudomonads produce siderophores such as pseudobactin and pyoverdine which chelate the iron available in the soil and make it unavailable to pathogen thus the pathogen dies for want of iron. In rice, seed treatment followed by root dipping and foliar spray with *Pseudomonas fluorescens* showed a higher induction of ISR against sheath blight pathogen, *R. solani.*

Plantibodies

Although plants have mechanisms to protect themselves against pathogen attack, in contrast to animals, there is no "immune system" *per se* in plants. With the advent of genetic engineering, plants can be engineered to express an antibody against a protein crucial for pathogenesis resulting in a high level of immunity or resistance to the pathogen (Martin, 1994). This promising approach is described under Plantibodies: an animal strategy imported to the plant kingdom to fight back pathogens.

Conclusion

Recently developed biotechnological tools provide promising new approaches for achieving control of diseases. Genetically engineering plants for resistance to virus, bacterial, fungal, and nematode pathogens has shown some success, but more research is needed. Also, techniques have been developed that allow researchers to trace specific genes and thus facilitate breeding resistant varieties. Biotechnology offers the exciting possibility of developing disease resistance in plants that cannot be accomplished through conventional breeding procedures (for example, by introducing genes for resistance from unrelated species).

REFERENCES

Baker, R. (1980). Pathogen-suppressive soils: in biocontrol of plant diseases. *Plant- Protection-Bulletin*, Taiwan, **22(2)**: 183-199.

Bar-Joseph, M. and Garnsey, S.M. (1981). In: *Plant disease and vector: ecology and epidemiology.* (Eds. Karl Maramorsch and Kerry F. Harris), Academic Press, Inc., New York, pp. 35-59.

Brough, E., Foster, J., Norton, G. (1995). Transgenic plants: a challenge to integrated pest management Australas. *Biotechnol.*, **5**, **6**: 360-364.

Clark, M.F. (1981). In: *Immunoassays for the eighties.* (Eds. A. Volley D.E. Bidwell and A. Bartlett), MTP press, Lancaster, England, pp. 481-494.

Donald, James Mac (1992). New fungal root disease detection kits may revolutionize yam disease control decisions. In: *Groner Talks*, Feb. pp: 73-79.

Martin, R.R. (1994). Genetic engineering of potatoes. *American potato Journal*, **71(5)**: 347-358.

McCartney, H.A., Foster, S.J., Fraaije, A. and Ward, E. (2003). Molecular diagnostics for fungal plant pathogens. *Pest-Management-Science,* **59(2)**: 129-142.

Mel-nikov, N.N. (1986). Current trends in the production of new pesticides. *Agrokhimia*, **(2)**: 119-129.

Mullis, K.B., F. Faloon, S. Scharf, R. Saiki, G. Horn and H. Erlich (1986). Specific enzymatic amplification of DNA *in vitro*: The polymerase chain reaction. *Cold spring Harber Symp., Quart Biol.*, **51**: 263.

2

Use of Indirect Competitive ELISA Technique for Detection of AflatoxinB$_1$ Contaminated in Chilli

K. AJITKUMAR[1], M.K. NAIK[1], F. WALIYGAI[2] & S.V. REDDY[2]

[1] *Department of Plant Pathology, College of Agriculture, Raichur – 584 101, India*
[2] *Division of Mycology, ICRISAT, Patancheru, Andhra Pradesh – 502 324, India*

ABSTRACT

Of late, aflatoxin contamination in chilli caused by *Aspergillus* spp. has become a serious problem since it influences the quality of the produce thereby affecting the export trade in global market. The stringent Sanitary and Phytosanitary Specifications (SPS) and Hazard Analysis Critical Control Point (HACCP) are imposed by the importing countries on chilli and their products. However, it is essential to monitor, detect and screen the aflatoxin contaminating fungi in chilli for confirming their toxigenic nature, which would be otherwise misleading. Infected chilli fruits were collected from the fields, storage and markets of Raichur, Gulbarga and Bellary districts of Karnataka, India. A total of sixteen isolates were collected from the samples and analyzed for aflatoxinB$_1$ (AfB$_1$) content by an indirect competitive ELISA at ICRISAT Asia Center, Patancheru, Andhra Pradesh. All the sixteen isolates produced aflatoxinB$_1$ at varying levels. The aflatoxin fungal isolates AFL10 and AFL14 produced 6.6 and 10.2 AfB$_1$ µg/kg respectively, which were in non-toxigenic range. The highest concentration of 1652.2 µg/kg was produced by AFL16 isolate followed by AFL8 with 1081.9 µg/kg. Overall, 88 per cent of *Aspergillus* isolates obtained from chilli fruits from the North- Eastern Karnataka region were proved to be toxigenic for AfB$_1$ aflatoxin. Awareness among the consumers about mycotoxins in chillies may induce producers as well as traders to market mycotoxin free chilli fruits.

Key Words: *Chillies, Spices, AflatoxinB$_1$, Indirect competitive ELISA.*

Aflatoxin contamination of agricultural commodities has gained global significance as a result of their deleterious effects on human as well as animal health and its importance to international trade. Aflatoxins are potent carcinogenic, mutagenic and immuno-suppressive agents, produced as secondary metabolites by the fungus *Aspergillus niger* Van. Tiegh. and *Aspergillus flavus* Link. Fres. on a wide range of food products. Many commodities are contaminated by aflatoxins, including cereals, oil-seeds and spices (Jelink *et al.*, 1989, Vasanthi and Bhat, 1998). Chillies (*Capsicum annum* L.) occupy a prominent place among the spices and form an ingredient of a variety of food preparations in India consumed by the majority of people, including those below the poverty line. India produces 946000 tonnes of chilli fruits from 957000 hectares. Over 50000 tonnes valued at more than 40 million US dollars are exported to various countries, including the UK and USA. In this paper we report on the estimation of aflatoxins in chilli fruits collected from different localities of North-Eastern Karnataka by an indirect competitive ELISA.

Material and Methods

Collection and Isolation of the Pathogenic Fungus

Fruit of *Capsicum annum* contaminated with aflatoxin fungus were collected from the field/storage, godown/market of different localities. The fungus was isolated by following standard tissue isolation method under aseptic condition using laminar flow. The infected tissues of the fruit were cut into small bits of 1-2 mm size and surface sterilized in 1 : 1000 mercuric chloride ($HgCl_2$) solution for one minute and washed repeatedly twice in sterile distilled water to remove the traces of mercuric chloride before transferring them to sterile potato dextrose agar (PDA) plates under aseptic conditions. The Petridishes were incubated at a temperature of 28° ± 1° C and observed for fungal growth. Then the pure culture of the fungus was obtained (Table 2.1).

Pathogenicity Test

Pinprick Method of Inoculation

For inoculation, pinprick method developed by Naik and Rawal (2002) was followed. Eight days old culture of *A. niger* was used for artificial inoculation. The fruits harvested at red ripened stage were surface sterilized with 0.1 per cent $HgCl_2$ and then washed in two charges of sterile water. Thereafter, the fruits were picked with pin bundles specially designed for pricking. The pinpricked fruits were then dipped in spore suspension (1×10^6 spores/ml) for one minute and then kept for incubation on a perforated tray under humid chamber. The humid chamber was prepared by keeping water in tray, which was placed below the perforated tray kept with inoculated fruits. The 3-4 wet cotton pieces were placed on the tray. The tray was covered with polythene sheet to maintain the relative humidity of over 90 per cent and then incubated at 28° ± 1° C.

Table 2.1: Source and Identity of Sixteen Isolates of *Aspergillus* spp.

Isolates	*Locality*	*Cultivar Source*	*Colour of Culture*	*Identity of Aspergillus spp.*
AFL 1	Shahpur	Guntur	Black	*A. niger*
AFL2	Chittapur	Mukthi	Black	*A. niger*
AFL3	Sedam	Jwala	Black	*A. niger*
AFL4	Raichur	Byadagi Kaddi	Green	*A. flavus*
AFL5	Raichur	Byadagi Kaddi	Black	*A. niger*
AFL6	Gulbarga	Wonder Hot	Black	*A. niger*
AFL 7	Devadurga	Byadagi Dabbi	Green	*A. flavus*
AFL8	Sindhanur	Byadagi Kaddi	Green	*A. flavus* var. *columnaris*
AFL9	Afzalpur	Jwala	Green	*A. flavus*
AFL 10	Bellary	Byadagi Kaddi	Black	*A. niger*
AFL11	Aland	Mukthi	Green	*A. flavus*
AFL12	Hagari	Byadagi Kaddi	Black	*A. niger*
AFL13	Shahpur	Mukthi	Green	*A. flavus*
AFL 14	Raichur	Byadagi Dabbi	Yellow	*A. niger*
AFL 15	Bellary	Byadagi Kaddi	Green	*A. flavus* var. *columnaris*
AFL16	Hagaribommanahalli	Byadagi Kaddi	Green	*A. flavus*

Antibody Production

For the production of polyclonal antibodies to AfB_1, commercially available aflatoxinB_1-bovine serum albumin (AfB_1-BSA) from Sigma chemical company was used. Five hundred gram of AfB_1-BSA in 0.25ml of 0.9 per cent NaCl was emulsified with an equal volume of Freud's complete adjuvant and injected intramuscularly at multiple sites into the hind leg of a New Zealand White inbred rabbit. For the subsequent four injections at weekly intervals, AfB_1-BSA was emulsified with Freund's incomplete adjuvant. The rabbit was bled and the titre of the antiserum was determined by an indirect competitive ELISA.

Sample Collection and Processing for ELISA

Dried graded chilli fruit samples were collected and samples of approximately 500g were drawn. About three each of 50g sub-samples were drawn from 500g samples. Dried fruits were cut into small pieces, thoroughly mixed and a 15g of analytical sample was drawn from each of 50g sub-samples. The analytical samples were ground in a wearing blender and then utilized for analysis. Each 15g of the chilli powder (from fruits as well as commercially prepared chilli powder) was extracted with 75ml of a solvent containing 70ml methanol + 30ml water + 0.5g KCl and blended for 2 min. This was followed by shaking in a rotary shaker for 30 min. at 250 rpm. The extract was filtered through Whatman No. 41 filter paper and diluted to 1 : 10 with 0.2 per cent bovine serum albumin (BSA) prepared in 0.05m PBS–Tween, pH–7.4, (PBST-BSA). Depending on the concentration of aflatoxin, the sample was either used directly or diluted further at 10 fold intervals prior to analysis by ELISA.

ELISA Procedure

An indirect competitive ELISA technique was used, as followed by previous workers. (Reddy *et al.*, 1988; Devi *et al.*, 1999 and Thirumala Devi *et al.*, 2000). The ELISA work was carried out at the International Crop Research Institute for the Semi and Tropics (ICRISAT) Asia Center, Patancheru, Andhra Pradesh, India. Prior to utilizing this procedure, concentrations of the various reagents required to give optimum results were determined. These include the concentration of AfB$_1$-BSA and dilution of polyclonal antiserum and goat antirabbit IgGs labelled with alkaline phosphate. Maxi–sorp (Nune A/s, DK-4000, Rorkilde, Denmark) ELISA plates were coated with 150 µl/well of AfB$_1$-BSA at a concentration of 125 µg/ml prepared in coating buffer (Hobbs *et al.*, 1987). At each step plates were incubated at 37° C followed by three washes with PBS–Tween. In the second step, plates were treated with PBST-BSA. AfB$_1$ standards ranging from 0.1 to 100 µg/ml were prepared in extracts (diluted to 10 per cent) from chillies not containing any aflatoxin. Chilli fruits (25g) free of aflatoxin, as determined by HPLC, were dried at 40° C for 2 days, powdered and extracted with 125ml of 70 per cent methanol containing 0.5 per cent KCl. The extract was filtered and diluted to 1 : 10 in PBST–BSA. This was used as diluents for preparing aflatoxin standards. Concentrations of the standards used were 100, 50, 25, 12.5, 6.25, 3.12, 1.56, 0.78, 0.39 and 0.097 µg/ml and each concentration were duplicated in two wells. Similarly, each test sample was duplicated in two wells. One hundred µl of each of test sample extract or standards was mixed with 50 µl of antiserum diluted to 1 : 10,000 in 0.2 per cent PBST-BSA. This step was followed by the addition of alkaline phosphatase labelled goat antirabbit IgG conjugate diluted to 1 : 2000 in PBST–BSA. The substrate was p-nitrophenyl phosphate prepared in 10 per cent diethanolamine. The plates were incubated at room temperature and then read in an ELISA reader. A maximum interval of 3 hours was allowed until optical densities from wells not containing any toxin reached 1.5–2.0 OD units at 405 nm. Log$_{10}$ values of concentration for aflatoxin standards were plotted on the X-axis and OD values were plotted on the Y-axis. Aflatoxin concentration in the sample extract was determined using the following formula (Reddy *et al.*, 2001):

$$\text{Aflatoxin Concentration } (\mu g/kg) = \frac{AfB_1\ (\mu g/ml) \times \text{ dilution} \times \text{Extract solvent (ml)}}{\text{Sample weight}}$$

Results and Discussion

The toxigenic nature of 16 isolates of *Aspergillus* spp. were detected and screened by using indirect competitive ELISA technique and the data pertaining to detection are given in the Table 2.2.

The maximum aflatoxinB$_1$ (AfB$_1$) production was recorded in AFL16 isolate (1652.2 AfB$_1$ µg/kg) indicating the most potent one followed by AFL8 isolate, producing 1081.9 AfB$_1$ µg/kg and AFL11 (864.3 AfB$_1$ µg/kg). On the contrary, AFL10 and AFL14 isolates produced lowest amount with 6.6 and 10.2 AfB$_1$ µg/kg respectively, which were in non-toxigenic range (Fig. 2.1). Whereas all other isolates produced AfB$_1$ above non-toxigenic range (above 30.00 AfB$_1$ µg/kg) indicating toxigenic nature.

The study of aflatoxin contamination caused by *A. niger* and *A. flavus* and its management, very essentially requires the detection and differentiation of toxigenic isolates from that of non-toxigenic ones and identification of their toxic nature should be based on the quantity of toxin production. Otherwise, a mere growth of these fungi on chilli fruits could be a misleading in the absence of such detection. The data from Table 2.2 and Fig. 2.1 indicates the availability of toxigenic and non-toxigenic strains in the same geographical region suggesting the existence of natural variation amongst the population of *Aspergillus* spp. This might, be attributed to several variability mechanisms such as asexual and parasexual cycle operating in *Aspergillus* spp. (Pontecorvo, 1956).

Table 2.2: Estimation of AflatoxinB_1 (AfB_1) from Sixteen Isolates of *A. niger* by Indirect Competitive ELISA Technique

Isolates	*Osmotic Density*	*Volume*	*Diluted (D)*	*Weight (G)*	*AfB_1 (g/kg)*
AFL1	1.105	100	10	20	57.9
AFL2	1.143	100	10	20	49.4
AFL3	0.9675	100	10	20	101.3
AFL4	0.4915	100	10	20	665.1
AFL5	1.0795	100	10	20	64.3
AFL6	1.212	100	10	20	36.9
AFL7	1.0865	100	10	20	62.4
AFL8	0.367	100	10	20	1081.9
AFL9	0.701	100	10	20	292.4
AFL10	1.5675	100	10	20	6.6
AFL11	0.4245	100	10	20	864.3
AFL12	0.749	100	10	20	241.9
AFL13	0.6965	100	10	20	297.6
AFL14	1.4895	100	10	20	10.2
AFL15	1.035	100	10	20	77.1
AFL16	0.2585	100	10	20	1652.2

A glance towards previous investigation supports the present findings. Reddy *et al.* (1988) employed potential ELISA technique to detect the mycotoxins produced by the fungus. Polyclonal antibodies and monoclonal antibodies were employed for detection of aflatoxin B_1 produced by *Aspergillus* spp. (Devi *et al.*, 1999 and Thirumala-Devi *et al.*, 2000). Reddy *et al.* (2001) used indirect competitive ELISA technique to detect the aflatoxinB_1 (AfB_1) in different grades of chilli in India. Direct antigen coating and protein A coating ELISA for detection of mycotoxins and plant viruses have been used (Hobbs *et al.*, 1987). Existence of aflatoxin fungal isolates with varied potency of toxigenic nature is a significant finding in the present investigation in the light of understanding the basics of genes responsible for such toxigenic or non-toxigenic strains. Further understanding of identification and isolation of gene responsible for toxin production is possible once such system of non-toxigenic and toxigenic is developed, and the present study has made stride in that direction. Overall, 88 per cent of *Aspergillus* isolates obtained from chilli fruits from the North-Eastern Karnataka region were proved to be toxigenic for AfB_1 aflatoxin. Awareness among the consumers about mycotoxins in chillies may induce producers as well as traders to market mycotoxin free chilli fruits.

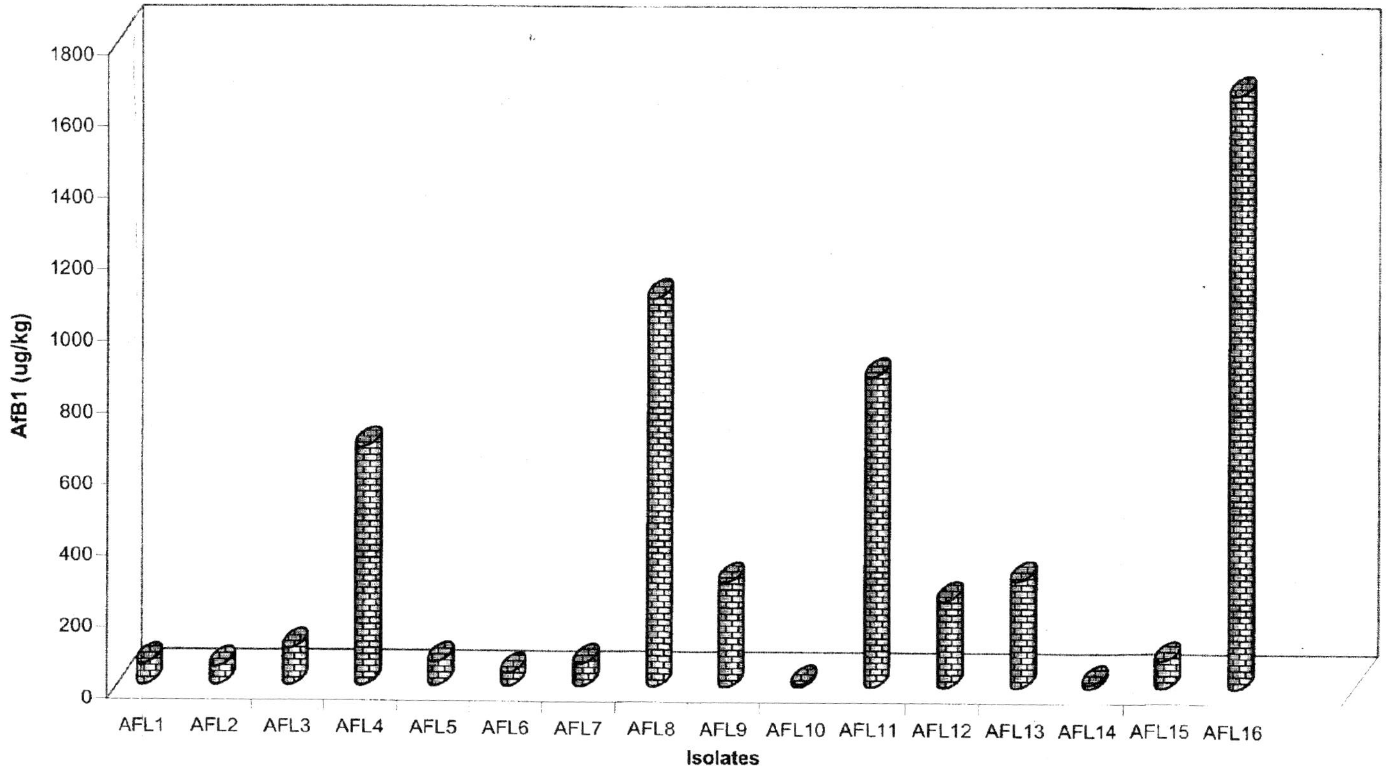

Fig. 2.1: Estimation of Afltoxin B$_1$(AfB$_1$) from Sixteen Isolates of *A. niger* by Indirect Competitive ELISA Technique

REFERENCES

Devi, K.T., Mayo, M.A., Reddy, K.L.N., Delfose, P., Reddy, G., Reddy, S. V. and Reddy, D.V.R. (1999). Production and characterization of monoclonal antibodies for aflatoxinB_1. *Letter in App. Microbiol.*, **29**: 284-289.

Hobbs, H.A., Reddy, D.V.R., Rajeshwari, R., and Reddy, A.S. (1987). Use of direct antigen coating and protein A coating ELISA procedures for detection of plant viruses. *Plant Dis.*, **71**: 747-749.

Jelinek, C.F., Pohland, A.E. and Wood, G. (1989). Worldwide occurrence of mycotoxins in food and feeds. *Journal of the Association of Official Analytical Chemists*, **72**: 223-230.

Naik, M.K. and Rawal, R.D. (2002). Disease resistance in horticultural crops, pp: 68-84. In: *Resources Management in Plant Protection*, Vol. 1 Eds. Sarath Babu *et al.*, Plant Protection Association of India, Hyderabad, p. 303.

Pontecorvo, G. (1956). The parasexual cycle in fungi. *Ann. Rev. Microbiol.*, **128**: 162-171.

Reddy, D.V.R., Nambiar, P.T.C., Rajeshwari, R., Mehan, V.K., Anjaiah, V. and McDonald, D. (1988). Potential of Enzyme-linked Immunosorbent Assay for detecting viruses, fungi, bacteria, MLO's, mycotoxins and harmones. *Biotechnology in Tropical Crop Improvement*, Proceedings of International Biotechnology Workshop, 12-15 January, 1987. International Crop Research Institute for the Semi Arid Tropics, Patancheru – 502 324, Andhra Pradesh, India, pp: 43-49.

Reddy, S.V., Kiran Mayi, D., Uma Reddy, M., Thirumala-devi, K.T. and Reddy, D.V.R. (2001). AflatoxinB_1 in different grades of chillies (*Capsicum annuum* L.) in India as determined by indirect competitive ELISA. *Food Additives and Contaminant*, **18(6)**: 553-558.

Thirumal-Devi, K.T., Mayo, M.A., Gopal Reddy, Reddy, S.V., Delfose, P., and Reddy, D.V.R. (2000). Production of polyclonal antibodies against Ochratoxin A and its detection by ELISA. *Journal of Agricultural and Food Chemistry*, **48**: 5079-5082.

Vasanthi, S. and Bhat, R.V. (1998). Mycotoxins in foods-occurrence, health and economic significance and food control measures. *Indian J. Med. Res.*, **108**: 212-224.

3

Studies on *in vitro* and *in vivo* Synthesis of Pectolytic and Cellulolytic Enzymes by the Leaf Spot and Fruit Rot Pathogen of Banana

M.M.V. BAIG[1] & D.S. MUKADAM[2]

[1] *Botany Research Laboratory & Plant Disease Clinic, Science College, Nanded, India*
[2] *Department of Botany, Dr. Babasaheb Ambedkar Marathwada University, Aurangabad, India*

ABSTRACT

Protopectinase-Macerating Enzyme (ME), Pectin methyl esterase (PME), endo-polygalacturonase (Endo PG), endo-poly methyl galacturonase (Endo-PMO), pectin trans eliminase (PTE), Cellulases-FPase, CMCase and -Glucosidase activities were investigated in the culture filtrates of *Dieghtoniella torulosum* (Syd) Ell. The enzymes were detected in the culture filtrate of the fungus grown on Pectin Peptone medium and leaf extract medium. They were also detected in extracts of naturally infected leaves and artificially inoculated Banana leaves. Only PME was detected in healthy (control) plants.

***Key Words**: Pectolytic enzymes, Cellulolytic enzymes, Dieghtaniella torulosum (Syd) Ell.*

Introduction

The Banana growing area of Nanded-Parbhani districts in Maharashtra, suffered a major setback in terms of yield due to blighting of leaves, early ripening and rotting of Banana fruits on the plants in field. Looking to the' extent of damage done to the crop and the involvement of *Dieghtaniella torulosum* (Sreeramulu, 1962; Hasija, 1962), a detailed study on production of cell wall degrading enzymes and toxins of *Dieghtaniella torulosum* was undertaken. In the preliminary studies carried out in this laboratory, the work indicated that toxins and enzymes produced by *D. torulosum* (Syd) Ell. (*Helminthosporium torulosum* (Syd) Ashby *Cercospora musarum* Zimm) played a role in pathogenesis.

The production of pectolytic and cellulolytic enzymes *in vitro* and *in vivo* by *Dieghtoniella torulosum* (Syd) Ell. and their possible role in the disease caused by the pathogen in Banana plants have been discussed in this study.

Materials and Methods

Host, Pathogen, and Inoculation

Pectin peptone Medium (Pectin: 10gm, Peptone: 2.5g, KH_2PO_4: 1g, $MgSO_4$: 0.5g, Distilled water: 1000ml, pH: 5.5) was used for the production of Pectolytic enzymes *in vitro* (Deshpande, 1967) whereas Carboxymethyl cellulose medium (Carboxymethyl cellulose: 10g, $CaNO_3$: 2.5g, KH_2PO_4: 1g, $MgSO_4$: 0.5g, Distilled water: 1000ml, pH: 5.5) was used for production of cellulolytic enzyme. Healthy leaves (100g) were extracted in 1000ml sterile distilled water. The extract was filtered through several layer of muslin cloth. To this extract 5g glucose was added and this medium was also used for enzyme production. The media were seeded with 7-mm disks cut from the edge of 5-days old colonies grown on potato dextrose agar (PDA) in petri dishes at room temperature at 27 ± 2°C.

For *in vivo* studies plants were grown for 30 to 60 days in controlled environment cabinets maintained at 27°C during a16-h light period with about 24000 lux (2200 ft-c) provided by cool white fluorescent tubes (Anonymous, 1993). They were inoculated by the conidial suspension of *D. torulosum* containing 1×10^5 conidia/ml and incubated in cabinets kept at 30°C, whereas other conditions were kept unchanged. All analyses of the enzymes were made in the leaves from inoculated and uninoculated plants.

Assays for Enzymes

Studies on enzymes production *in vitro* of *D. torulosum* was carried by distributing 50ml of medium in 250ml Erlenmeyer flask and the pH was adjusted before autoclaving to 4.5 with 0.1 N HCl or to pH 6 with 0.1 N NaOH. Cultures were incubated at 30°C in the dark for 8 to 15 days. After incubation the cultures were filtered through Whatman no 1 and the filtrate was centrifuged at 7800g for 20 min. The supernatant was dialyzed against 50 volumes of 0.01 M citrate or Tris-HCL buffer at 5°C from 15 to 22h (Hancock, 1966).

Cultures were also prepared on leaf extract mediums *in vitro* eeded with mycelial disks and incubated in the same way as the liquid cultures. After incubation the cultures were homogenized in a waring Blender in 100ml of 0.25 M NaCl for pectin methyl esterase (PME) and in 0.1 M citrate buffer for polygalacturonase (PO), FPase, CMCase and glucosidase assays. They were then filtered, centrifuged, and dialyzed in the same way as the liquid cultures.

Determinations of *in vivo* enzyme production were made after 30 to 50 days old Banana plant leaves 15 days after inoculation. The tissues were harvested and immediately frozen at –15°C. Uninoculated healthy tissues were prepared in same ways.

They were then homogenized in the appropriate buffers (Hancock 1966), filtered through four layers of muslin cloth, centrifuged, and dialyzed in the same way as the leaf extract medium and liquid cultures. Similarly determinations of enzymes were also made in naturally infected leaves following method prescribed by Bisen and Agarwal (1980). All determinations were made similarly on uninoculated plant (control), and on boiled or autoclaved culture filtrates.

Macerating Enzyme (Protonectinase)

Macerating activity was determined by the method suggested by Brown (1915) with slight modifications. Cylindrical plugs of 8 mm diameter were cut from healthy potato tubers with cork borer. The plugs were infused with distilled water under vacuum for 1 hr. Discs of 4 mm diameter were cut with sliding hand microtome in watch glass. 10 discs were blotted dry and added to enzyme solution in watch glass. The reaction mixture contained enzyme solution, 0.01 M phosphate buffer and water. At 5 min. interval each disc was subjected to needle test. The discs were pulled apart by mounting needles for loss of coherence till all the discs lose coherence. The mean time for loss of coherence in all discs was recorded and taken as reaction time (RT) in min. ME (Macerating enzyme activity) was determined as 1000/RT (Mukadam & Gangawane, 1982).

Endopoly Galacuturonase

Endo-PG activity was measured by a modification of the viscometric method of Bell *et al.* (1955). The substrate used was a 1.2 per cent solution of sodium polypectate in 0.05 M citrate buffer, pH 4.5. The relative enzyme activity (RA) was calculated by a formula described by Hancock *et al.* (1964).

Endopoly Methyl Galacturonase

Endo-PMG activity was measured by a modification of the viscometric method. The substrate used was a 1.0 per cent solution of Pectin in 0.05 M citrate buffer, pH 4.5. The relative enzyme activity (RA) was calculated.

Pectin Methyl Esterase

A modification of the continuous titration technique of Kertsez (1951, 1955) was used to measure PME activity. A 1.2 per cent solution of pectin in water or in 0.25 *M* NaCl, adjusted to pH 5.5 with 1 *N* NaOH, was used as substrate.

At 0 time the pH was determined immediately after addition of 10ml of dialyzed culture filtrate to 30ml of substrate. For *in vivo* tests, 6ml of dialyzed plant tissue extract was added to 20ml of substrate.

The mixture was incubated at 30°C. At intervals the mixture was titrated with 0.02 *N* NaOH to bring the pH back to that recorded at 0 time, and the resultant values (micro equivalents) were used to express enzyme activity (Hancock *et al.*, 1964).

Pectin Trans-Eliminase

Pectin trans-eliminase (PTE) activity was measured by the viscometric and TBA. The viscometric method was similar to that used for endo-PMG, except that filtrate or extracts were dialyzed at pH 8.5.in 0.01 M Tris-HCI buffer, then added to a 1.2 per cent solution of pectin in 0.05 M tris-HCI buffer at pH 8.5.In the TBA method (Albersheim *et al.*, 1960; Starr and Moran, 1962) 2ml of 0.2 per cent solution of Pectin in 0.05 M Tris HCI buffer pH 8.5, 2ml of dialyzed culture filtrates or tissue extract, and 2 drops of toluene were incubated in stoppered test tubes at 30°C for 0, 3, 12, 18, and 24 h(culture filtrates) and, 0, 12, 24, 36 and 48 h for *in vivo* extracts. After incubation, 17.5ml of TBA reagent was added; after reaction at boiling point, colour intensity was measured at wavelengths between 547 nm.

Cellulolytic Enzymes

Assay for Cellulases was done following method of Berghem and Petterson (1973) with slight modifications.

For FPase

100mg of Filter paper dust was suspended in 1ml of 0.01 M Sodium acetate buffer of pH 4.8 was incubated with 2ml of crude enzyme solution.

For CMCase

2ml of crude enzyme solution was mixed with 4ml of CMC and 1ml of the 0.01 M Sodium acetate buffer of pH 4.8 and were incubated at 27 ± 2°C. Aliquots were drawn from the mixture at regular time interval and the release of glucose due to the enzyme activity was assayed by 3,5 Dinitrosalicylic acid method (Miller, 1959) using D glucose as standard.

For Glucosidase

Activity of glucosidase was assayed by the method of Eberhart *et al.* (1963) using p-nitro glucoside as substrate. The reaction mixture consist of 50mg of p-nitro glucoside in 2ml of 0.01 M acetate buffer at 4.8 pH and 1ml crude enzyme preparation incubated at 27 ± 2°C. At regular time intervals aliquot were drawn and added with 0.1 *N* NaOH and the release of p-nitrophenol was from substrate was estimated for absorbance at 420 nm in a spectrophotometer.

Soluble reducing sugars (equivalent to glucose) released from filter paper was estimated and the Cellulase activity was expressed in enzyme units. Enzyme activity is expressed in Units (U). 1 enzyme unit is equivalent to amount of enzyme required to release 1µ mole of D glucose (for C_1 and C_x) or p-nitro phenol (For glucosidase) per minute from respective substrate.

Results and Discussion

Macerating Enzyme (Protopectinase)

ME was synthesized by *D. torulosum* and activity was recorded from the culture filtrates of PP medium, leaf extract medium, naturally infected as well in inoculated Banana leaves where as it was not recorded from the healthy uninoculated leaves extract. Maximum activity was recorded in the filtrate of leaf extract medium followed by inoculated leaves (Table 3.1).

Table 3.1: Pectolytic Enzyme Produced by *Dieghtoniella torulosum*

Enzymes	*In vitro*		*In vivo*		*Healthy Plant Leaves*
	PP Medium	*Leaf Extract Medium*	*Artificially Inoculated Leaves*	*Naturally Inoculated Leaves*	
ME	40.0	44.5	80	86	ND
Endo PG	56	64	12	10	ND
Endo PMG	48	54	10	12	ND
PME	0.024	0.065	0.084	0.072	0.011
PTE	10	18	15	08	06

Endo Polygalacturonase

Endo-PG activity of filtrate from PP medium was 56 RA/ml, from leaf extract medium filtrate was 64 RA/ml, from inoculated banana leaf extracts medium filtrate was 12 RA/ml and from extract of naturally infected leaves was 10 RA/ml (Table 3.1). Endo-PG was not detected in healthy tissues. Because it is very effective in breaking down the pectic acid constituent of the middle lamellae and primary cell walls, endo-PG is considered to have an important function in the pathogenesis of *D. torulosa,* particularly in the early stages of infection.

The cleavage of polygalacturonate by endo-PG (Demain and Phaff, 1957; Jansen and MacDonnell, 1945) would result in the accumulation of oligogalacturonides and subsequently into galacturonic acid which is utilized by the pathogen. Endo poly methyl galacturonase Endo-PMG activity was detected in culture filtrate of PP medium, leaf extract medium and extracts from naturally infected leaves and from inoculated leaves as well (Table 3.1).

Pectin Methyl Esterase

PME activity was detected in filtrates of PP medium and Leaf extract medium of *D. torulosa,* and also in extracts of healthy and inoculated Banana leaves (Table 3.1). Activity was higher in the extracts from diseased than from healthy tissues. *D. torulosum* produced PME in culture, but much of the activity in extracts from diseased plants may have been of host origin. PME has been reported to enhance the activity of PME (Jansen *et al.,* 1945), the part played by PME in the pathogenesis of *D. torulosum* difficult to determine.

Pectin Trans-Eliminase

Pectin Trans-Eliminase activity determined by the viscosimetric method was 10 RA/ml for cultures filtrate of PP medium where as it was 18 RA/ml in filtrate of leaf extract medium. 15 RA/ml was detected in the extracts from inoculated leaves and 8 RA/ml in extracts of naturally infected leaves. Mixtures prepared by the TBA method showed maximum light absorption between 545 and 550 mμ.

PGTE activity was detected in infected leaves and in culture filtrates but found PGTE activity occasionally in host tissues infected by *Sclerotinia sclerotiorum,* but not in culture filtrates (Hancock, 1966).

Cellulases

D. torulosum synthesizes C_1, C_x and -Glucosidase, and the activity was observed in culture filtrate of CMCN medium, leaf extract medium, inoculated as well naturally infected leaves. However no cellulolytic activity was detected in health uninoculated leaves (Table 3.2). Enzyme activity in culture filtrate of leaf extract medium was higher than culture filtrate of CMC medium.

Table 32: Cellulolytic Enzyme Produced by *Dieghtoniella torulosum*

Enzymes	*In vitro*		*In vivo*		*Healthy Plant Leaves*
	CMC Medium	*Leaf Extract Medium*	*Artificially Inoculated Leaves*	*Naturally Inoculated Leaves*	
FPase	0.04	0.06	0.01	0.01	ND
CMCase	0.06	0.07	0.03	0.02	ND
glucosidase	0.06	0.08	0.02	0.02	ND

All the enzymes except PME were found to be produced on PP medium, CMC medium or leaf extract medium and in diseased host tissue of the pathogen. It is clear that each of them plays a role in the development of leaf blight in banana caused by *D. torulosa,* from the early stages of infection.

Acknowledgements

We are grateful to Dr. Mrs. Usha Deshpande, Nanded for her constant encouragement and to Prof. L.V.Gangawane, Department of Botany, Dr. Babasaheb Ambedkar Marathwada University, Aurangabad for critical advice on analytical procedures.

REFERENCES

Albershiem, P., H. Neukom and H. Deuel (1960). Uber die Bilidung von ungesattgiten Abbauprodukten durch ein pektinabbauendes Enzym. *Helv. Chim. Acta.,* **43**: 1422-1426.

Anonymous (1993). Tissue Culture propagation of Banana. Tech. Report. No. **11**. IIHR, Bangalore.

Baig. M.M.V. (1997). Role of cell wall degrading enzymes in induction of resistance in plants. Ph.D. Thesis. Swami Ramanand Teerth Marathwada University, Nanded.

Bergham L.E.R. and L.G. Petterson (1973). The mechanism of enzymatic cellulose degradation, purification of cellulolytic enzymes from *Trichoderma viride* active on highly ordered cellulose. *Eur. J. Biochem.,* **37**: 38-40.

Bisen P.S. and G.P. Agarwal (1980). *In vitro* production of pectolytic enzymes by *Aspergillus niger* van Tieghem causing soft rot in apples. *Phytopathology Z.,* **97**: 317-326.

Brown. W. (1915). Studies in the physiology of parasitism I: the action of *Botrytis cinerea. Ann. Botany* (London), **29**: 313-348.

Demain, A.L. and H.J. Phaff (1957). Recent advances in the enzymes hydrolysis of the pectic substance. *Wallerstein Lab. Common,* **20**: 119-139.

Deshpande, K.S. (1967). Physiology of Helminthosporium species from Marathwada. Ph.D. Thesis, Marathwada University, Aurangabad.

Eberhart, B., D.F. Cross and L.R. Chase (1963). glucosidase system of *Neurospora crassa* 1. glucosidase and Cellulase activities of mutant and wild type strains. *J. Bacteriol.,* **87**: 761-770.

Hancock J.G., R.L. Millar and J.W. Lowbeer (1964). Pectolytic and Cellulolytic enzymes produced by *Botrytis cinerea, B. allii,* and *B. squamosa in vitro* and *in vivo. Phytopathology,* **54**: 928-931.

Hancock, J.G. (1966). Degradation of pectic substance associated with pathogenesis by *Sclerotinia sclerotiorum* in sunflower and tomato stems. *Phytopathology,* **56**: 975-979.

Hasija, S.K. (1962). Addition to fungi of Jabalpur (M.P.) VI *Mycopath. et Mycol. Appl.,* **23**: 33-41.

Jansen, E.F., and L.R. MacDonnell (1945). Influence of methoxyl content of pectic substance on the action of polygalacturonase. *Arch. Biochem.,* **8**: 97-112.

Jansen, E.F., L.R. MacDonnell, and R. Jang. (1945). Simultaneous actions of polygalacturonase and pectinesterase on pectin. *Arc. Biochem.,* **8**: 113-118.

Kertesz, Z.I. (1951). *The pectin substances*. Interscience publishers, Inc., New York.

Kertesz, Z.I. (1955). Pectin enzymes. In: *Methods in enzymology*. Vol. I. Edited by S.P. Colowick and N.O. Kaplan. Academic Press, Inc., New York, pp. 158-166.

Miller, G.L. (1959). Use of Dinitrosalicylic acid reagent for determination of reducing sugars. *Anal Chem.*, 426-429.

Mukadam, D.S. and L.V. Gangawane (1982). *Methods in Experimental Plant Pathology*. Marathawada University Press, Aurangabad.

Naaz, R. (1987). Studies on cellulolytic enzymes of *Alternaria carthami.* Chow. Ph.D. Thesis, Marathwada University, Aurangabad.

Sreeramulu T. (1962). Some observation on the *Dieghtoniella* fruit and Leaf spot disease of Banana. *Curr. Sci.*, **31**: 258-259.

Starr, M.P. and F. Moran (1962). Eliminative split of pectic substances by phytopathogenic soft- rot bacteria. *Science*, **135**: 920-921.

4

Strategies for the Management of Groundnut Diseases

M.P. GHEWANDE & VINOD KUMAR

National Research Centre for Groundnut, Ivnagar Road, P.O. Box No. 5, Junagadh, Gujarat – 362 001, India

ABSTRACT

Diseases are the limiting factors for the successful production of groundnut crop in India. Control of groundnut diseases has been mostly through the use of synthetic fungicides. This has created a lot of environmental hazards and economic problems besides the development of resistance by some pathogens to fungicides. Therefore, hunt for nature-based products as alternative and researches aimed at finding products that mirror the natural plant mechanisms, narrowing the differences between chemical and biological products is highly desirable. This chapter describes the integrated management practices for the major diseases of groundnut and aims at bringing a review of the subject including the basic scientific data. It covers the disease management approaches like host plant resistance, cultural methods, use of botanicals, biopesticides and biotechnological approaches like tissue culture and genetic engineering including molecular strategies. The chemical method, though indispensable is least emphasized. The information on integrated disease management approach is also provided.

Introduction

Groundnut is one of the five most important oilseed crops of the world and is grown in about 26 million ha with a production of 34.5 million tonnes of nuts in shell. In India, about 7.6 m ha area is under groundnut, with a production of 7.8 m tonnes. Although India is the largest grower and second largest producer after China, the average yield is low around 1000 kg/ha, as against worlds average yields of 1304 kg/ha or on farm yields obtained in developed countries (2500 kg/ha). This may be attributed to the rainfed nature of cultivation of this crop coupled with attack by a variety of diseases and insect pests. The role of groundnut diseases in reducing yield has been clearly demonstrated (Ghuge *et al.*, 1981; Ghewande *et al.*, 1983; Subramaniyam *et al.*, 1985). More than 55 pathogens including viruses have been reported to affect groundnut. Diseases render the crop uneconomical in the rainy season, which is the major period of groundnut cultivation.

Major Diseases of Groundnut

Foliar Fungal Diseases

The major foliar fungal diseases, which are economically important, are early leaf spot, late leaf spot and rust. Early leaf spot is caused by *Cercospora arachidicola* Hori (Perfect stage- *Mycosphaerella arachidis* Deigton) and late leaf spot is caused by *Phaeoisariopsis personata* Berke & Curt (Perfect stage- *Mycosphaerella berkeleyi* Jenkins) both commonly called 'tikka disease'. These diseases occur wherever the groundnut crop is grown. However, the incidence and severity of each disease varies between localities and seasons and there can be both short and long term fluctuations in their relative proportions (Mc Donald *et al.*, 1985). The magnitude of yield losses caused by these diseases is very high and ranged from 10 to 70 per cent all over the world, but vary considerably from place to place and between seasons (Ghewande, 1983; 1985; 1990a; Subramaniyam and Mc Donald, 1983). Lesions caused by *C. arachidicola* are sub-circular and from 1 to over 10 mm in diameter. They are dark brown on the upper surface of leaflet where most sporulation occurs and a lighter shade of brown on the lower surface of leaflet. Lesions caused by *P. personata* are usually smaller, almost circular and darker in color than those of *C. arachidicola*. On the lower surface, where most sporulation occurs, the lesions are black with a slight rough appearance. The color of the lesion on the lower surface of leaflet is light brown for early leaf spot and black for late leaf spot. The sporulation occurs on the upper surface of leaflet in case of early leaf spot while in late leaf spot it occurs on lower surface. On severity of infection, leaflets first become chlorotic then necrotic, lesions often coalesce, and leaflets are shed. Temperatures in the 25°C to 30°C range and high relative humidity favour infection and disease development. The pathogen survives from season to season on volunteer groundnut plants and infected crop debris. Long distance distribution of the pathogens may be by airborne conidia, by movement of the infected crops debris or by movement of pods or seeds that are surface contaminated with conidia. Early and late leaf spot pathogens are also soil borne (Mc Donald *et al.*, 1985) Rust is caused by *Puccinia arachidis* Speg. It is readily recognized as orange colored pustules (uredinia) that appear on the lower surface of leaflets and rupture to expose masses of reddish brown urediniospores. The pustules are usually circular and range from 0.5 to 1.4 mm in diameter. Infected leaves become necrotic and dry up by end to remain attached to the plant. Groundnut rust is known to perpetuate, spread and produce severe disease out-breaks by means of its urediniospores. Groundnut crop or volunteers are available in one or the other parts of the country enabling the survival of uredial stage round the year. Long distance dissemination for rust may occur via airborne urediniospores, by movement of infected crop debris, or by the movement of pods or seed surface contaminated with uredinospores. An optimum temperature of 20°C, the presence of liquid water on the leaf surface, and high relative humidity, favour infection and subsequently disease development. Spread of the disease within growing crops is facilitated by wind movement, by rain splash, and also by insects (Subramaniyam and Mc Donald, 1987).

Seed and Soil Borne Fungal Diseases

Seed and soil borne diseases may develop from fungi already established in the seeds before sowing or may result from direct invasions of seeds or seedling by soil fungi. Of seed and soil borne diseases, collar rot/seedling blight (*Aspergillus niger* van Tieghem), stem rot/Sclerotium wilt (*Sclerotium rolfsii* Sacc) dry wilt or dry root rot (*Macrophomina phaseolina* (Tassi) Goid = *Rhizoctonia bataticola* (Taub., Butler) have been recognized to be the major diseases. These diseases cause severe seedling mortality resulting in patchy crop stand mostly in sandy loam soil and reduce the yields from 25-50 per cent (Chohan, 1974; Ghewande, 1985). In India, stem rot occurs in all groundnut growing states, particularly more severe in Gujarat, Maharashtra, Madhya Pradesh, Orissa and Tamil Nadu, where approximately over 50, 000 ha of groundnut fields are infected with *S. rolfsii*. About 27 per cent or more yield loss due to this disease has been reported from India (Chohan 1974). Mayee and Datar (1988) have reported yield losses of over 25 per cent in Maharashtra. Collar rot is prevalent in almost all groundnut growing areas of the country and reported to cause 40 per cent yield loss (Chohan, 1973). It is particularly serious in the sandy loam and medium black soil of the Punjab, Tamil Nadu, Uttar Pradesh, Rajasthan and Haryana. Since the disease causes considerable seedling mortality in early stages of crop growth, crop yield is directly affected by reduction in the stand of the crop. Dry root rot and stem rot are sporadic in occurrence. Dry rot is particularly serious in Rajasthan, U.P., Tamil Nadu, A.P. and Maharashtra. Stem rot is becoming more prevalent in T.N., A.P., Karnataka, M.P. and Gujarat. Collar rot disease is spread by the spores and mycelium adhering to the seedpods from season to season. In addition, collar rot pathogen is saprophytic, common soil dwellers in soil humus and plant debris and capable of initial infection. The main carryover from season to season is in plant debris rotting in the soil and on the seeds harvested from infected fields. *A. niger* can tolerate low soil-moisture and develops best at temperature between 30 to 35^{0}C. Soil moisture to the extent of 40-45 per cent of water holding capacity and temperature between 29-32^{0}C has been found optimum for the development of stem rot. Dry root rot has wide host range. The pathogen *M. phaseolina* is facultative saprophyte and soil dweller. Infected soil, plant debris and pods serve as sources of inoculum. The optimum temperatures for seedling infection and pods invasion are 29 to 35^{0}C and 26 to 32^{0}C. The dissemination of the fungus is by sclerotia via plant debris and soil and also infected pods, shell and kernel.

Virus Diseases

The economically important virus diseases of groundnut are peanut bud necrosis, peanut mottle, Indian peanut clump and peanut stem necrosis disease. Peanut bud necrosis disease caused by peanut bud necrosis tospovirus, is wide spread with a wide host range and is transmitted by *Thrips palme* (Kendre *et al.*, 2000). The incidence of the disease was observed to be as high as 92 per cent on Spanish bunch varieties of groundnut during 1985 (Summer) in Jamnager district of Saurashtra region of Gujarat (Ghewande, 1987). It has been reported to cause yield loss of up to 50 per cent in early infected crop of groundnut in India (Ghanekar, 1980). The typical symptoms include chlorotic rings, terminal bud necrosis, severe stunting and proliferation of axillary shoots. Necrosis of the bud may spread to the petiole and stem, sometimes leading to death of the plants. In late infections pod size is reduced. Seeds are often shrivelled and have mottled and discolored testa (Reddy *et al.*, 1991). Peanut mottle caused by peanut mottle (PMV) potyvirus is reported to occur mostly on *rabi*/summer groundnut mainly in the states of Andhra Pradesh, Maharashtra and Gujarat. The disease can cause up to 30 per cent loss in yield. (Kuhn and Demski, 1975). Newly formed leaves show mild mottling and vein clearing, whereas older leaves show upward curling and interveinal depression with dark green islands. Infected plants are not severely stunted and older plants seldom show typical disease symptoms. Some pods from plants infected with PMV may be smaller than normal and have irregular, green to brown patches. Seeds from such pods are discoloured. Yield losses of 15-45 per cent have been recorded in India (Reddy *et al.*, 1978). The virus is transmitted through the seed to the extent of 8.5 per cent. (Adams and Kuhn, 1977). Indian peanut clump caused by peanut clump furovirus has been reported from Punjab, Gujarat, Andhra Pradesh, Rajasthan and Uttar Pradesh. Yield losses up to 60 per cent have been reported from fields infected with peanut clump virus. (Ghanekar, 1980). Symptoms are severe stunting, mosaic mottling and peeling of root epidermis. The virus has a wide host range and several weeds commonly growing in groundnut fields are also infected by the virus. Indian peanut clump virus isolates are transmitted through seed (Nolt and Reddy, 1984). A soil-borne fungus (*Polymyxa graminis*) also transmits the virus (Ratna *et al.*, 1991). Recently a new virus disease, peanut stem necrosis has been reported from Andhra Pradesh (Reddy *et al.*, 2002). This disease caused nearly 300 crores rupees worth of crop losses to groundnut in Andhra Pradesh during *kharif* 2002. It is caused by an isolate of tobacco streak virus (TSV). The affected plants show necrotic lesions on young quadrifoliates, which coalesce to cover entire leaflets followed by necrosis of entire stem. The entire young plant is often necrosed. The virus survives on weeds such as *Parthenium* and also on hosts like cowpea, black gram, marigold and sunflower. Transmission occurs via thrips.

Nematode Diseases

Although more than 70 species of plant parasitic nematodes have been reported in association with groundnut disease (Sharma and Mc Donald, 1990) only a few species are known to cause economically important disease (Subrahmaniyam, *et al.*, 1990). Pod and root knot disease caused by four species of *Meloidogyne viz. M. arenaria* (Neal) Chitwood, *M. hapla* Chitwood, *M. incognita* Kofoid & White and *M. javanica* (Treub) Chitwood is the most important disease of groundnut. *M. arenaria* is the most wide spread. Recently nematode (*Tylenchorynchus brevilineatus*) induced disease locally known as "Kalahasti malady" was observed in Chittoor and Nellore district of Andhra Pradesh, India. The symptoms appear as small brownish yellow lesions on the pegs, pod stalks and on young developing pods. Pod stalks are much reduced in length and in advance stage of the disease the entire pod surface become discolored. Affected plants are stunted and greener than normal foliage.

Components of Integrated Disease Management

The changing production system scenario demands for cost effective, easily adaptable and eco-friendly tools for the efficient management of the major diseases. Diseases management requires judicious adoption of the several management tools. The important among them are discussed below.

Host Plant Resistance

Use of disease resistant cultivars is one of the best means of reducing crop losses. Sources of disease resistance have been identified both in the primary gene pool and in the wild relatives of *A. hypogaea*. Hundreds of land races, advanced breeding lines, cultivars and wild species have been screened for resistance against foliar fungal diseases by several workers and a large number of resistant sources have been identified (Chahal and Sandhu, 1972; Chiteka *et al.*, 1988; Ghewande *et al.*, 1992; Mc Donald *et al.*, 1985; Moraes and Godoy, 1985; Subramaniyam and Mc Donald, 1987; Subramaniyam *et al.*, 1989). Source of multiple disease resistance especially for late leaf spot and rust are available from ICRISAT and NRCG. These are EC76446 (292), NCAc17133 (RF), PI 259747, PI 350680, PI 389595, USA 63, CS-19 and NCAc 343. Recently 5 multiple resistant germplasm accessions namely NCAc17149, NCAc927, NCAc17133 (RF), PI 393646 and PI 341879 have been identified (Ghewande *et al.*, 1992) which were found to be resistant to early leaf spot, rust and *Alternaria* leaf spot. Genetic resistance to foliar diseases caused by *P. personata* and *C. arachidicola* has usually been associated with low yields and late maturity (Pixley *et al*., 1990). In field trials the genotypes, ICGV 86252 and JL 24-3 were found resistant to leaf spot (Reddy *et al*., 1997). Bera *et al*. (1999) found that leaf spot resistant genotypes were capable of maintaining a higher level of chlorophyll, soluble sugars, acid proteins, RNA and total phenol content in their leaves. National Agricultural Research Systems and ICRISAT laid a lot of emphasis on development of disease resistant varieties. Efforts have been made to breed foliar disease resistant genotypes using conventional breeding methods as well as interspecific hybridization. In general, it is very difficult to get promising sources of resistance to many of the seed and soil borne pathogens such as *Sclerotium rolfsii*, *Macrophomina phaseolina* and *Aspergillus niger* as they are not very specialized pathogens and they have a very wide spectrum of host attacking mechanism. However, a few genotypes have been identified as tolerant to these pathogens. Recently an advanced breeding line, CS-19 at NRCG had been found resistant to stem rot and incorporation of resistance from this to agronomically suited variety is in progress (Ghewande *et al*., 2003). At ICRISAT, ICGV 86029, ICGV 86030, ICGV 86031, ICGV 86032, ICGV 86033 and ICGV 86558 were identified to possess field tolerance to peanut bud necrosis (PBNV) and also good agronomic traits (Reddy *et al*., 1991). Several aspects of PBNV including resistant entries have been examined by Buiel *et al*. (1993). *Arachis chacoense* and *A. pussilla* were found to be resistant to peanut mottle virus (Reddy, 1988). Two genotypes *viz*. EC-76446 (292) and NCAc 17133 RF have not shown any seed transmission of peanut mottle virus (Ghanekar, 1980) and PI 261945 and PI 261946 were tolerant to PMV (Kunj *et al*., 1978). Four genotypes JSP 1, ICGMS 2, NCA X and CGC 4 have been reported to be resistant to root knot disease caused by *M. arenaria* (Grewal *et al*., 1986). The released varieties resistant/tolerant to major diseases are listed below which continues to be used in endemic areas.

Diseases	*Resistant/Tolerant Varieties*
Early leaf spot, late leaf spot, rust	ALR 1, ALR 2, ALR 3, Girnar 1, ICGV 86590, ICGV 87160, ICGV 86325, CSMG 84-1, OG-52-1, RSHY 1, DRG 12, DRG 17, TAG 24, BSR 1, VRI 5, CO 4
Collar rot and aflaroot	J 11, JCG 88 and OG-52-1
Stem rot	OG-52-1, Dh-8, and ICGV 86590
Peanut bud necrosis disease (PBND)	ICGS 11, ICGS 44, ICGS 37, Kadiri 3, ICGV 86325, K 134, DRG 12, *R* 8808, JCG 88, CSMG 884, Chandra

Source: Ghewande *et al*., 2002.

Cultural Practices

Several cultural practices like removal of volunteer plants, removal or burial of infected crop debris, crop rotation, organic amendments of soil, intercropping etc. are important for the management of diseases. Removal of volunteer groundnut plants, removal or burial of infected crop debris, crop rotation and early sowing of the crop can reduce intensity of leaf spot and rust. Altering plant population could also reduce foliar disease severity (Ghewande *et al*., 1983, Kodmelwar and Ingle, 1989). Intercropping groundnut with pigeon pea, cowpea, black gram, pearl millet, sorghum and other crops have been reported to reduce the intensity of foliar diseases. Deep ploughing of fields and rotation of groundnut with gram and wheat are useful in reducing the collar rot disease incidence (Sathiyanarayanmurthy *et al*., 1988). Early sowing of groundnut in June reduces collar rot incidence. The crop planted with a spacing of 30 cm between the rows had lowest root rot incidence than when planted with a row spacing of 45 cm or 60 cm. Closer spacing reduced incidence of collar rot and stem rot in early sown crop. Cultural practices such as deep covering or burial of organic matter before planting, non-durting cultivation by avoiding movement of soil up around the base of plant, prevention of development of organic debris are useful for reducing the incidence of stem rot. Cotton is a suitable rotation crop for management of *S. rolfsii* (Rodriguez and Morgan, 1987). Other rotation crops like maize, sorghum, onion and garlic are also useful for the management of stem rot. Plant density can be increased to decrease the proportion of infected plants of bud necrosis (Reddy *et al.*, 1983). Groundnut intercropped with pearl millet and early sowing showed less incidence of bud necrosis disease as compared with sole crop at ICRISAT (Ghanekar, 1980). Cultivation of groundnut away from soybean, cowpea, navy beans could reduce peanut mottle virus considerably as in these crops the virus is seed transmitted and serve as foci of inoculum (Demski and Khun, 1983). Groundnut crop grown during *rabi* escapes the clump disease (Reddy, 1988). Soil borne pathogens can be effectively managed through deep ploughing during summer months, increased use of farmyard manure, deoiled cakes, other organic amendments and soil solarization. Application of neem cake and farmyard manure to soil gave good control of collar rot (Karthikeyan, 1996). In pot experiment the best reduction of pre- and post- emergence death of seedlings infected by *S. rolfsii* was given by amendments of safflower oil cake and sun hemp which was more effective than seed dusting with captan (Kulkarni *et al.*, 1995). Rago *et al.* (1997) observed that there was no sclerotial production on lucerne residues. Germination of sclerotia produced on rye, love grass, sorghum, soybean, groundnut, oat and maize was >70 per cent, 45 per cent on wheat and 6 per cent on sunflower. Kanzaria and Patel (1994) observed maximum mycelial growth and oxalic acid production at 30^0 C. Wheat straw application reduced *Sclerotina* blight in 1992 and 1993 but not in 1994 compared with unamended plots (Ferguson, 2001). The crops like sesamum, castor and cotton are good crops to include in rotations for reducing the incidence of *M. arenaria* (Rodriguez and Morgan, 1987; Rodriguez *et al.*, 1991). Poultry manure and neem cake amendments give good control of Kalahasti malady (*Tylenchorhynchus brevilineatus*) (Naidu *et al.*, 2000). Nematode problems in groundnut can be effectively managed through deep ploughing during summer months, increase use of F.Y.M. and other organic amendments and soil solarization. For the management of foliar diseases especially leaf spots and rust, deep burying of crop residues, destruction of crop debris by burning, removal of affected groundnut plants, early planting and wider row spacing (40-45 cm) and intercropping of pearl millet, sorghum and pigeon pea have been advocated (Ghewande *et al.*, 2002). Early sown crops suffered least from leaf spot and rust due to low inoculum potential whereas late sown crops suffered more because of ready availability of inoculum built in early sown crops (Naidu and Chandrika, 1997). Further they also suggested that late *Kharif* sowing might be utilized for screening groundnut germplasm for disease resistance due to the existence of maximum inoculum load on the late sown crop. Similarly, Ashtaputre *et al.* (1994) found that early sowing preferably in June helped in the reduction of late leaf spot incidence across genotypes when staggered planting at four dates of sowing at fortnightly interval commencing from June 13 when Dh-3-30, JL-24, and TMV-2 varieties were used.

Biological Control

Biological control agents are microbes with proven antagonistic activity against other microbes. A critical analysis of the literature on biological control reveals a skewed research towards soil-borne pathogens, as the response has been more positive in this area as compared to foliar pathogens. The most often used biocontrol agents are either fungi or bacteria. The biocontrol agents have to be produced in bulk and applied with the onset of the disease to get the desired level of control. Mass production of biocontrol agents using low cost technology is the basic requirement for successful exploitation of biocontrol agents.

Among the fungi, *Trichoderma* has been extensively used to manage a variety of plant pathogens. The genus *Trichoderma* has nine species aggregates with a lot of strains. The mode of biocontrol is by mycoparasitism, competition, production of hydrolases, and antimicrobial chemicals. Whereas some strains possess more than one mode of action and effective against many pathogens, some strains are specific to some pathogens. Several strains of *Trichoderma* have been commercially exploited and are available for use by farmers. *T. harzianum* reduced seedling mortality due to collar rot and soil drenching was more effective than seed treatment (Kulkarni *et al.*, 1994). *T. harzianum* and *T. viride* applied as seed dressing, effectively reduced mortality of groundnut seedlings due to stem rot and collar rot (Nagaraju and Urs, 1998). Seed rot and collar rot of groundnuts were reduced in pot cultures when *Trichoderma* isolates were incorporated into the soil (Prabhu and Urs, 1998). The spores of *Trichoderma* remained viable up to nine weeks of storage at room temperature on seeds when the seeds were coated and reduced the stem rot incidence (Biswas *et al.*, 2000). Biswas *et al.* (2000) reported significant reduction in stem rot incidence by two strains of *Trichoderma* either as seed dressing (33-50 per cent) or soil application in pot trials (72-83 per cent). Their studies on formulation of the strains showed that spores of T8 and T10 isolates showed better longevity after 13 weeks (1.3×10^3 cfu/seed) and upto 15 weeks (1.0×10^4 cfu/seed) respectively, when the seeds coated with spores were dipped in 2 per cent carboxy methyl cellulose and pelleted with bentonite at 25 g/kg seed. Bacterial isolates *viz.* BCB-135, AF-52, *Pseudomonas fluorescens*-2 and *Bacillus* isolates were effective in reducing the population of *A. flavus*. *Bacillus* spp. could reduce the population by 53.08 per cent followed by *P. fluorescens* (47.94 per cent), AF-52 (40.60 per cent) and BCB-135 (36.56 per cent). These isolates also showed plant growth promotion activity, (Desai *et al.*, 2002).

Rust was significantly reduced by *T. harzianum* on detached groundnut leaves. Pre-treatment with *T. harzianum* conidia resulted in a greater inhibition of germination percentage and germ tube growth of *P. arachidis* than mixed/post-treatment. On detached groundnut leaves, both pre- and mixed treatments with the antagonist showed significant reduction in the number of rust uredosori/leaflet and number of uredospores/pustules as compared to post-treatment. *T. harzianum* colonized better on uredosori (16 day old) which were post treated with the antagonist than on pustules developed on pre-treated and simultaneously treated leaves. A phenol-like folin positive antifungal compound was separated from the 36 hr old germinating *T. harzianum* (Govindasamy and Balasubramanian, 1989). The mycoparasites like *Acremonium persicinum*, *Eudarluca caricis*, *Penicillium islandicum*, *Tuberculina costaricana* and *Verticillium lecanii* and their culture filtrates inhibited urediniospore germination and reduced rust development by varying degrees. Sprays of culture filtrates of *V. lecanii* and *P. islandicum* significantly reduced rust disease under field conditions (Ghewande, 1990b). By monitoring the fermentation conditions, biocontrol agents against late leaf spot and rust such as *P. islandicum* and *V. lecanii* were mass multiplied and formulated using cheap and simple substrates (Desai and Bagwan, 2002). Another antagonist *Fusarium chlamylosporum* has been found to rapidly colonize the rust pustules. Mathivanan and Murugesan (2000) found that uredospore greatly lost its ability to germinate and also result into bursting and degradation of cell walls of uredospores. A significant reduction (20-38 per cent) was recorded due to spraying of culture filtrates along with uredospores.

Extra cellular chitinolytic enzymes of microorganisms have a potential to suppress the activities of the pathogens by degrading the chitin in their cell walls and thus protect the plants from disease. A 40 kDa extra cellular chitinase was detected from *M. verrucaria*, an antagonist of *P. arachidis*, and the purified chitinase inhibited uredospore germination (Govindsamy *et al.*, 1998). Krishna Kishore *et al.* (2003) found that *Serratia marcescens* GPS-5 and *Bacillus circulans* GRS-242 had potential in terms of chitinolysis and inhibition of *in vitro* conidial germination of *P. personata* when these strains were used as foliar spray in controlled environment, reduction in late leaf spot was non significant but supplementation of foliar spray with 1 per cent colloidal chitin resulted in effective control. Mycoparasites, *Dicyma pulvinata* and *V. lecanii*, and their culture filtrate inhibited the *in vitro* spore germination (33-75 per cent) and reduced *in vivo* development of *P. personata* (Ghewande, 1989b). Among bacterial antagonists, species of *Pseudomonas* are known to be highly potential. Fluorescent pseudomonads have been reported to be antagonistic against many soil-borne and foliar pathogens. Podile and Prakash (1996) have studied in detail the mode of action of *B. subtilis* AF1 strain, effective against *A. niger*. The bacterial cells adhered to the fungal mycelium, multiplied *in situ*, colonized and lysed mycelial surface. Groundnut seeds bacterized with *B. subtilis* showed a reduced incidence of crown rot in *A. niger* infested soil. Patil *et al.* (1998) reported that two strains of *Pseudomonas fluorescens viz.* FPD- 10 and FPD- 15 inhibited sclerotial germination of *S. rolfsii in vitro* as well as in soil. The strain FPD- 15 was more 'ecologically fit' with greater multiplication and survival in the soil. Among six carriers tried for the soil application of a native isolate of *Pseudomonas fluorescens* for the management of *A. niger*, disease incidence was the lowest (23.33 per cent) in the peat soil followed by the farmyard manure and gobar gas effluent (Sheela *et al.*, 1998). Bacteriarization with species of *Bacillus* resulted in the effective biological control of stem rot of groundnut and enhanced plant growth/yield (Sakthivel *et al.*, 1998). *B. subtilis* AF1 not only inhibited *A. niger* causing collar rot but also possessed plant growth promoting activity (Sailaja *et al.*, 1998). Hence, the isolate could be used for management of *A. niger*.

Botanicals

Green plants appear to be the reservoirs of effective therapeutants and can provide renewable sources of useful biofungicides. For example, aqueous leaf extracts (2 per cent) of neem (*Azadirachta indica*) and mehandi (*Lawsonia inermis*) were found to be effective and economical in controlling leaf spots and rust diseases of groundnut (Ghewande, 1989a). Neem has been extensively used for the management of diseases of crop plants. Research on other plant products is still at its infancy as compared to progress made with neem products. However, in groundnut use of neem and its products has not been very extensive. In the field trials conducted at the NRCG during *kharif* 1985-86 and 1989-90, three sprays of aqueous neem leaf extract (2 per cent) at fortnightly interval starting from 4-5 weeks reduced severity of early leaf spot by 13.6 per cent, late leaf spot by 22.2 per cent and rust by 30.6 per cent. The treatment also increased the pod yield by 26.1 per cent with a benefit/cost ratio of 2.26: 1. Aqueous neem leaf extract also reduced growth, seed infection and colonization by *A. alternata*, *A. tenuissima*, and *A. raphani* (Ghewande, 1995). Field experiments for the integrated management of ELS and LLS diseases of groundnut using 2 per cent aqueous leaf extract of neem as one of the components were conducted during 1988 to 1990 rainy seasons at the NRCG. Results showed that groundnut intercropped with pigeon pea + one spray each of neem leaf extract, fungicides and cell-free culture filtrate of *P. islandicum* at 40, 55 and 70 DAS, respectively, effectively controlled ELS and LLS and gave significantly higher net monetary returns (Ghewande *et al.*, 1993). Ganapathy and Narayanaswamy (1990) reported that water extracts of neem leaf, neem cake and neem oil (1 per cent) were useful in inhibiting the LLS and rust pathogens under laboratory conditions. They also reported that 1 per cent neem oil reduced the incidence of LLS and rust and increased pod yields by 62.3 per cent of groundnut over control. Soil application of 55 kg/ha potash plus 1 per cent foliar application of potash coupled with 2 per cent neem seed kernel extract significantly reduced the incidence of tikka leaf spot and increased pod and haulm yield (Chandrasekhar *et al.*, 1994). On the contrary, Lokhande *et al.* (1998) have found that chemical spray was superior to neem products. Such differences in the performance of plant products is not unusual as the efficacy of broad spectrum pesticides such as neem depend on the initial inoculum load of the pathogens, the weather prevailing at the time of spray and efficacy of the formulation. Usually, when the disease levels are already high, such pesticides show little influence in disease control. *A. niger*, *M. phaseolina*, and *A. flavus* were effectively controlled by the extracts of *Polyalthia longifolia* which was of course next to fungicides (Sobti *et al.*, 1995). Aqueous leaf extract (2 per cent) of neem (*Azadirachta indica*), neem oil (1 per cent), neem kernel extract (3 per cent) and Nerium leaf extract (10 per cent) reduced the incidence leaf spot and rust diseases of groundnut (Sudheendra-Ashtaputre, 1999). Narain *et al.* (1981) found that soaking groundnut seed in *Vinca rosea* leaf extract was not as effective as that of Carbendazim and thiram to reduce leaf spot incidence. Ethanol and essential oil extracts of *O. sanctum* inhibited growth and multiplication of *A. niger* and *A. flavus* and increased seed germination of groundnut. In greenhouse studies ethanol extracts gave the best control of both diseases (Mahapatra *et al.*, 1994). Bhore (1995) reported that crude extract of *C. lanceolatus* was effective against *A. flavus* and an acetone extract of the same was effective against *C. lunatus*. A 1 per cent solution of Indiara, commercial formulation, was effective in inhibiting spore germination, suppressing collar rot and minimising the leaf spots caused by the fungi. Spraying of 2.5 g mancozeb + 1.0 g carbendazim at 50 DAS followed by spraying of *C. procera* 5 per cent leaf extract at 70 DAS proved to be highly significant in reducing the leaf spot disease and in increasing the yield. (Srinivas *et al.*, 1997). Leaf extracts of coconut, sorghum, neem and *Parthenium* reduced the late leaf spot incidence and increased the yield, neem leaf extract being the most effective. (Sudheendra-Ashtaputre, 1999). Application of mustard cake @ 2 per cent enhanced the growth of *Trichoderma* spp., where as germination of sclerotia of *S. rolfsii* were inhibited at the same concentration (Desai, *et al.*, 2003). Growth of *S. rolfsii* was effectively inhibited by all the concentrations tested of the extracts of *Agave americana*, *Cassia occidentalis* and *Azadirachta indica*, the maximum inhibition being at 10 per cent concentration (Seshakiran and Adiver, 2003). Two years field experiments (2001-2002) showed that application of fresh leaves of *Parthenium hysterophorus*, neem, *Pongamia glabra* and wild sorghum @500 kg/ha in furrow at the time of sowing gave higher yield and effective control of stem rot as compared to control treatment (Ghewande, 2003)

Chemical Methods

Several systemic and non-systemic fungicides were tried for reducing the severity of major foliar diseases *viz.* early leaf spot, late leaf spot and rust in India and elsewhere (Kolte, 1984; Mc Donald *et al.*, 1985; Subramaniyam and Mc Donald, 1987). However, no single fungicide was found effective against all the foliar diseases. Efforts were made to evaluate new chemicals released from time to time for their efficacy against groundnut pathogens. However, as seen from the overall analysis, most of the studies found that Carbendazim was effective against leaf spots and Mancozeb to be effective against rust over time. This also probably gives an indication that there is not considerable variability in the pathogens. Singh and Naik (1977) reported efficacy of 2 sprays of Carbendazim and Benomyl in managing leaf spots and in increasing yield. In field trials at 2 locations, application of 3 foliar sprays of 0.2 per cent Mancozeb or Kocide (Copper hydroxide) or 25 kg sulphur dust/ha at 15-day intervals beginning 45 days after sowing, was effective against *P. arachidis* and increased yields of unshelled nuts from 459 to 764-996 kg/ha; Mancozeb was the most effective, followed by sulphur dust and Kocide (Durairaj and Mohan, 1978). Three years of study revealed that three sprays of tridemorph sprays at 0.07 per cent, beginning 45 days after planting and continuing at 10-day intervals, gave effective control of rust and increased pod yield (Mayee *et al.*, 1979). For the management of leaf spots and rust, benlate at 0.5g/l and Duter at 2.00g/1 were found to be most effective in reducing the disease intensity and also increased the yield significantly (Alam *et al.*, 1980). In field trials, four sprays of Carbendazim at 150 g in 250 L water/ha, at fortnightly intervals, substantially decreased leaf spots and increased yield. However, it was also suggested that to avoid development of resistance, the last spray should be made with a different, non-systemic fungicide (Sekhon and Dhillon, 1981; Rattan and Kang, 1984). Three sprays of Carbendazim (0.1 per cent), Benomyl (0.1 per cent), Copac (0.1 per cent), Tridemorph (0.05 per cent), Mancozeb (0.2 per cent) and Propiconazole (0.1 per cent) reduced intensity of leaf spots and rust and increased the pod yield significantly, (Rahman *et al.*, 1989). However, Carbendazim and Tridemorph were best against leaf spots and rust, respectively. Plantavax, BAS-3170 F, Mancozeb, RH-124 and Dithane Z-78 were significantly effective in reducing rust infection (Wangikar *et al.*, 1985). Adiver and Anahosur, (1995) reported Tebuconazole, Cyproconazole, Propiconazole and Difenoconazole were effective against late leaf spot. Best control of leaf spots and rust was achieved with Hexaconazole and Difenoconazole, and these treatments also gave best pod and fodder yields. Seed treatment with thiram at 3g/kg or Carbendazim at 1g/kg in combination with 3 sprays of 0.1 per cent Carbendazim gave the most effective control of leaf spots (Mittal, 1996). Spraying of a mixture of Mancozeb and Carbendazim at 50 days after sowing followed by spraying of *C. procera* 5 per cent leaf extract at 70 DAS proved to be highly significant in reducing the leaf spot disease and in increasing the yield (Srinivas *et al.*, 1997). Leaf spots and rust were effectively managed when three sprays of Carbendazim (0.05 per cent) + Mancozeb (0.2 per cent) were given at 15 days interval commencing from the first appearance of symptoms (Patil *et al.*, 1979; Naidu and Rao, 1997). Johnson *et al.* (1998) reported spray treatments of Carbendazim + Mancozeb gave good control of foliar diseases with an increase in yield by 20 per cent, giving an additional income of Rs. 2250 per hectare. Rust and leaf spots were substantially reduced by spraying a mixture of Carbendazim (0.025 per cent) and Tridemorph (0.4 per cent) five times at fortnightly intervals commencing 35 days after sowing during summer. The Hexaconazole treated plots gave a 71 per cent higher pod yield and an 87 per cent higher fodder yield than an untreated control with the highest net return of Rs 9793/ha (Jadeja *et al.*, 1999). Instead of taking up spray schedules based on crop age, Das *et al.* (1997) used weather dependable spray schedule (WDSS) to minimize application schedule to only 2 sprays for management of leaf spots and rust. Such schedule would be highly useful in areas where the disease is endemic and the variety sown is susceptible.

For management of seed and soil-borne diseases only seed treatment has been practical as soil application of chemicals has not only been expensive but also not practical. Carbendazim @ 1g/kg of seed was recommended for the control of collar rot. Seed treatment with either Thiram @ 3g/kg seed or Carbaendazim @ 2g/kg seed is effective in controlling seed and soil-borne diseases. Among 12 fungicides screened, Propiconazole and Iprodione at 5-1000 ppm inhibited the mycelial growth and also controlled spore production of *A. niger* (Woothisak *et al.*, 1991). Tok-E-25 was the most effective treatment in controlling *A. niger* and *A. flavus, in vitro* (Siddaramaiah *et al.*, 1981). Tebuconazole and Cyproconazole were effective against *S. rolfsii*, which inhibited ergosterol biosynthesis (Adiver and Anahosur, 1995).

In addition to traditional antifungal chemicals, efforts were made to use non-pesticides to induce resistance in the plant systems. The chemicals used were simple salts or chitosan etc. Collar rot was controlled using seed treatment with 9 non-conventional chemicals (barium sulfate, barium nitrate, copper sulfate, copper chloride, zinc sulfate, zinc chloride, manganese sulfate, ferric chloride and mercuric chloride (Dasgupta *et al.*, 1995). Dasgupta *et al.* (1998) found that seed soaking in chitosan (0.1 per cent) is economical in controlling these diseases. Chitosan treatment of groundnut leaves before inoculation reduced the number of lesions, lesion diameter and sporulation of *P. arachidis* (Sathiyabama and Balasubrahmaniam, 1998). Further it was observed that chitosan-treated leaves showed an increase in endogenous salicylic acid and intercellular chitinase and glucanase activity which are associated with induced resistance responses. A mixture of Di-xiu-na (a.i. sodium sulphanilate) and colloidal sulphur controlled infection by rust and leaf spot pathogens, increasing yield by 30-40 per cent. Addition of 0.1 per cent copper sulphate to the spray improved efficiency. To avoid phytotoxicity, concentration of Di-xiu-na should not exceed 1: 500, or 1: 600-700 at high air temp. The mixture controlled rust well even in the rainy seasons (Lin and Cheng, 1980).

Biotechnological Approaches

Host resistance forms the pivotal point in any IPM strategy as it is not only cheap and effective but, ecofriendly also. The rate of success in conventional disease resistance breeding in groundnut, at least in India, has been very low. Though, some good sources of resistance to foliar diseases are available (Subramaniyam *et al.*, 1995), a very high degree of resistance could not be transferred to the high yielding background mainly because of the complexity of inheritance of resistance and undesirable linkages (Miller *et al.*, 1990). Wild relatives of groundnut possess resistance to the level of even immunity. But, the interspecific hybridization has not been highly successful in introgressing the desirable traits where desired due to several inherent breeding barriers, (Bandyopadhyay *et al.*, 1992).Though a moderate degree of resistance against aflatoxigenic fungi is available in the cultivated gene pool, success in breeding has been almost non-existent. Relatively little efforts have been made so far in breeding for resistance to the soil-borne pathogens. In addition, the breeders have been using a very limited stock of the primary gene pool (Stalker, 1991). To add to this problem, isozyme and restriction fragment length polymorphism (RFLP) have shown that variability at the molecular level is low in cultivated groundnut (Kochert *et al.*, 1991). This limitedness of variability may explain to some extent why the groundnut yield is not going over the plateau. Given, these conditions, it becomes imperative to explore at least the alternative possibilities of building up host resistance as a part of IDM.

Through the aids of non-conventional breeding methods like transformation and protoplast fusion, disease resistance could be introduced into an otherwise desirable genetic background. The success of these non-conventional methods have started generating good hopes in the case of virus diseases where coat-protein-mediated resistance has been used. For a successful non-conventional gene transfer programme, the basic requirements are standard, and easily reproducible *in vitro* regeneration and recombination techniques. In groundnut, protocols have been amply standardized for regeneration (Ozias-Akins *et al.*, 1993; Bandyopadhyay *et al.*, 1995). Chitinases and glucanases are two potential classes of enzymes, which can hydrolyze the cell walls of the pathogenic fungi (Bama and Balasubramaniam, 1991; Buchala *et al.*, 1991). The chitinase III, isolated from rust infected leaves, inhibited germination, germ tube growth and appressorium formation of *P. arachidis* uredospores (Govindaswamy and Balasubramaniam, 1994). Infection of groundnut leaves with the early leaf spot pathogen *C. arachidicola* lead to a marked increase in activity of three extra cellular -1, 3 glucanases, which when acting together, were capable of degrading the pathogen cell wall *in vitro* (Roulin and Buchala, 1995). In groundnut, chitinase genes form a small multigene family. Herget *et al.* (1991) isolated four chitinase cDNAs (*chit* 1-4) from cultured cells. Expressions of individual *chit* genes were assayed by the polymerase chain reaction followed by RFLP analysis. They concluded that *chit* 2 gene expression may be controlled by pathogen-specific regulatory elements as the expression was strongly activated by the cell wall components of *Phytophthora megasperma*. Kellmann *et al.* (1996) characterized, cloned and sequenced two classes II chitinases from groundnut cv. NC4 (*A.h.Chi2; 1* and *A.h.Chi2; 2*). Transgenic tobacco plants containing the complete peanut *A.h.Chi2; 1* gene exhibited essentially the same expression pattern in leaves as observed in peanut cell cultures. Transcriptional fusion of a 1.2 kb 5'-upstream region of *A.h.Chi2; 1* to the GUS reporter gene conferred its expression in leaves of transgenic tobacco plants after challenging with *Phytophthora megasperma* elicitor or *Botrytis cineria* spores. Thus, the chitinase gene system offers, apparently, a way for building up of resistance to fungal pathogens of groundnut. Such resistance, if successful, may be durable being non-specific in nature. Fertile transgenic plants of groundnut cv.TMV-2 expressing tobacco (*Nicotiana* spp.) chitinase gene were generated. The transgene stably integrated in the genome of peanut plant and inherited in offsprings. Small-scale field tests indicated increased ability of these plants to resist the fungal pathogen *C. arachidicola* (Rohini and Rao, 2001). Somaclonal varients of groundnut having higher biomass and field tolerance to leaf spot and rust disease were selected in fourth generation from shoot tip regenerants of peanut variety TAG-24. Field testing of the regenerants revealed that the characters are stable and heritable (Eapen *et al.*, 1999)

Against soil-borne fungal pathogens of groundnut, relatively less work has been done mainly due to lack of information on nature of disease resistance against pathogens such as *Sclerotium rolfsii*, *Aspergillus* spp., *Macrophomina phaseolina* and *Rhizoctonia solani*. The role of phytoalexins in groundnut disease resistance has been recently reviewed (Strange and Subba Rao, 1994; Daniel and Purkayastha, 1995). This information can help to track down the phytoalexin production metabolism so that the genes for the synthesis of these phytoalexins could be isolated and used. It has been established that the biocontrol agents possess differing modes of action against plant pathogens and it is not necessary that all strains of biocontrol agents possess all these modes of actions for effective biocontrol ability. *Trichoderma* spp. are the most studied among the various biocontrol agents studied so far. However, they have not been successful as a part of integrated disease management for several reasons. The absence of more than one mode of action in a single strain may be one of them. Efforts were thus made for cataloguing of the strains of *Trichoderma* for various desired traits to introgress the desirable traits into an otherwise desirable strain of the biocontrol agent, (Desai, 1996). However, Stasz *et al.* (1989) found limited vegetative compatibility between and within the species of *Trichoderma* following protoplast fusion. Sivan and Harman (1991) developed intraspecific hybrids of *T. harzianum*. Strain 1295-22, derived from fusing protoplasts of auxotrophic mutants of the prototrophic strains T12 and T95, was more effective in colonizing the middle sections of the roots than either of the parental strains. Protoplast fusion in the genus *Trichoderma* gives rise to great variability, (Stasz *et al.*, 1988; Stasz and Hannan, 1990). To enable marker-assisted selection, protocols were standardized for transformation of spores of spp. *Trichoderma* by electroporation for the introduction of marker characters (Desai, 1996). Similar efforts were made in the past also by Sivan *et al.* (1992) to introduce *Hyg*B gene conferring resistance to hygromycin B in *T. viride* and *T. harzianum* including the details of selection protocols, regeneration of transformants and chromosomal integration of *Hyg*B gene. Lorito *et al.* (1993) compared efficiency of biolistic and protoplast-mediated procedures to transform strains of *Gliocladium virens* and *T. harzianum*. The biolistic procedure was technically simpler and increased transformation frequency and genetic stability in the progeny as compared with the protoplast-mediated transformation. Electrofusion protocols were also standardized for intra- and inter- specific hybridization of protoplasts with a view to introgress the desirable traits to enhance biocontrol ability (Desai, 1996). The detection of biocontrol agents (Gowdu and Balasubramaniam, 1991) helps in assessing the natural load of the biocontrol agent, and thus to plan a strategy to enrich the conditions for the multiplication of these mycoparasites.

Cole *et al.* (1988) hypothesized that phytoalexins were involved in resistance to moisture-deficit-stress induced preharvest aflatoxin contamination of immature groundnuts and this resistance might have evolved in an ecosystem that included seasonal periods of moisture-deficit-stress. Cooksey *et al.* (1988) identified a stilbene (3 isopentadienyl 1-4, 3′, 5'-trihydroxystilbene) as the major anti fungal component elicited by slicing imbibed kernels. The compound was inhibitory to both spore germination and hyphal extension of *A. flavus*. The invasion of groundnut kernels by *A. flavus* can occur under many conditions, but aflatoxin contamination does not occur until kernels lose the capacity to produce stilbenes as a result of moisture-deficit-stress induced dehydration (Cole and Dorner, 1993). The biosynthesis of stilbenes was due to the activity of two enzymes *viz.* stilbene synthase, which catalyses a one-step reaction to phytoalexins from precursors present in all plant cells, and a reductase, which in co-action with [naringenin-] chalcone synthase channels the metabolite flow into the biosynthesis of isoflavonoid phytoalexins (Schroder *et al.*, 1993). The preliminary evidence of the possibility of stilbene-gene amplification has been obtained. Fisher and Hain (1994) generated transgenic tobacco plants expressing (trishydroxyl) stilbene synthase genes from groundnut and grapes resulting in pathogen-inducible resveratrol synthesis. Glutathione-S transferase (GSH) is a constitutive enzyme of the plants. It has a major role in imparting herbicide resistance to the plants. Aflatoxins are oxidatively metabolized by the oxygenase system localized on endoplasmic reticulum forming an unstable but highly reactive AFB 1-8, 9 epoxide. The levels of GSH in mice are high enough to convert 12 times more AFB1-8, 9 epoxide to AFB1-GSH and hence are resistant to the carcinogenic effects of AFB1 compared with male rats (O'Brien *et al.*, 1983; Ramsdell and Eaton, 1990). The genes for GSH have already been cloned in maize. Attempts could be made to produce transgenic with GSFI genes in groundnut also. This could be another possible approach to manage aflatoxin contamination (Keenan and Savage, 1995). ICRISAT has developed specific monoclonal antibodies for aflatoxin estimation. Besides this transgenic groundnut against Indian peanut clump virus (IPCV) using coat protein (*Cp*) gene has been produced. The field testing of IPCVcp transgenics is underway. cDNA copies of the *Cp* gene of (IPCV)-H were introduced into cells of *Nicotiana benthamiana* or *Escherichia coli* by transformation with vector based on pROK II or pET respectively. The IPCV *Cp* was expressed and assembled to form virus-like particles (VLP). When transgenic plants expressing IPCV-H *Cp* were inoculated with IPCV-L, a strain that is serologically distinct from IPCV-H, the virus particles that accumulated contained both types of *Cp* (Bragard *et al.*, 2000).

Integrated Disease Management

Integrated disease management (IDM), a systems-approach, is a concept on which the modern crop disease management is based. IDM includes use of host plant resistance, biocontrol agents, fungicides of plant origin, cultural practices including cropping systems, and synthetic fungicides in optimal blends. IDM thus can further be integrated with weed and insect-pests management to form Integrated Pest Management package.

Gorbet *et al.* (1990) successfully used host plant resistance along with fungicides for the control of leaf spots. Barnes *et al.* (1990) integrated fungicides, genotypes, cultural practices (irrigation) and environment to control *Rhizoctonia* limb rot and stem rot. They reported that the application of difenoconazole every 14 days at 0.28 kg a.i./ha on Florunner and New Mexico Valencia cultivars with reduced irrigations resulted in maximum reduction of the limb rot. Combining Southern runner, a moderately resistant variety to late leaf spot, *S. rolfsii* and tomato spotted wilt virus with applications of chlorothlonil can provide late leaf spot control (Culbreath and Brenneman, 1992). At NRCG, various control measures were integrated for the management of early and late leaf spot. Early and late leaf spot control was achieved where groundnut was intercropped with pigeon pea and sprayed twice (55 and 70 DAS) with a mixture of fungicides (carbendazim 0.05 per cent + mancozeb 0.2 per cent) or where groundnut was intercropped with pigeon pea and received one spray each of neem leaf extract (2 per cent), fungicide mixture, and cell-free culture filtrate of *P. islandicum* at 45, 55 and 70 DAS respectively. These treatments, and groundnut intercropped with pigeon pea and sprayed separately with neem leaf extract and *P. islandicum* inoculum at 55 and 70 DAS gave significantly higher net monetary returns (Rs 9569-11561/ha) with cost: benefit ratio ranging from 1:2.99 to 1:3.63 for Girnar 1 and 1:2.67 to 1:3.48 for JL-24 cultivars (Ghewande *et al.*, 1993). An integration of date of sowing (10th or 20th June), seed treatment with thiram @ 5 g/ha and soil application of phorate granules reduced the plant mortality and gave higher returns (Ghewande *et al.*, 1993). Karthikeyan (1996) reported the ability of organic amendments (neem cake, mahua cake, castor cake, farm yard manure, sheep manure and poultry manure), the biocontrol agent (T. *harzianum*) and fungicides (carbendazim and thiram), to control seed and collar rot (*A. flavus* and *A. niger*) of groundnut. The lowest disease incidence was recorded when carbendazim was applied as a seed treatment. A soil application of *T. harzianum* was also effective. Of the organic amendments, neem cake and farmyard manure gave good disease control. These treatments also increased the yield of groundnut crop. Srinivas *et al.* (1997) found that spraying of 2.5 g mancozeb +1.0 g carbendazim at 50 DAS followed by spraying of *Calotropis procera* 5 per cent leaf extract at 70 DAS was highly significant in reducing the leaf spot disease and increasing the yield. Seed treatment with *P. fluorescens* @ 10 g/kg seed + soil application of neem cake @ 160 kg/ha before sowing was found to be the best treatment in reducing collar rot incidence, however, seed treatment with *T. viride* @ 4g/kg seed +soil application of neem cake gave higher pod yield (1849 kg/ha) which was next best treatment for reducing incidence of collar rot (Sheela and Packiaraj, 2000). Bag *et al.* (2000) reported that soil application of NPK at 15:35:15 kg/ha in combination of 6.5 t/ha FYM as basal dose and 4 sprays of Mancozeb (Dithane M-45) at 2 g/L of water 25 days after sowing at 15 days intervals was most effective and economical in controlling early and late leaf spots. On-station experiments on integrated disease management (IDM) at ICRISAT-Patancheru, India, have clearly demonstrated that when moderate levels of host plant resistance are combined with seed treatment and affordable levels of chemical control, expected yields and economic returns are higher than obtained with chemical control of susceptible genotypes (Pande *et al.*, 2001). Soil application of castor cake @500 kg/ha + commercial preparation of *T. viride* (Monitor-S) 62.5 kg/ha in furrow at the time of sowing reduced the incidence of collar rot and stem rot of groundnut considerably and gave higher gross monetary return of Rs 29865/ha than farmers practice (Rs 23685/ha) under the IVLP-TAR programme (Dr. M.P. Ghewande, personal communication).

It is evident from the literature that there is a great scope for integration of suitable and compatible disease control measures for the development of a holistic IDM package and in turn a holistic IPM package for the management of all biotic stresses including diseases, insect-pests and weeds.

REFERENCES

Adams, D. B. and Kuhn, C. W. (1977). Seed transmission of peanut mottle virus in peanut. *Phytopathology*, **67**: 1126-1129.

Adiver, S. S. and Anahosur, K. H. (1995). Efficacy of some triazole fungicides against late leaf spot of groundnut and their subsequent effects on *Sclerotium rolfsii*. *Indian Phytopath.*, **48(6)**: 459-462.

Alam, K. B., Bakr, B. A. and Talukdar, M. J. (1980). Effect of fungicides on the control of 'tikka' disease of groundnut [Bangladesh]. *Bangladesh Journal of Agricultural Research*, **5**: 17-20.

Ashtaputre, S. A., Srikant Kulkarni, Anahosur, K. H. and Kulkarni, S. (1994). Influence of sowing dates on the incidence of late leaf spot of groundnut (*Arachis hypogaea* L.). *Karnataka Journal of Agricultural Sciences*, 7: 178-180.

Bag, S. S., Raj, S. K. and Das, S. (2000). Management of leaf spots of groundnut by nutrition and fungicide: An integrated Approach. *International Arachis Newsletter*, **20**: 51-53.

Bama, M. S. and Balasubrahmaniam, R. (1991). Chitinases and glucanases in *Arachis hypogaea* L. in detached leaves infected with *Puccinia arachidis* Speg. *Zeitschrift fuer Pflanzenkrankheiten und Pflanzenschutz*, **98**: 305-316.

Bandyopadhyay, A., Murthy, T. G. K. and Reddy, P. S. (1992). Breaking the yield barriers in groundnut: some possible approaches. Invited lecture: *Special session on rabi/summer groundnut*, XLI *Rabi*/ summer Oilseed Workers' Group Meeting. 17.8.92, PKV, Akola.

Bandyopadhyay, A., Radhakrishnan, T., Murthy, T. G. K. and Desai, S. (1995). Somatic embryogenesis in groundnut (*Arachis hypogaea* L.). Paper presented at the review meeting of the NARP Basic Research Sub-Project Field Crop Biotechnology, IARI, New Delhi, August 7-8, 1995.

Barnes, J. S., Csinos, A. S. and Hook, J. E. (1990). Effects of fungicides, cultivars, irrigation and environment on *Rhizoctonia* limb rot of peanut. *Plant Dis.*, **74**: 671-675.

Bera, S. K., Dash, P., Singh, S. P. and Dash, M. M. (2002). Bacterial pod rot: A new threat to groundnut under rice-based cropping system in Orissa. India. *International Arachis Newsletter*, **22**: 40-41.

Bhore, S. T., Thakare, C. S. and Kadam, V. C. (1995). Efficiency of some plant extracts and commercial plant products against test fungi. *Journal of Maharashtra Agricultural Universities*, **20**: 251-254.

Biswas, K. K., Chitreshwar, Sen and Sen, C. (2000). Management of stem rot of groundnut caused by *Sclerotium rolfsii* through *Trichoderma harzianum*. *Indian Phytopath.*, **53(3)**: 290-295.

Bragard, C., Duncan, G. H., Wesley, S. V., Naidu, R. A. and Mayo, M. A. (2000). Virus-like particles assemble in plants and bacteria expressing the coat protein gene of peanut clump virus. *Journal of General Virology*, **81(1)**: 267-272

Buchala, A. J., Roulin, S., Coquoz, J. L. and Meier, H. (1991). Host pathogen interactions in the system *Arachis hypogaea-Cercospora arachidicola*. In: *Advances in Molecular Genetics of Plant-Microbe Interactions*. Volume I (Eds., Hennecke, H. and Verma, D.P.S.), Kluwer Academic, The Netherlands, pp. 284-287.

Buiel, A. A. M., Jacobs, T. and Parlevliet, J. E. (1993). Resistance in groundnut to peanut bud necrosis virus. Durability of disease resistance, 207-210. In: Cur*rent Plant Science and Biotechnology in Agriculture*. Vol. 18, ed. Jacobs, T., Kluwer Academic Publishers, Dordrecht, Netherland.

Chahal, A. S. and Sandhu, R. S. (1972). Reaction of groundnut varieties against *Cercospora personata* and *C. arachidicola*. *Pl. Dis. Reptr.*, **56**: 601-603.

Chandrasekhar, V., Narayanaswami, R. and Ramabadran, R. (1994). Effect of foliar spray of potash and neem seed extract on the tikka leaf spot of groundnut. *Indian Phytopath.*, **47(2)**: 188-189.

Chiteka, Z. A., Gorbet, D. W., Knauft, D. A., Shokes, F. M., Kucharek, T. A. and Norden, A. J. (1988). Components of resistance to late leaf spot in peanut (*Arachis hypogaea* L.). Proceedings, American Peanut Research and Education Society, **19**: 25.

Chohan, J. S. (1972). Final Progress Report, ICAR. Scheme for research on important diseases of groundnut in the Punjab for the period 1957-67. Department of Plant Pathology, Punjab Agricultural University, Ludhiana, India, p. 117.

Chohan, J. S. (1973). Biological control of seed borne pathogens of groundnut. *Indian J. Mycol. Plant Pathol.*, **4**: 80.

Chohan, J. S. (1974). Recent Advances in diseases of groundnut in India. In: *Current Trends in Plant Pathology*, Uttar Pradesh, India, Lucknow University Press, pp.171-178.

Cole, R. J. and Dorner, J. W. (1991). Peanut Phytoalexins. In: *Mycotoxins and Phytoalexins* (Eds., Sharma, R.P. and Salunkhe, D.K.). CRC Press, Boca Raton, USA, pp. 501-509.

Cole, R. J., Dorner, J. W., Blankenship, P. D. and Sanders, T. H. (1988). Potential role of phytoalexins in aflatoxin contamination of peanut. In: *Biotechnology for Crop Protection* (Eds. Hedin, P.A., Menn, J.J. and Hollingsworth, R.M.). ACS symposium Series No. 379, American Chemical Society, Washington, D.C., USA, pp. 73-81.

Cooksey, C.J., Garrett, P.J., Richards, S. E. and Strange, R. N. (1988). A dienyl stilbene phytoalexin from *Arachis hypogaea*. *Phytochemistry*, **27**: 1015-1016.

Culbreath, A.K. and Brenneman, T. B. (1992). Combining Centre Pivot irrigation applications of clorothalonil with moderately resistant cultivar for control of late leaf spot in peanut. *Plant Dis.*, **76**: 29-32.

Culver, J.N., Sherwood, J.L. and Melouk, H.A. (1987). Resistance to peanut stripe virus in *Arachis* germplasm. *Plant Dis.*, **71**: 1080-1082.

Daniel, M. and Purkayastha, R. P. (1995). *Handbook of phytoalexin metabolism and action*. Marcel Dekker, New York, USA. 627pp.

Das, N. D., Srivastava, N. N. and Rao, B. V. R. (1997). Comparative efficacy of spray schedule against leaf spot/tikka diseases of groundnut. *Annals of Plant Protection Sciences*, **5**: 103-107.

Dasgupta, S., Raj, S. K. and Das, S. (1998). Management of collar rot disease of groundnut by seed treatment with growth regulators-an alternative approach. *International Arachis Newsletter* **18**: 21-22.

Dasgupta, S., Raj, S. K., Srikanta, Das and Das, S. (2000). Control of collar rots disease of groundnut caused by *Aspergillus niger* by seed soaking with non-conventional chemicals. *Indian Phytopath.*, **53**: 227-229.

Demski, J. M. and Khun, C. W. (1983). Peanut mottle In: Peanut Disease Compendium Eds. D.M.Porter, D.H. Smith and R.Rodriguez-Kabana, St. Paul, Minesota. The U.S.American Phytopathological Society, pp. 45-46.

Desai, S. (1996). Strain improvement of *Trichoderma* for enhanced biocontrol ability. In *Abstract of the Poster Sessions, 2nd International Crop Science Congress*, November 17-24.1996, New Delhi, India, pp.433.

Desai, S. and Bagwan, N. B. (2002). Mass multiplication, and formulation of fungal biocontrol agents against late leaf spot pathogen of groundnut in the Saurashtra region of Gujarat, Asian Congress of Mycology and Plant Pathology, Mysore, (Karnataka), October 1-4, 2002: ISMPP.

Desai, S., Bagwan, N. B. and Yeole, R.D. (2003). Effect of mustard cake extract on *Sclerotium rolfsii* causing stem rot of groundnut and *Trichoderma*, a common biocontrol agent at National Seminar on "*Stress Management in Oilseed for Attaining Self-Reliance in Vegetable Oils*" January 28-30, 2003: ISOR.

Durairaj, P. and Mohan, R. (1978). Evaluation of fungicide for the control of groundnut rust. *Madras Agricultural Journal*, **65**: 692.

Eapen, S., Kale, D. M., Murty, G. S. S., Leela George, George, L. and Kishor, P. B. K. (1999). Somaclonal variation in peanut. *Plant tissue culture and biotechnology: emerging trends*. Proceedings of the Symposium held at Hyderabad, India, 29-31 January 1997, pp. 240-243., University Press (India) Ltd. Hyderabad, India.

Feakin, S. D. (1971). Pest control in groundnuts. In: PANS Manual (3rd Ed.) London, U.K. Center for Overseas Pest Control Foreign and Commonwealth Office, pp. 19-81.

Ferguson, L. M. and Shew, B. B. (2001). Wheat straw mulch and its impact on three soil borne pathogens of peanut in microplots. *Plant Dis.*, **85**: 661-667.

Fiesher, R. and Hain, R. (1994). Plant disease resistance resulting from the expression of foreign phytoalexins. *Current Opinion in Biotechnology*, **5**: 125-130.

Ganpathy, T. and Narayanasamy, P. (1990). Effect of plant products on the incidence of major diseases of groundnut. *International Arachis Newsletter*, **7**: 20-21.

Ghanekar, A. M. (1980). Groundnut Virus Research at ICRISAT. Proceedings of the International Workshop on groundnuts, 13-17 October 1980. Eds. R. W. Gibbons and J. V.Martin, International Crops Research Institute for the Semi- Arid Tropics, Patancheru, India, pp. 211-216.

Ghewande, M. P. (1980). Management of foliar diseases of groundnut (*Arachis hypogaea* L.) using plant extracts. *Indian Journal of Agricultural Science*, **59**: 133-134.

Ghewande, M. P. (1983). Effect of cultural practices on the disease (bud necrosis, collar rot, stem rot) incidence and yield of groundnut. *Indian Botanical Reporter*, **2**: 176-177.

Ghewande, M. P. (1985). Seed and seedling diseases of groundnut deserve attention for better productivity. *Seeds and Farms*, **11**: 9-12 & 19.

Ghewande, M. P. (1987). Virus diseases of groundnut and their management. In: Symposium on Diseases of Oilseed and Pulse Crops and their Control., Eds. S.C. Vyas and V.N. Shroff, J.N.K.V.V. Indore, (M.P.) pp. 62-67.

Ghewande, M. P. (1989a). Management of foliar diseases of groundnut (*Arachis hypogea*) using plant extracts. *Indian J. Agric. Scis.*, **59(2)**: 133-134.

Ghewande, M. P. (1989b). Biological control of late leaf spot (*Phaeoisariopsis personata* Berk. Curt.) V. Arx. of groundnut. *Indian J. Agric. Scis.*, **59(3)**: 189-190.

Ghewande, M. P. (1990a). Diseases of groundnut and their management. *J. Oil seeds Res.*, **7(1)**: 78-97.

Ghewande, M. P. (1990b). Biological control of groundnut (*Arachis hypogaea* L.) rust (*Puccinia arachidis* Speg.) in India. *Tropical pest Management*, **36(1)**: 17-20.

Ghewande, M. P. (1995). Management of foliar fungal diseases using aqueous leaf extract of neem (*Azadirachta indica A. juss*) In: *"Neem" for the Management of crop Diseases*. 1995. (Ed. V. Mariappan) Associated Publication Co., New Delhi, pp. 37-42.

Ghewande, M. P., Desai, S. and Basu, M. S. (2002). Diagnosis and management of major diseases of groundnut, *NRCG Technical Bulletin*, 3-12 pp.

Ghewande, M. P., Desai, S., Prem Narayan and A. P. Ingle (1993). Integrated management of foliar diseases of groundnut (*Arachis hypogaea* L.) in India. *International Journal of Pest Management*, **39(4)**: 375-378.

Ghewande, M. P., Desai, S., Prem Narayan and Kamble, S. D. (1992). Some sources of resistance to early leaf spot of groundnut (*Arachis hypogaea* L.) in India. *Tropical Agriculture* (Trinidad), **69**: 284-286.

Ghewande, M. P., Savaliya, S. D., Hingrajia, H. M. and Padavi, R.D. (2003). Management of stem rot (*Sclerotium rolfsii* Sacc.) through organic soil amendments in groundnut. In: Proceedings of the National Symposium on *Recent Advances in Management of Plant Diseases, Technology Development and Applications*, held at Pune, India, 20-21 December, 2003, pp. 9-10.

Ghewande, M. P., Shukla, A. K. Pandey, R. N. and Mishra, D. P. (1983). Losses in groundnut yields due to leaf spots and rust at different intensity levels. *Indian J. Mycol. & Pl. Pathol.*, **13(1)**: 123-127.

Ghewande, M. P., Shukla, A. K., and Pandey, R. N. (1986). Management of foliar diseases of groundnut through agronomic practices. *Indian Botanical Reporter*, **5**: 179-181.

Ghewande, M. P., Shukla, A. K., Pandey, R. N. and Mishra, D. P. (1983). Losses in groundnut yields due to leaf spots and rust at different intensity levels. *Indian J. Mycol. and Plant Pathol.*, **13**: 125-127.

Ghuge, S. S., Mayee, C. D. and Godbole, G. M. (1981). Assessment of losses in peanut due to rust and tikka leaf spots. *Indian Phytopath.*, 34: 179-182.

Gorbet, D. W., Knauft, D. A. and Shokes, F. M. (1990). Response of peanut genotypes with differential levels of leaf spot resistance to fungicide treatment. *Crop Science*, **30**: 529-533.

Govindasamy, V. and Balasubramaniyam, R. (1989). Biological control of groundnut rust, *Puccinia arachidis*, by *Trichoderma harzianum*. *Zeitchrift fuer Pflanzenkrankheiten und Pflanzenschutz*, **96**: 337-345.

Govindasamy, V. and Balasubramaniyam, R. (1994). Purification and properties of apoplastic chitinases from rust infected leaves of *Arachis hypogaea* L. *Botanical Helvetica*, **104**: 79-86.

Govindasamy, V., Gunaratna, K. R. and Balasubramaniyam, R. (1988). Properties of extra cellular chitinase from *Myrothecium verrucaria*, an antagonist to the groundnut rust *Puccinia arachidis*. *Canadian Journal of Pathology*, **20**: 62-68.

Gowdu, B. J. and Balasubramanium, R. (1991). Detection of hyperparasites in rust (*Puccinia arachidis*) infected groundnut leaves by ELISA. *Indian Journal of Microbiology*, **31**: 153-156.

Grewal, P. S., Chhabara, H. K. and Kaul, V. K. (1986). Screening of groundnut germplasm against root knot nematode *Meloidogyne arenaria*. *Indian Journal of Nematology*, **17**: 151-152.

Herget, T., Schell, J. and Schreiber, P. H. (1990). Elicitor-specific induction of one member of the chitinase gene family in *Arachis hypogaea*. *Molecular and General Genetics*, **224**: 469-476.

Jadeja, K.B., Nandolia, D. M., Dhruj, I. U. and Khandar, R. R. (1999). Efficacy of four triazole fungicides in the control of leaf spots and rust of groundnut. *Indian Phytopath.*, **52(4)**: 421-422.

Johnson, M. Rao, M. M. and Meenakumari, K. V. S. (1998). Chemical control of groundnut late leaf spot in Anantpur district of Andhra Pradesh. *Indian Phytopath.*, **51**: 382-384.

Kanzaria, K.K. and Patel, V.A. (1994). Influence of environmental factors on growth and oxalic acid production by *Sclerotium rolfsii*. *Groundnut News*, **6**: 7.

Karthikeyan, A. (1996). Effect of organic amendments, antagonist *Trichoderma viride* and fungicides on seed and collar rot of groundnut. *Plant Dis. Res.*, **11(1)**: 72-74.

Keenan, J. L., and Savage, G. P. (1994). Mycotoxins in groundnuts, with special reference to aflatoxin. In: *The groundnut crop–a scientific basis for crop improvement*. (Ed. Smartt, J.), Chapman & Hall, London, UK, pp. 509-551.

Kellmann, J. W. Kleinow, T., Engelhardt, K., Phillipp, C. D., Schell, J. and Schreier, P. H. (1996). Characterization of two class II chitinase genes from peanut and expression studies in transgenic tobacco plants. *Plant Molecular Biology*, **30**: 351-358.

Kendre, M.S., Patange, N.R., Neharkar, P.S. and Telang, S.M.(2000). Occurrence of *Thrips palmi*, a vector of bud necrosis disease of groundnut in Marathawada region. *Journal of Soil and Crops*, **10(2)**: 226-230.

Keshwal, R. L., Kulkarni, S. N. and Jain, A. C. (1976). Occurrence of new strain of *Pseudomonas solanacearum* E.F. Smith in Madhya Pradesh infecting groundnut (*Arachis hypogaea* L.) and Chilli (*Capsicum annum* L.) *J N K V V Research Journal* (India), **10**: 89.

Kochert, G., Halward, T., Branch, W. D. and Simpson, C. E. (1991). RFLP variability in (*Arachis hypogaea* L.) cultivars and wild species. *Theoretical and Applied Genetics*, **81**: 565-570.

Kodmelwar, R.W. and Ingle, A.Y. (1989). Effect of sowing dates, spacing and meteorological factors on the development of tikka and rust of groundnut. *Indian Phytopath.*, **42**: 274.

Kolte, S. J. (1984). Diseases of annual edible oilseed crops. Peanut diseases Vol.1 Florida, CRC Press Incorporated, USA, p. 29.

Krishna Kishore, G., Pande, S., Naryana, R. J. and Podile, A. R. (2003). Evaluation of chitinolytic strains of *Serratia marcescens* and *Bacillus circulans* for biological control of late leaf spot of groundnut. *ISOR National Seminar: Stress Management in Oilseeds*, Jan. 28-30, 2003, pp. 19-20.

Kuhn, C. W. and Demski, J. W. (1975). The relationship of peanut mottle virus to peanut production. Georgia Agriculture Experiment Station Research Report, **213**: 19.

Kuhn, C. W., Paguio, D. R. and Adams, D. B. (1978). Tolerance in peanut to peanut mottle virus. *Pl. Dis. Reptr.*, **62**: 365-368.

Kulkarni, S. A., Srikant Kulkarni and Kulkarni, S. (1994). Biological control of *Sclerotium rolfsii* Sacc.– a casual agent of stem rot of groundnut. *Karnataka Journal of Agricultural Sciences*, **7**: 365-367.

Kulkarni, S. A., Srikant Kulkarni and Kulkarni, S. (1995). Effect of organic amendments and green manuring on the survival of collar rot fungus of groundnut. *Current Research University of Agricultural Sciences*, Bangalore, **24(8)**: 135-136.

Lin, K. H. and Cheng, Z. (1980). Control of groundnut leaf rust with a mixture of di-xiu-na and colloidal sulphur. *Journal of the South China Agricultural College* , **1**: 73-85.

Lokhande, N. M., Lanjewar, R. D. and Newaskar, V. B. (1998). Effect of different fungicides and neem products for control of leaf spot of groundnut. *Journal of Soils and Crops*, **8(1)**: 44-46.

Lorito, M., Hayes, C. K., Pietro, A. di., Harman, G. E. and Di-Pietro, A. (1993). Biolistic transformation of *Trichoderma harzianum* and *Gliocladium virens* using plasmid and genomic DNA. *Current Genetics*, **24**: 349-356.

Mahapatra, T. K. and Tewari, S. N. (1994). *Ocimum sanctum* L. leaf extract toxicity against collar rot (*Aspergillus niger*) and yellow root (*A. flavus*) diseases of groundnut. *Allelopathy Journal*, **1**: 114.

Mathivanan, N. and Murugesan, K. (2000). *Fusarium chlamydosporum*, a potent biocontrol agent to groundnut rust, *Puccinia arachidis*. *Zeitschrift-fur-Pflanzenkrankheiten-und-Pflanzenschutz*, **107(3)**: 225-234.

Mayee, C. D. and Datar, V. V. (1988). Disease of groundnut in the tropics. *Review of Tropical plant Pathology*, **5**: 169-198.

Mayee, C. D., Patil, M. A., Godbole, G. M., Kide, D. S. and Patil, F. S. (1979). Fungicidal control of groundnut rust. *Pesticides*, **13**: 13-14.

Mc Donald, D., Subrahmaniyam, P., Gibbons, R. W. and Smith, S. D. H. (1985). Early and late leaf spots of groundnut. *Information Bulletin No.* **21**. International Crops Research Institute for the Semi-Arid Tropics, Patancheru, A.P., India.

Millar, I. L., Norden, A. J., Knauft, D. A. and Gorbet, D. W. (1990). Influence of maturity and fruit yield on susceptibility of peanut to *Cercosporidium personatum* (late leaf spot pathogen). *Peanut Science*, **17**: 52-58.

Moraes, S. A. and Godoy (1985). Evaluation of resistance to *Cercosporidium personatum* in *Arachis hypogaea* genotypes. *Summa Phytopathologica*, **11**: 140-151.

Nagarju, P. and Urs, S. D. (1998). Comparative efficiency of fungicides and bio-agents against *Aspergillus niger*, a casual agent of collar rot of groundnut. *Current Research University of Agricultural Sciences*, Bangalore, **27**: 137-139.

Naidu, P. H. and Rao, A. S. (1997). Relative efficacy of number of sprays of fungicides on the control of late leaf spot in groundnut. *Indian Journal of Plant Protection*, **25**: 45-47.

Naidu, P. H., Mosas, G. J. and Sitaramaiah, K. (2000). Control of groundnut Kalahasti malady (*Tylenchorhynchus brevilineatus*) through organic and inorganic soil amendments. *J. Mycol. Plant Pathol*, **30(2)**: 180-183.

Naidu, P. H. and Chandrika, V. (1997). Effect of dates of sowing on the occurrence of tikka late leaf spot and rust on groundnut in southern zone of Andhra Pradesh. *Journal of Oilseeds Research*, **14(2)**: 238-240.

Narain, A., Sahu, K. C. and Swin, N. C. (1981). Comparative efficacy of selected fungicides and a plant extract on tikka disease of groundnut. *Pesticides*, **15**: 27-28.

Nolt, B. L. and Reddy, D. V. R. (1984). Peanut Clump. In: *Compendium of Peanut Diseases* Eds.D.M.Porter, D.H., Smith and R. Rodriguez-Kabana. American Phytopathological Society, pp. 73.

O'Brien, K., Moss, E., Judah, D. and Neal, G. (1983). Metabolic basis of the species difference to aflatoxin B_1 induced hepatoxicity. *Biochemical and Biophysical Research Communications* **114**: 813-821.

Ozias, Akins, P., Schnall, J. A., Anderson, W. F., Singsit, C., Clemente, T. E., Adang, M. J. and Wessinger, A. K. (1993). Regeneration of transgenic peanut plants from stably transformed embryonic callus. *Plant Science*, **93**: 185-194.

Pande, S., Rao, J. N., Upadhyaya, H. D. and Lenne, J. M. (2001). Farmers' participatory integrated management of foliar diseases of groundnut. *International Journal of Pest Management*, **47(2)**: 121-126.

Patil, M. B., Wani, P. V., and Patil, G. D. (1979). Evaluation of fungicides for the control of groundnut leaf rust (*Puccinia arachidis* Speg.). *Pesticides*, **13**: 24-25.

Patil, R. P., Jagadeesh, K. S. and Kulkarni, J. H. (1998). Isolation of fluorescent pseudomonds associated with roots of different plants and their *in vitro* antagonism against groundnut collar rot pathogen *Sclerotium rolfsii* Sacc. *Karnataka Journal of Agricultural Sciences*, **11**: 45-4.

Pixley, K. V., Boote, K. J., Shokes, F. M. and Gorbert, D.W. (1990). Growth and partitioning characteristics of four peanut genotypes differing in resistance to late leaf spot. *Crop Science*, **30**: 796-804.

Podile, A. R., and Prakash, A. P. (1996). Lysis and biological control of *Aspergillus niger* by *Bacillus subtilis* AF1. *Canadian Journal of Microbiology*, **42**: 533-538.

Prabhu, K. S. and Urs, S. D. (1998). Efficiency of bioagents for management of collar rot of groundnut caused by *Aspergillus niger*. *Current Research University of Agricultural Sciences*, Bangalore, **27**: 114-115.

Rago, A., March, G. J. and Marinelli, A. (1997). Effect of plant residues on Sclerotia production by *Sclerotium rolfsii*. *Fitopatologia*, **32**: 121-125.

Rahman, M. A., Ahmed, H. U. and Alam, K. B. (1986). Studies on the efficacy of fungicides and the date of commencing of spray in controlling tikka and rust of groundnut. *Bangladesh Journal of Pathology*, **2**: 61.

Ramsdell, H. S. and Eaton, D. L. (1990). Species susceptibility to aflatoxin B1 carcinogenesis: Comparative kinetics of microsomal biotransformation. *Cancer Research*, **50**: 615-620.

Ratna, A. S., Rao, A. S., Nolt, B. L., Reddy, D. V. R., Vijayalakshmi, M. and Mc Donald, D. (1991). Studies on the transmission of Indian peanut clump virus disease by *Polymyxa graminis*. *Annals of Applied Biology*, **118**: 71-78.

Rattan, G. S. and Kang, M. S. (1984). Efficiency of different systemic and non-systemic fungicides against 'tikka' disease of groundnut and their effects on host physiology, *Pesticides*, **18**: 30-33.

Reddy, A. S., Prasada Rao, R.D.V.J., Thirumala-Devi, K. Reddy, S.V., Mayo, M.A., Roberts, I., Satyanarayana, T., Subramaniam, K. and Reddy, D.V.R. (2002). Occurrence of Tobacco streak virus on peanut (*Arachis hypogaea* L.) in India. *Plant Dis.*, **86**: 173-178.

Reddy, D. V. R., Iizuka, N., Ghanekar, A. M., Murthy, V. K., Kuhn, C. W., Gibbons, R. W. and Chohan, J. S. (1978). The occurrence of peanut mottle virus in India. *Pl. Dis. Reptr.*, 62: 978-982.

Reddy, C. D. R., Srinivas, T. and Reddy P.N. (1997). Evaluation of advanced groundnut lines for resistance to early and late leaf spots. *International Arachis Newsletter*, **17**: 13-15.

Reddy, D. V. R., Amin, P. W., Mc Donald, D. and Ghanekar, A. M. (1983). Epidemiology and control of groundnut bud necrosis and other diseases of legume crops in India caused by tomato spotted wilt virus. In: *Plant Diseases Epidemiology*, Eds.R.T. Plumb and J.M. Tresh, J.K., Blackwell, Oxford, pp. 93-102.

Reddy, D. V. R., Wightman, J. A., Bashear, R. J., Highland, B., Black, M., Sreenivasalu, P., Dwivedi, S. L., Demski, J. W., Mc Donald, D., Smith, Jr. J. W. and Smith, D. J. (1991). Bud necrosis: a disease of groundnut caused by tomato spotted wilt virus. Information Bulletin no. 31. International Crops Research Institute for the Semi-Arid Tropics, Patancheru, A.P., India.

Reddy, P. S. (1988). Genetics, breeding and varieties. In: *Groundnut*, Ed. P.S. Reddy, pp. 508-525.

Rodriguez-Kabana, R. and Morgan-Jones, G. (1987). Novel rotations and organic materials show promise for management of nematode. Alabama Agricultural Experiment Station. *Highlights of Agricultural Research*, **34**: 13.

Rodriguez-Kabana, R., Roberson, D. G., Wells, L., Weaver, C. F. and King, P. S. (1991). Cotton as a rotation crop for the management of *Meloidogyne arenaria* and *Sclerotium rolfsii* in peanut. *Supplement of Journal of Nematology*, **23**: 652-657.

Rohini, V. K. and Rao, K. S. (2001). Transformation of peanut (*Arachis hypogea* L.) with tobacco chitinase gene: variable response of transformants to leaf spot disease. *Plant Science*, **160(5)**: 889-898.

Roulin, S. and Buchala, A. J. (1995). The induction 1, 3-ß-glucanases and other enzymes in groundnut leaves infected with *Cercospora arachidicola*. *Physiological and Molecular Plant Pathology*, **46**: 471-489.

Sailaja, P. R., Podile, A. R. and Reddanna, P. (1998). Biocontrol strain of *Bacillus subtilis* AF1 rapidly induces lipoxygenase in groundnut (*Arachis hypogaea* L.) compared to crown rot pathogen *Aspergillus niger*. *European Journal of Plant Pathology*, **104**: 125-132.

Sakthivel, N., Sivamani, Anuratha, C. S., Savithiry, S., Gnanamanickam, S. S. and Mahadevar, A. (1988). Beneficial bacteria for plant disease management. Proceeding of the National Symposium on Phytobacteriology held at the University of Madras, Madras, India during March, 14-15, 1986, 213-220.

Sathiyabama, M. and Balasubrahmanium, R. (1998). Chitosan induces resistance components in *Arachis hypogaea* against leaf rust caused by *Puccinia arachidis* Speg. *Crop Protection*, **17(4)**: 307-313.

Sathiyanarayanmurthy, G. V., Ramasubbiah, K., Rama Subba Rao, K. and Krishnamurthy, M. M. (1988). Efficacy of certain synthetic pyrithroids and in combination against thrips and leaf miner of groundnut. *Journal of Agricultural Research*, Andhra Pradesh Agricultural University, **16**: 16-19.

Schroder, J., Schanz, S., Tropfg, S., Karcher, B. and Schroder, G. (1993). Phytoalexin biosynthesis: stilbene synthase and co-action of a reductase with chalcone synthase. In: *Mechanism of Plant Defense Responses* (Fritig, B., [editor Legrand, M.). Kluwer Academic, The Netherlands, pp. 257-267.

Sekhon, I.S., and Dhillon, A.S. (1981). Studies on chemical control of 'tikka' disease of groundnut. *Pesticides*, **15**: 19-21.

Seshakiran, K. and Adiver, S. S. (2003). Efficacy of plant extracts against *Sclerotium rolfsii*- the incitant of stem rot of groundnut (*Arachis hypogea* L). *ISOR National Seminar: Stress Management in Oilseeds*, Jan. 28-30, 2003. pp. 13-15.

Sharma, S. B. and Mc Donald, D. (1990). Global status of nematode problem of groundnut, pigeon pea, chickpea, sorghum and pearl millet and suggestion for future works. *Crop Protection*, **9**: 453-458.

Sheela, J. and Packiaraj, D. (2000). Management of collar rot by *Pseudomonas fluorescens*. *International Arachis Newsletter*, **20**: 50-51.

Sheela, J., Sivaprakasam, K. and Seetharaman, K. (1998). Effect of soil application of *Pseudomonas fluorescens* using different carriers on groundnut collar rot disease. *Madras Agricultural University Journal*, **85**: 191-192.

Siddaramaiah, A. L., Desai, S. A., and Hegde, R. K. (1981). Effect of few herbicides on crown rot and afla- root disease of groundnut. *Mysore Journal of Agricultural Sciences*, **15**: 44-47.

Singh, S. D. and Naik, S. M. P. (1977). Field evaluation of modern fungicides for the control of tikka disease of groundnut. *Pesticides*, **11**: 49-50.

Sivan, A. and Harman, G. E. (1991). Improved rhizosphere competence in a protoplast fusion progeny of *Trichoderma harzianum*. *Journal of General Microbiology*, **137**: 23-29.

Sivan, A. Stasz, T. E., Hemmat, M., Hayes, C. K. and Harman, G. E. (1992). Transformation of *Trichoderma* spp. with plasmids conferring hygromycin B resistance. *Mycologia*, **84**: 687-694.

Sobti, A. K., Sharma, O. P. and Bhargava, A. K. (1995). A comparative study of fungicidal compounds and plant extracts against three pathogens of *Arachis hypogea*. *Indian Phytopath.*, **48(20)**: 191-193.

Srinivas, T., Rao, M. S., Reddy, P. S. and Reddy, P. N. (1997). Integrated management of leaf spot of groundnut (*Arachis hypogaea* L.) with botanicals and chemicals. *Zeitschrift fuer Pflanzenkrankheiten und Pflanzeschutz*, **104**: 528-530.

Stalker, H. T. (1991). Utilizing wild species for crop improvement. In: Scientific management of germplasm: characterization, evaluation and enhancement. *IBPGR training courses: lecture series*, **2**: 139-161.

Stasz, T. E. and Harman, G. E. (1990). Nonparental progeny resulting from protoplast fusion in *Trichoderma* in the absence of parasexuality. *Experimental Mycology*, **14**: 145-159.

Stasz, T. E., Harman, G. E., and Gullino, M. L. (1989). Limited vegetative compatibility following intra- and inter- specific protoplast fusion in *Trichoderma*. *Experimental Mycology*, **13**: 364-371.

Strange, R. N., and Subba Rao, P. V. (1994). The phytoalexin response of groundnut and its role in disease resistance. *Oleagineux*, **49**: 227-233.

Subrahmaniyam, P. and Mc Donald, D. (1983). Rust diseases of groundnut. Information Bulletin No.13. International Crops Research Institute for the Semi-Arid Tropics, Patancheru, A.P., India.

Subrahmaniyam, P. and Mc Donald, D. (1987). Groundnut rust disease: Epidemiology and control. ICRISAT (International Crops Research Institute for the Semi-Arid Tropics), 1987. Groundnut rust disease proceedings of Discussion Group Meeting, 24-28, September 1984. ICRISAT Center, Patancheru, A.P., India.

Subrahmaniyam, P., Mc Donald, D., Waliyar, F., Reddy, L. J., Nigam, S. N., Gibbons, R. W., Ramanatha Rao, V., Singh, A. K., Pande, P. M. and Subba Rao, P. V. (1995). Screening methods and sources of resistance to rust and late leaf spot of groundnut. *Information Bulletin No.* **47**, ICRISAT, Patancheru, A.P., India, 20pp.

Subrahmaniyam, P., Rao, V. R., Mc Donald, D., Moss, J. P. and Gibbons, R. W. (1989). Origin of resistance to rust and late leaf spot in peanut (*Arachis hypogaea* L.). *Economic Botany*, **43**: 444-445.

Subrahmaniyam, P., Reddy, D. V. R., Sharma, S. B., Mehan, V. K. and Mc Donald, D. (1990). A world list of groundnut diseases. Legumes Pathology Progress Report **12**, Patancheru, A.P., India.

Subrahmaniyam, P., Reddy, L. J., Gibbons, R. W. and Mc Donald, D. (1985). Peanut rust: a major threat to peanut production in the semi arid tropics. *Plant Dis.*, **69**: 813-819.

Sudheendra, Ashtaputre, Srikant Kulkarni, Ashtaputre, S. and Kulkarni, S. (1999). Effect of plant extracts and chemicals on late leaf spot of groundnut. *Karnataka- Journal of Agricultural Sciences*, **12**: 195-196.

Wangikar, P. D., Buldeo, A.N. Wankar, B.N., Wadibhasme, S. S. and Khune, N.N. (1985). Chemical control of rust of groundnut. *PKV Research Journal* (India), **9**: 89-94.

Woothisak, Butranu, Sudthi, Suriya, and Montien, Somabhi. (1991). Chemical control on peanut crown rot (*Aspergillus niger* L.) 10th Seminar on Thailand National Groundnut Meeting for Rayong (Thailand). 16-19 Oct 1991.

5

Biotechnological Approaches for Integrated Management of Plant Diseases

C.D. MAYEE* & P.K. CHAKRABARTY

Central Institute for Cotton Research, Shankar Nagar, Nagpur – 440 010, India

Plants are constantly under attack by a variety of pathogens including bacteria, viruses, fungi, insects and nematodes. The percentage of crop losses caused by plant pathogens, insect pests and weeds worldwide has been estimated to 42 per cent, accounting for $ 500 billion worth of damage (Oerke *et al.*, 1994). Out of this, global losses caused by pathogens alone are estimated to be 13 per cent of the potential crop production (FAO 1993). Recent survey has shown that the world population has almost doubled in past 50 years, while the food production has increased meagerly by 25 per cent per head. To meet the food requirements for 8 billion population by 2020, the production has to be raised by 30 per cent (Anonymous, 2000). One possibility to achieve this demand is by striving to prevent damage caused due to pest and diseases. Presently most of the control measures rely on the use of conventional methods including use of chemical pesticides. Biological control measures through deployment of antagonists have been used with limited success to control soil-borne diseases caused

* Agriculture Commissioner, Ministry of Agriculture, Department of Agriculture and Cooperation, Krishi Bhawan, New Delhi-110 001.

by species of *Fusarium, Rhizoctonia* and other pathogens involved in seedling disease complex. Even though sanitation, quarantine, seed treatment and chemical control measures have been used with several limitations, deployment of resistant cultivars selected or bred through conventional breeding are considered the most common and dependable method of controlling plant diseases. The biotic losses to the tune of 42 per cent occur despite the use of all these conventional management practices.

With advancement of research in molecular biology and genetic engineering, the traditional research needs to be coupled with novel technologies to optimize management of diseases in crop plants. Breeding a new variety with improved resistance requires considerable time and efforts, particularly when the source of resistance is a poorly adapted wild species that requires a lengthy backcrossing/breeding programme to eliminate deleterious genes. Often pathogens can develop virulence even before a new variety is released leading to boom and bust cycles whereby varieties possessing effective resistance when grown under a vast area (boom) selects for the evolution of matching virulence which spreads within the pathogen population (bust) overcoming the resistance (Pink and Puddephat, 1999). Plant transformations through development of transgenic offers efficient means to introduce specific genes, determining disease resistance into crop species without the problem of introducing deleterious genes and the linkage drag that occurs in resistance breeding. Genetic engineering thus has become a highly potential tool for crop improvement and lines expressing genes conferring resistance against viruses, insects, herbicides have all found commercial applications (Birch, 1997).

Several exciting approaches and candidates are now available for developing designer plants capable of providing effective resistance against diseases. This article presents some of the potential biotechnological strategies that could hold a great promise for plant disease management through deployment of transgenic resistance and other novel approaches.

Engineering Universal Resistance with Genes from Plant Kingdom

First group of genes that have been exploited in agriculture from time immemorial through conventional breeding are the plants own defense genes. Advances in molecular biology has made it possible to isolate and clone these *R* genes and use them for engineering resistance against different groups of pathogens, for several exciting reasons mentioned in subsequent sections.

Marker-aided Selection of Resistance (*R*) Genes

Bacterial blight induced by *Xanthomonas* is the only major bacterial disease of cotton. It reduces lint yield and fibre quality in all the cotton growing zones of the world (Blasingame and Patel, 2000). At least twenty major genes, which can each, confer either complete or partial resistance to one or more races of the pathogens have been identified (Brinkerhoff, 1970 and Brinkerhoff *et al.*, 1984). These genes were derived from number of sources and backgrounds of cotton. These include B_1, B_2, B_3, B_4, B_{sm}, B_7, B_{In}. De Feyter and Gabriel (1991) cloned six *Avr* genes from *X. c. pv. malvacearum*, each of which conferred avirulence towards one or more of Hunter and Brinkerhoff's resistant lines (Hunter and Brinkerhoff, 1961). De Feyter *et al.* (1993) observed that some of the above resistance genes (B_4, B_6, B_7, $B_{In.}$) were effective against single *Avr* genes in gene- for -gene manner whereas, some like B_1, B_2 *and* B_{in3} appeared to recognize multiple *Avr* genes. Molecular characterization of several resistant genes from different plant species led to identification of several mechanistic classes and provides opportunity for transgenic disease control strategies for generating resistant plants having different specificities. With the advent of molecular techniques breeding program based on plant disease resistance genes have to be optimized by incorporating molecular marker techniques and biotechnology to speed up the deployment of resistance genes (Rommens and Kishore, 2000). Number of molecular markers linked to resistance genes has been identified using a variety of approaches (Michelmore, 1995). The markers are specifically useful to locate resistance genes in F2 or F3 generations when disease screenings are difficult or unreliable. Identification of markers (viz RFLP, RAPD, AFLP etc) tightly linked to resistance locus greatly facilitate the marker-aided selection, map-based cloning of resistance gene, germplasm characterization and exploitation, fingerprinting and identification of cultivars.

An early success of the CSIRO cotton-breeding program was the incorporation of bacterial blight resistance in commercial cultivars (Rungis *et al.*, 2002). The source of resistance used was the so-called immune lines carrying B_2, B_3, B_7 and B_{sm} genes. However, resistance to race 18 in an Australian cultivar CS 50 segregated as a single dominant locus. Use of RFLP markers to map this resistance locus established that the resistance gene was not located on chromosome 20 near *B2* or *B3* but co-segregated with a marker on chromosome 14 linked to broad spectrum *B12* gene for resistance. AFLPs and SSRs were also used to search for novel markers linked to the *X. axonopodis*. pv. *malvacearum* resistance locus to facilitate introgression of this trait into *G. barbadense* through a program of marker-assisted selection. Although overall level of polymorphism between the two species was low, an RFLP marker was found linked to the resistance locus. Such novel markers found linked to resistance locus can help in marker aided selection of resistance genes. RAPD markers were developed and used for identification of 28 upland cotton cultivars currently grown in Greece (Linos *et al.*, 2002).

Cloning and Characterization of *R* Genes

More than 30 genes have been cloned and characterized from different plant species both monocots and dicots, that provide resistance against a variety of different pathogens (Hulbert *et al.*, 2001). Structures of most of these genes are consistent with a role in pathogen recognition and defence response signalling because they encode protein with motifs found in receptor and signal transduction proteins. Such genes are abundant in plant genomes and most belong to tightly linked gene families. No function other than disease resistance has yet been ascribed to them. Various *R* genes cloned so far from different plant species have many structural and functional features like, NBS domain and C-terminal LRR (Penuela *et al.*, 2002) in common, regardless of the plants from which they were isolated or the pathogens against which they are effective *viz.*, bacteria, fungi, viruses or nematodes, implying a common related mechanism of resistance (Table 1). Genetic analyses of plant pathogen interactions have revealed that resistance reactions in most of these cases are controlled by two dominant loci: an avirulence (*Avr*) gene in the pathogen (Chakrabarty and Raj 2002) and the corresponding resistance (*R*) gene (Hulbert *et al.*, 2001) in the plant. The absence of either member of the gene pair results in the development of disease. The interactions have been genetically defined by the gene-for-gene theory (Flor 1971). It is envisaged that *R* gene interacts with specific *Avr* gene of the pathogen. *R* genes are either involved in detoxifying or neutralizing virulence factors produced by the pathogens or they recognize pathogen legends (Jones 1996; Dangl 1995). Recognition of specific legends (possibly the products of *Avr* genes of pathogen) triggers a cascade of signalling, involving multiple signal transduction pathways that eventually lead to manifestation of resistance, characterized by hypersensitive response (HR) or by other means. HR involves the death of first cell or cells infected and the local accumulation of antimicrobial compounds. Many *R* genes cloned are components of signal transduction pathways. Isolation and characterization of *R* genes from various sources have indicated that these genes have remarkable structural similarities, indicating that plants evolve and use similar signal transduction pathways to trigger resistance against a large number of unrelated pathogens. The NBS and LRR motifs in most of these *R* genes are known for protein-protein and protein-nucleic acid interaction between the host and pathogens.

R genes are the potential candidates for cloning in plants through genetic engineering. Most of them are cloned by map-based, positional cloning or T-DNA tagging (Hulbert *et al.*, 2001). Using degenerate PCR primers directed towards conserved motifs of the *R* genes it is now possible to isolate, resistance gene analogues (RGAs) and then test them for their association with *R* genes (Meyers *et al.*, 1999). Employing this approach RGAs have been amplified and cloned from a number of diverse crop species including corn (Collins *et al.*, 1998), flax (Dodds *et al.*, 2001), soybean (Graham *et al.*, 2000), citrus (Deng *et al.*, 2000) wheat and barley (Penuela *et al.*, 2002), lettuce (Shen *et al.*, 1998) and several other plants including cotton. RGAs amplified from plants species can be used as DNA probes to clone *R* genes from the BAC and cDNA libraries of plant genes. Using similar approach RGAs were cloned from cotton possessing different bacterial blight resistance genes (Chakrabarty, unpublished). Twelve isogenic lines of cotton each having an individual blight resistance gene, *viz.*, *B1*, *B2*, *B3*, *B4*, *B5*, *B6*, *B7*, *Bin3* were analysed for polymorphism of *R* genes. Four sets of degenerate primers based on conserved nucleotide binding motifs of *R* genes were designed to amplify resistance gene analogues from the isogenic lines including the bacterial blight immune IM216 parent line. Southern hybridisation of the isogenic lines with four RGA probes was done. Polymorphism was detected in case of *Bin3* line with only one of the four probes tested. No discernible polymorphism was detected when other three RGAs were used as probes. Several probes need to be generated and tested by designing primes directed towards different regions of the resistance genes, to delineate variability amongst different BBR genes. Cloning and characterization of *R* genes having broad-spectrum resistance against bacterial blight of cotton is contemplated using a BAC library prepared from the resistant line. Besides, RGAs cloned from different plants (using same sets of primers) can also be used as potential candidates for developing resistant plants with different specificities by conventional breeding or through plant transformation. Transformation of indigenous cotton hybrids NHH-44 and DCH 32 with -Gus was demonstrated recently by Agrobacterium mediated and particle bombardment methods (Banerjee *et al.*, 2002). High Gus expression was observed in callus tissues as well as in somatic embryo-like structures in shake cultures.

Deployment of Cloned *R* Genes Through Genetic Engineering

Molecular characterization of several resistant genes from different plant species led to identification of several mechanistic classes and provides opportunity for transgenic disease control strategies for generating resistant plants having different specificities. Transgenic resistance thus can be deployed through inter-specific transfer of *R* genes in various plant species (Hulbert 2001).

The first plant resistance gene that was cloned was the maize *Hm1* gene (Table 5.1) that conferred resistance against the leaf spot fungus *Cochliobolus carbonum*. Unlike most of the characterized *R* genes, which are implicated in signal transduction or defence signalling pathway, *Hm1* codes for an enzyme HC-toxin reductase that detoxifies the toxin made by race 1 of *C. carbonum* (Johal and Briggs 1992). This gene could be useful for engineering resistance in other crops affected by this pathogen or against other related toxin producing pathogens (Mitra 2001). *Pto* is a serine-threonine kinase gene of tomato that confers resistance against *Pseudomonas syringae* pv. *syringae* strains containing *avrPto*, in gene-for-gene manner (Martin *et al.*, 1993). The transgenic tobacco containing *Pto* showed race-specific resistance against *P. syringae* pv. *tabaci* harbouring the avirulence gene *avrPto* (Rommens *et al.*, 1995). A barley *R* gene *mlo*, confers durable field resistance to powdery mildew fungus *Erysiphe graminis* fsp. *hordei* (Jorgenson 1992). This is an unusual gene where the allele conferring resistance is functionally recessive. The gene probably works through activation of host's inherent defence system. Most other *R* genes cloned recently possess NBS-LRR sequences, which are known for protein-protein and protein-DNA interactions and are the components of defense signalling cascades.

N gene governs resistance against TMV and possesses NBS-LRR motifs (Whitham *et al.*, 1994). In addition to these two conserved motifs, which it shares with most of the cloned *R* genes, *N* also possesses a TIR domain, homologous to the *Drosophila Toll* and mammalian *interleukin-1* receptor at N terminus. Several species of tobacco that possess *N* gene are resistant against TMV. Transformation of this gene in susceptible tobacco as well as tomato rendered the plants, resistant against TMV (Whitham *et al.*, 1996). Several other genes (Table 5.1) governing resistance against fungal, viral, bacterial and nematode pathogens share the same conserved motifs characteristics of *R* genes (Hulbert *et al.*, 2001; Tolin 2001; Creusot 1999).

Table 5.1: List of *R* Genes Characterized to Govern Pathogen Resistance

Pathogen Group	Genes	Host	Effective Against	Protein Motifs
Fungi	*L*	Linseed	*Melampsora lini*	TIR-NBS-LRR
	M	Linseed	*Melampsora lini*	TIR-NBS-LRR
	P	Linseed	*Melampsora lini*	TIR-NBS-LRR
	RPP1	Arabidopsis	*Peronospora parasitica*	TIR-NBS-LRR
	RPP5	Arabidopsis	*Peronospora parasitica*	TIR-NBS-LRR
	Dm3	Lettuce	*Bremia lactucae*	NBS-LRR
	I2	Tomato	*Fusarium*	NBS-LRR
	Mla	Barley	*Blumeria*	NBS-LRR
	Pib	Rice	*Magnaporthe grisea*	NBS-LRR
	Pi-ta	Rice	*Magnaporthe grisea*	NBS-LRR
	Rp1	Maize	*Puccinia*	NBS-LRR
	RPP8	Arabidopsis	*Peronospora parasitica*	NBS-LRR
	RPP13	Arabidopsis	*Peronospora parasitica*	NBS-LRR
	Cf-2/5	Tomato	*Cladosporium fulvum*	LRR-TM
	Cf-4/9	Tomato	*Cladosporium fulvum*	LRR-TM
	Rpw8	Arabidopsis	*Erysiphe*	Unique
	mlo	Barley	*Blumeria*	Membrane protein
	Hm1	Maize	*Cochliobolus*	HC-Toxin reductase
		Bean	*Colletotichum lindemuthianum*	NBS-LRR
Bacteria	*RPS4*	Arabidopsis	*Pseudomonas*	TIR-NBS-LRR
	Bs2	Pepper	*Xanthomonas*	NBS-LRR
	Prf	Tomato	*Pseudomonas*	NBS-LRR
	RPM1	Arabidopsis	*Pseudomonas*	NBS-LRR
	RPS5	Arabidopsis	*Pseudomonas*	NBS-LRR
	Xa1	Rice	*Xanthomonas*	NBS-LRR
	Pto	Tomato	*Pseudomonas*	Protein Kinase
	Xa21	Rice	*Xanthomonas*	LRR-TM-Kinase
Virus	*N*	Tobacco	*TMV*	TIR-NBS-LRR
	Rx1	Potato	*PVX*	NBS-LRR
	Rx2	Potato	*PVX*	NBS-LRR
	Sw-5	Tomato	*Tospovirus*	NBS-LRR
Nematode	*Gpa2*	Potato	*Globodera rostrochinesis*	NBS-LRR
	Mi	Tomato	*Meloidogyne*	NBS-LRR
	HS1	Beet	*Heterodera*	Unique

Bacteria belonging to genus *Xanthomonas*, is a highly destructive pathogen as it can infect atleast 124 monocot and 268 dicot plant species, causing variety of diseases including leaf spots and streaks, wilts and blights, rots and cankers (Starr 1981; Hayward 1993). *Xa21* gene is the first *Xanthomonas R* gene that was characterized at the molecular level (Song *et al.*, 1995). Transgenic plants expressing the cloned *Xa21* gene conferred resistance to 31 strains of *X. oryzae* pv. *oryzae* collected from all over the world (Wang *et al.*, 1996). Several research groups in different countries have tried and introduced the cloned *Xa21* gene into elite rice varieties and evaluated its field performance (Zhang *et al.*, 1998; Tu *et al.*, 1998). *Xa21* transgenic rice proved excellent in terms of their yield and broad-spectrum resistance against different strains of *X. o.* pv. *oryzae*. Other cloned *R* genes against *Xanthomonas* include the rice gene *Xa1* conferring resistance to Japanese race 1 of *X. o.* pv. *oryzae* (Yoshimura *et al.*, 1998) and the pepper gene *Bs2* controlling resistance to strains of *X. campestris* pv. *vesicatoria* containing *avrBs2* (Tai *et al.*, 1999). *Bs2* functions both in pepper and tomato. These *R* genes of rice and pepper are the potential candidates for cloning into citrus, cotton, beans and other crops which too are invaded by Xanthomonads possess the highly homologous members of *Xanthomonas avr/pth* gene family (Chakrabarty & Gabriel 1996; Chakrabarty *et al.*, 1997a; 2002; Chakrabarty & Raj 2002; Gabriel 1999). Inter-specific transfer of *R* genes is expected to be effective because of several reasons:

1. Many plants *R* genes are known to interact with avirulence genes from different pathogens, inciting HR responses.
2. Several *R* genes have been demonstrated to function after transfer as transgenes to different but related plant species (Hulbert *et al.*, 2001). *Bs2* gene cloned from pepper (Tai *et al.*, 1999) can interact with several members of *Avr* gene family widespread amongst pathovars of *X. campestris* (Gabriel 1999) including pathovar *malvacearum* that infects cotton (Chakrabarty *et al.*, 1997a).
3. *R* genes cloned from a number of plant species and effective against diverse array of pathogens including viruses, bacteria, fungi, nematodes (Baker *et al.*, 1997) show significant similarities in conservation of their structural motifs.

Merits of Deployment of *R* Genes

It is possible to pyramid resistance genes with each gene effective against different strains of a single pathogen or different pathogens. Transgenic approaches overcome two objections to pyramiding resistance genes by classical methods.

1. Unlike sequential introduction of multiple genes through traditional back crossing, several *R* genes can be introduced at one time through molecular approach. Such introduction of transgenes will allow different types of resistance or mechanistic classes of resistance genes to be combined.
2. Transfer of *R* genes by transformation is an efficient substitute to back crossing and will overcome the possible linkage drag of deleterious genes.

Despite significant theoretical feasibility it remains to be seen if *R* gene with a known function in one species will have a role in different species if the recipient does not share the same pathogen.

Deployment of Novel Genes other than *R* Genes

R genes are known to be the initial players in signal transduction pathway of disease resistance. Their function is believed to intercept the pathogens possessing corresponding *Avr* genes and transcend signal to trigger plant's inherent defence system. In the absence of cognate *Avr* genes, pathogens go unrecognised leaving the plants unchallenged and defence responses untriggered, resulting in manifestation of disease. Since specific '*R*' genes in plants confer resistance to pathogen containing specific '*Avr*' genes, the resistance of the host plant can also be compromised if the pathogen sheds or mutates the cognate *Avr* gene (Hirt 2002). Therefore, besides *R* genes, which act mostly in gene-for-gene specific manner, there are host of other defence genes and non-host *R* genes (Chisholm 2000), which can confer general resistance that is independent of interaction with pathogen *Avr* genes. Transformation of plants with novel genes that act downstream *R* gene signal transduction pathway appear to be a potential strategy for developing disease resistant transgenic plants. Asai *et al.* (2002) described flagellin, a mitogen activated protein kinase (MAPK) signaling cascade as a potential tool for engineering crops with broad-spectrum resistance against bacterial and fungal pathogens. Although the defence genes in plants and animals differ considerably, the signal transduction mechanism connecting the activation of host receptors to induction of defence responses shares similarities between the kingdoms. In both animal and plants, kinases of MAPK class are activated by pathogen factors. The MAPKs are the ideal intracellular mediators of information between cytoplasm and nucleus where they act upon their targets- the transcriptional factors that in turn trigger expression of defense genes/responses. Transient over-expression of MAPKs and other related kinases conferred resistance to *Aradiopsis thaliana* leaves upon infection by the bacterial and fungal pathogens (Asai *et al.*, 2002). Expression of these novel proteins like MAPKs and other receptor like kinases in plants thus offers a challenging strategy for disease control. Twenty major bacterial blight resistance genes including *B1, B2, B3, B4, Bsm, B7, Bln* etc have been cloned (Essenberg, 2002). Some *R* genes are effective against single *Avr* genes in gene-for-gene manner whereas, others like *B1, B2, B3* etc. recognize multiple *Avr* Genes (De Feyter *et al.*, 1993). The ability to engineer one common signaling pathway in cotton that can confer resistance to various pathogens would be highly beneficial allowing creation of host with resistance against a wide spectrum of pathogens. This could either be attempted by classical breeding method assisted by molecular markers or by direct engineering of crop plants with variants of active MAPKs.

Engineering broad-spectrum durable resistance in plants by over-expression of signalling components and defence response genes that act downstream of *R* genes (Rommens and Kishore, 2000) in signal transduction pathway can provide effective resistance. Plant intrinsic responses that can be engineered to attain a wider, more durable resistance include systemic acquired resistance (SAR) response. The potential candidate in this category is the *NPR1* gene, which manifests resistance in plants by regulation of SAR. *NPR1* is the first SAR-regulating gene cloned from *A. thaliana* and is also known as *NIM1* and *SA1* gene (Cao *et al.*, 1994; Delaney *et al.*, 1995; Glazebrook *et al.*, 1996; Shah *et al.*, 1997; Cao *et al.*, 1997; Ryals *et al.*, 1997). Mutations in this *Arabidopsis* gene resulted in plants that were no longer responsive to SAR induction. PR gene expression and disease resistance were both abolished in the *npr1* mutants. Over-expression of this gene in *Arabidopsis* increased resistance to virulent isolates of *P.syringae* and *Peronospora parasitica* in a dose-dependent manner (Cao *et al.*, 1998). Interestingly, over expression of NPR1 did not lead to constitutive PR gene expression prior to pathogen challenge, indicating that the NPR1 protein requires activation to be functional. This active-only-upon-induction characteristic of NPR1 makes it favourable target for engineering broad-spectrum disease resistance. Also no obvious detrimental effects to the plants like constitutive PR gene expression or spontaneous cell death occurs. The gene has been over-expressed successfully in wheat and rice and as such is an exciting candidate for development of disease resistant crop plants. Although a tangible approach to boost plant's resistance through deployment of various components of resistance in holistic manner is yet to be achieved, individual components of signal transduction pathways for resistance are being tried with different degree of success.

Engineering Resistance Against Fungal Pathogens Using Antimicrobial Proteins

Genetic engineering of plants with traits like over-expression of anti-microbial proteins (*e.g.* chitinases, glucanases etc.) or induction of key plant defence pathways through signalling molecules (*e.g.* salicyclic acid, jasmonic acid or ethylene) are other novel strategies for manipulation of resistance. Genes encoding hydrolytic enzymes such as chitinases, chitosanase and glucanase, which can degrade fungal cell wall components, have been specifically used for fungal disease resistant transgenic plants (Mitra 2001). The cDNAs from cotton encoding PR protein, GaPR-10 (Zhou *et al.*, 2002), chitinases and 1, 3- glucanase (Hudspeth *et al.*, 1996) are the potential candidates for expression of enhanced resistance in plants. Expression of these genes from infection inducible promoters in transgenic plants would provide ingenious and effective method of resistance. Broglie *et al.* (1991) demonstrated for the first time that transgenic tobacco and canola plants expressing a chitinase gene were highly resistant to a root pathogen *Rhizoctonia solani* in green house tests. Chitin is a major component of fungal cell walls and degradation of chitin has been suggested to have protective effects against chitin containing, plant pathogenic fungi. Chitinases hydrolyze the -1, 4 linkage of the N-acetyl–D-glucosamine polymer chitin. Several chitinase genes have been cloned from insects and rice plants (Lin *et al.*, 1995). Transgenic rice plants constitutively expressing these chitinase genes showed a high degree of resistance to the sheath blight fungus, *R. solani*. Co-expression of chitinase and glucanase in transgenic crops heightened resistance to a number of fungal pathogens including species of *Alternaria* and *Cercospora* (Zhu *et al.*, 1994). This strategy will be more useful in case of the diseases caused by polyphagous pathogens (like species of *Rhizoctonia* and *Fusarium* as well as other pathogens that cause root rot and seedling diseases) against which the inherent resistance is scanty. Gene encoding a strong antifungal endochitinase from *Trichoderma harzianum* provided broad-spectrum fungal resistance in transformed tobacco and potato plants. Transgenic plants were highly resistant to *Alternaria alternata*, *A. solani*, *Botrytis cinerea* and *R. solani* (Lorito *et al.*, 1998). Apple scab caused by the fungus *Venturia inaequalis* is one of the greatest limitations to successful cultivation of this fruit in India as well as in the whole world. The pathogen drastically affects the productivity and quality of the affected fruits. Growers almost exclusively rely upon multiple applications of fungicides to control this disease during the growing season. Transgenic 'McIntosh' apple trees expressing either the endo- or exochitinase gene or both genes have increased resistance to apple scab (Bolar *et al.*, 1999, Bolar 1999). These results suggest the potential broad usage of chitinase transgenes to control fungal diseases of plants.

A novel root specific gene, MIC3 was cloned from nematode resistant cotton (Zhang *et al.*, 2002). The localized expression of MIC3 in roots of resistant cotton plants and its enhanced expression in nematode induced immature galls of resistant cotton M-249 suggested that MIC-3 might play a critical role in the resistance against root knot nematode. The molecular manipulation of this gene may result in enhanced resistance against root-knot nematode in cotton and other crops.

Osmotin, a thaumatin–like protein isolated from plants demonstrated antifungal activity against *Phytophthora infestans*, a destructive pathogen of potato and several other solanaceous plants. (Liu *et al.*, 1994). Nakajima *et al.* (1997) expressed a human lysozyme in tobacco to obtain resistance against Powdery mildew causing fungus *Erysiphe cichoracearum*.

The production of active oxygen species like superoxide anions, hydroxy radicals and hydrogen peroxide, H_2O_2, have been observed in many plant-pathogen interactions and are known to play an important role in plant defence. Plants have been engineered to continuously produce active oxygen species. Transgenic potatoes constitutively expressing the glucose oxidase gene, responsible for generating H_2O_2 and cloned from *Aspergillus niger* showed apoplastic accumulation of peroxide ions. The production of the free oxygen species in transgenic plants resulted in enhanced resistance against diseases such as late blight (*Phytophtora infestans*), wilt (*Verticillium dahliae*), and early blight (*Alternaria solani*) of potato (Wu *et al.*, 1997). Transgenic tobacco plants expressing a ribosome inactivating protein (RIP) from barley were shown to be resistant against *R. solani*. RIPs inhibit eukaryotic protein synthesis by modification of 28 S RNA of ribosomes. These proteins show specific characteristics in that they are inactive in the species from which they are isolated but can inactivate non-self ribosomes (Longemann *et al.*, 1992). Jach *et al.* (1992) expressed three anti-fungal genes in tobacco individually or in tandem. Expression of a barley type II chitinase, a -glucanase and a type I, RIP resulted in synergistic disease resistance in transgenic tobacco inoculated with *R. solani*. Transgenic tomato plants expressing yeast–9 desaturase gene have shown to have enhanced resistance against *Erysiphe polygoni*, the powdery mildew pathogen of tomato. Resistance was attributed to the increased levels of fatty acids in transgenic tomato (Wang *et al.*, 1998).

Expression of genes encoding hydrolytic enzymes from infection inducible promoters in transgenic plants would provide ingenious and effective method of resistance. Production of hydrolytic enzymes, anti-nutrients, enzymes for phytoalexin synthesis and viral components have been shown to protect plants against pathogens that have been typically difficult to control (Michelmore, 1995), thereby building durable resistance.

Engineering Resistance Against Bacterial Pathogens

Application of genetic engineering to combat bacterial diseases is in the stage of infancy. Most phytopathogenic bacteria thrive and multiply in the intercellular spaces in plants. Therefore signal peptide fusions have been used to direct transgene products to intercellular spaces. A variety of strategies have been tried to engineer bacterial resistance in transgenic plants, namely expression of non-plant antibacterial proteins, detoxification or inhibition of bacterial virulence factors, expression of antibacterial proteins of plant origin and over expression of cloned resistance genes in transgenic plants.

Expression of Non-plant Antibacterial Proteins

Several natural and synthetic polypeptides capable of disrupting bacterial membranes and eventually killing the bacterium have been expressed in transgenic plants. Lytic peptides are one of them and are small proteins with an amphipathic -helical structure, which makes pores in membranes resulting in the lysis of the bacterial cell membrane (Boman 1991). Cecropins are antibacterial lytic peptides produced in the hemolymph of the giant silk moth, *Hyalophora cecropia*. Transgenic tobacco plants expressing cecropins conferred increased resistance to *P. syringae* pv. *tabaci*, the causal agent of the devastating wild fire disease (Huang *et al.*, 1994; Huang & McBeath 1997). Similarly the bacterial blackleg of potato caused by *Erwinia carotovora* subsp. *atroseptica* is a potential menace to potato production worldwide. The disease can cause up to 30 per cent loss in field as well as in storage despite the use of chemicals and resistant varieties. Synthetic lytic peptide analogs, Shiva-1 and SB-37, produced from transgenes in potato plants reduced bacterial infection caused by *E. carotovora* subsp. *atroseptica* in transgenic potato plants (Wilson & Conner 1995; Arce *et al.*, 1999). Similarly, fire blight of apple caused by *E. amylovora* is a difficult disease to control by conventional management strategies. Limited success of antibiotics, development of antibiotic-resistant strains and lack of commercially acceptable fire blight resistant varieties makes the conventional management strategies less effective. Transgenic apple expressing the SB-37 lytic peptide analog showed increased resistance to *E. amylovora* in field tests (Norelli *et al.*, 1998).

Attacins are a group of antibacterial proteins produced by *H. cecropia* pupae (Hultmark *et al.*, 1983). Proteins of this group act by inhibiting the synthesis of outer membrane protein of Gram-negative bacteria (Carlsson *et al.*, 1998). Transgenic potato expressing Attacin exhibited enhanced resistance to bacterial infection by *E. carotovora* subsp. *atrospetica* (Arce *et al.*, 1999). Attacin gene expressed in transgenic pear and apple also lead to significant enhancement of resistance against *E. amylovora* (Norelli *et al.*, 1994; Norelli *et al.*, 1999; Reynoird *et al.*, 1999; Ko *et al.*, 2000). Targetting attacin to intercellular spaces- a site of multiplication of *E. amylovora* in transgenic apple significantly reduced the incidence of fire-blight incidence even in apple plants with low attacin production levels (Ko *et al.*, 2000).

Lactoferrin, another antibacterial protein is an iron binding glycoprotein, highly effective against both Gram- positive as well as Gram-negative bacteria. Expression of a mammalian lactoferrin protein in transgenic tobacco plants resulted in delayed symptom expression in plants inoculated with *R. solancearum*. Quantification of lactoferrin in transgenic plants indicated a positive correlation between lactoferrin accumulation and level of disease resistance (Zhang *et al.*, 1998). Lactoferrin is effective at a very low concentration and affects bacterial cells by altering the permeability of cell membranes resulting in disruption of bacterial cells (Montville and Bruno 1993). Thus this antimicrobial metabolite qualifies to be a novel and potential gene for engineering resistance against bacterial diseases. Tachyplesin is another antimicrobial peptide produced by Asian horseshoe crabs. Expression of Tachyplesin gene in transgenic potato resulted in resistance against soft rot disease caused by *Erwinia* spp. Lysozyme is a common antibacterial protein that causes lysis of cell walls through cleavage of the murein layer of bacterial peptidoglycan. Lysozyme genes from diverse sources, from bacteriophages (Jaynes *et al.*, 1987), plants (Broekaert *et al.*, 1997), animals (Vunnam *et al.*, 1997), and humans (Mitra and Zhang 1994, Nakajima *et al.*, 1997) have been cloned and can serve as the potential transgene for cloning into plants. Expression of T4L lysozyme from T4 bacteriophage in transgenic potato plants demonstrated significant reduction in tuber tissue maceration by soft rot pathogen *E. carotovora* (During *et al.*, 1993). Similarly transgenic apple with T4L lysozyme showed significant resistance to fire blight disease caused by *E. amylovora* (Ko 1999). Lysozyme genes have also been used to confer disease resistance against plant pathogenic bacteria in transgenic tobacco plants (Trudel *et al.*, 1995, Kato *et al.*, 1998. Human lysozyme transgenes have conferred disease resistance in tobacco through inhibition of fungal and bacterial growth, suggesting the possible use of the human lysozyme gene for controlling plant disease (Nakajima *et al.*, 1997).

Another effective gene that seems to hold promise in development of transgenic plants not only resistant to fungal disease as described earlier but also to bacterial diseases, is the glucose oxidase gene cloned from *Aspergillus niger*. The gene when expressed in transgenic potato imparted resistance against the bacterial soft rot pathogen *E. carotovora*. Expression of the transgene led to enhancement in the levels of hydrogen peroxide, which in turn activated the cellular defence genes and triggered a programmed cell death (Wu *et al.*, 1995).

Expression of Antibacterial Plant Proteins

Plants are the sources for many antimicrobial proteins, which are usually small, cationic, amphipathic and have open chain forms. The amphipathic structure with an -helix and an anti-parallel -sheet is highly conserved among plants, insects and animals. Most of these antimicrobial peptides contain cysteine residues, which are likely to form disulfide bonds leading to a rigid and compact structure. Antimicrobial peptides from plants possess 3-4 disulfide bonds such as thionin from barley (Reimann-Philipp *et al.*, 1989) and plant defensin from radish (Thomma *et al.*, 2002). Various kinds of antimicrobial peptides called defensins have been identified in plants including thionins (Bohlmann 1994), maize zeamatin (Malehorn *et al.*, 1994), coffee circulin (Tam *et al.*, 1999) and wheat puroindoline (Krishnamurthy *et al.*, 2001). Radish defensins (AFP1, 2, 3, 4) are small, 45-54 amino acid cysteine-rich peptides (Broekaert *et al.*, 1995) conserved in several plant species including members of Brassicaceae and inhibit growth of a broad range of microbes. Transgenic rice expressing AFP1 homologs of radish defensin cloned from *Brassica* spp. conferred effective resistance against paddy blast and blight caused by *Magnaporthe grisea* and *X. oryzae*, respectively (Kawata *et al.*, 2003). Simlarly expression of barley thionin gene resulted in significant reduction in growth of *P. syringae pv. syringae* and *P. syringae pv. tabaci*. Extracts of the transgenic plants when bioassayed against *Clavibacter michiganensis* pv. *sepedonicus* exhibited high antibacterial activity (Carmona *et al.*, 1993).

Expression of Bacterial Genes

Toxins are virulence factors produced by many plant pathogens and are responsible for the damage or disease caused by these pathogens (Bender *et al.*, 2002; Chakrabarty *et al.*, 2003). Because these toxins are deleterious to pathogens in addition to their host plants, the bacteria produce compounds that detoxify these toxins or produce toxin resistant target enzymes. Phaseolotoxin produced by the phytopathogenic bacteria *P. syringae* pv. *phaseolicola* is rendered inactive by a peptidase octase, both of plant and bacterial origin. Expression of octase of *P. syringae* pv. *phaseolicola* in transgenic tobacco confers resistance against the same pathogen (Fuente-Martinez 1992). Tabtoxin is a dipeptide toxin produced by *P. syringae* pv. *tabaci* the causal agent of Wild-fire disease of tobacco. The toxin acts by inhibiting the activity of gutamine synthetase in plants. Expression of a tabtoxin-inactivating enzyme, tabtoxin acetyl transferase, from *P. syringae* pv. *tabaci* enhanced resistance against the same bacterium in transgenic tobacco (Anazi *et al.*, 1989).

Exploring Prospects of Pathogen Derived Resistance

Although mostly effective against viral diseases, the pathogen derived resistance strategy appears to be a potential approach for engineering bacterial blight resistant cotton plants expressing the sequence of pathogenicity genes of *X. a.* pv. *malvacearum*. In *Xam* the genes governing pathogenicity were found to be plasmid-borne (Chakrabarty *et al.*, 1995, Chakrabarty and Gabriel, 1995). Two plasmid-borne pathogenicity genes were cloned from a highly virulent African strain of Xam (Chakrabarty *et al.*, 1997a & b). Characterization of these genes revealed that the sequences are conserved in several Xanthomonads (Chakrabarty *et al.*, 1997a; Gabriel 1999) all of which belonged to *Xanthomonas avr/pth* gene family. The sequence analysis also revealed several characteristic motifs that are involved in interaction with their hosts during pathogenesis (Chakrabarty and Gabriel, 1996; Chakrabarty *et al.*, 1997a&b; Gabriel *et al.*, 1997). The cloned genes have also been exploited for developing molecular diagnostic tools (Chakrabarty 1999; Chakrabarty *et al.*, 2001; Chakrabarty 2003 a, b & c) for detection of this pathogen in infected seeds or other sources. It would be interesting to see if transgenic cotton expressing *pthN*, the bacterial pathogenicity gene, can impart resistance against bacterial blight disease.

Engineering Resistance Against Viral Pathogens

Pathogen Derived Resistance

Transformation of plants with pathogen derived resistance (PDR) factors has become a potential tool for engineering resistance against diseases especially those that are caused by viruses. The term used for transformation of plants with portion of pathogen genome, frequently gives rise to lines of plants that are resistant to pathogens from which the sequence was derived. The nature of resistance obtained is variable and can either be RNA or protein mediated. RNA mediated resistance protects against very high levels of inoculum but is highly specific, whereas, protein mediated resistance in general offers low levels but broader spectrum of resistance (Pang *et al.*, 1993, Lomonosoff 1995, Baulcombe 1996a, Dawson 1996).

The number of plant species amenable to transformation and regeneration has increased greatly in recent years and it will likely be possible to produce fertile transformation of most if not all, significant crops in near future. Expression of literally any part of a plant virus gene can confer PDR (Lomonosoff, 1995). Most common are the transgenic plants expressing viral coat protein. Coat protein-mediated protection has been reported for tobacco mosaic virus, TMV (Nelson *et al.*, 1988), tomato mosaic virus (Sanders *et al.*, 1992), cucumber mosaic virus, CMV (Namba *et al.*, 1991, Quemada *et al.*, 1991), alfalfa mosaic virus (Loeshc-Fries *et al.*, 1987, Tumer *et al.*, 1987), potato virus X (Hemenway *et al.*1988), potato virus Y (Perlak *et al.*, 1994) and potato leaf roll virus, PLRV (Kaniewski *et al.*, 1993). Primers having engineered restriction enzyme sites were designed based on the conserved sequence of the coat protein gene of geminiviruses. These primers served as highly specific diagnostic tool in detecting CLCuV in infected cotton plants by amplification of a 770 bp coat protein gene (Chakrabarty 2003c). The gene is contemplated for use in developing transgenic cotton resistant to CLCuV. Besides coat-protein mediated resistance, resistant plants have also been engineered by expression of viral replicases, proteases, movement proteins, satellite virus RNA, viral antisense RNA and untranslated regions of viral genes (Gosal and Gosal 2000; Chawla 2000).

Examples of successful deployment of virus resistant transgenic plants that have been released in commercial scale include two cultivars of papaya ('Sunrise' and 'Rainbow') that are resistant to Papaya ring spot virus (PRSV) and a cultivar of squash (Freedom II), resistant to Zucchini yellow mosaic and watermelon mosaic viruses. PRSV is a highly destructive viral pathogen of papaya in several countries of the world (Lius *et al.*, 1997) and had been the biggest threat to the cultivation of papaya in Hawaii for several decades (Gonsalves 1998 a & b). The papaya industry was revived during 1990s by release of PRSV transgenic cultivars developed through coat protein mediated resistance. Freedom II, a virus resistant squash, is another transgenic cultivar that has been developed through coat-protein mediated PDR approach and has been commercially released at large scale (Fuchs and Gonsalves 1995). Similarly, Transgenic tomato engineered to resist CMV, a virus causing severe stunting and yield reductions throughout the world, showed high levels of resistance under field conditions (Fuchs *et al.*, 1996). Research is underway to generate transgenic tomato plants that are resistant to different strains of tomato spotted wilt virus and different tospoviruses, a group of viruses that seriously limit tomato production in the glasshouse and field, by combining virus transgenes with the native plant resistance gene Sw-5 (Gubba, 2000).

RNA Based Methods for Symptom Attenuation

Double stranded RNA interference mediated silencing of the viral gene has recently become very potential mechanism of development of disease resistant transgenic plant. The method is based on cloning a part of the viral gene in sense and antisense orientation in such a way that a double stranded RNA interference (dsRNAi) construct is formed. Transformation of plants with dsRNAi construct makes the plant resistant to subsequent infection by the virus having sequence complementary to the construct. This strategy is being exploited widely to engineer resistance in plants against RNA viruses (Hannon, 2002; Tenllado & Diaz-Riaz 2001). Most recently this phenomenon has been demonstrated successfully to engineer resistance against the bipartite DNA containing *Vigna mungo* yellow mosaic gemini virus from mung bean (Pooggin *et al.*, 2003).

Genetic engineering is proving to be highly effective for controlling virus diseases in a wide range of crops grown worldwide (Wilson, 1993; Gonsalves and Slightom, 1992). Compared to conventional breeding for virus resistance, genetic engineering provides a quicker and more precise technology to obtain plants that are resistant to viruses, although most transgenic virus-resistant plants are still under laboratory development. There have been concerns regarding a potential risk of recombination between the invading virus and the transgene leading to development of virus strains with altered specificities like wide host range and higher virulence.

Plantibodies: A Novel Strategy to Defend Plants Against Pathogens

Production of antibodies in response to invasion of an antigen (proteins or antigenic components of pathogen origin) was considered a domain of animal immune system. The remarkable potential of the recombinant DNA technology has made it possible for the plants to express the antibodies against pathogen proteins, which in turn enable them, defend against the target pathogen. The expression of pathogen specific antibody in plants, termed plantibody (Smith 1996; Gibbs 1997) is a promising approach for defending plants against potential pathogens (Schots *et al.*, 1992). The strategy for this unique approach includes fusion of the antibody molecules with carrier protein followed by its cloning and expression in transgenic plants. The plantibodies produced in the cell cytosol are expected to interact with their targets rendering them inactive (Zhang and Wu 1998). Most encouraging results so far have come from the cases of plantibodies generated against human pathogens, such as oral vaccines (Tacket and Mason1999).

Drawing a clue of the potential of plantibodies in combating human disease, plant scientists are now geared to extend this remarkable technology for plant disease control. The technology has been contemplated to control Tomato Spotted Wilt Virus, TSWV (Franconi *et al.*, 1999) and root knot nematodes caused by *Meloidogyne* spp. (Baum *et al.*, 1996). TSWV is a destructive pathogen of tomato and is one of the limitations to its successful cultivation. The TSWV monoclonal antibody producing mammalian gene, cloned and expressed from the transgenic plants are expected to inactivate the biological activities of series of viral proteins including N (important for viral replication), G1/G2 (uptake by thrips) and NSm (spread through the plant) (Franconi *et al.*, 1999). In the pathogenesis of root knot nematodes *Meloidogyne incognita* and other species of *Meloidogyne* cellulases are known to play important role. Anti-cellulase plantibodies expressed from transgenic plants are expected to interfere with pathogenesis of these nematodes by inactivating cellulases produced by them, thereby affecting their migration inside the plant roots (Baum *et al.*, 1996). In addition antibodies against cell cycle proteins are being generated to block initiation of giant cell formation through which these nematodes feed (Vrain 1999). Stabilising motifs have been found to ensure significant expression levels of single chain antibody fragments in the cytosol of plant cells. Although research on this novel approach of plant defence mechanism has been initiated, their practical application and commercialisation is yet to be realised.

Induced Resistance in Plant Health

Introduction of the first reliable chemical ASM (acibenzolar-s-methyl), an activator of broad-spectrum disease resistance in crops provides an additional tool of induced resistance in crop health management (Tally *et al.*, 1999). This chemical induction of plant inherent resistance strategy can be integrated with available genetic, cultural and biological methods judiciously to optimize their benefits and devise a sustainable disease control strategy (Staub, 2001). The strategy is based on the principle that of many natural defence mechanisms plants have evolved to survive in nature, only few are inducible and can be triggered by biological or chemical agents without any deleterious side effects and in a sufficiently controlled way for practical use. The best inducible defence mechanism is the systemic acquired resistance (SAR) which is triggered by localized infection by pathogen and non-pathogen including viruses, bacteria or fungi. The activation of SAR results in manifestation of broad-spectrum resistance against a wide array of pathogens. ASM is the first commercially available product that activates SAR in plants like other biological inducers with same spectrum of protection against pathogens (Ruess *et al.*, 1996; Staub *et al.*, 2001; Tally *et al.*, 1999). The spectrum and duration of SAR activation are crop-specific. The role of ASM in induction of defence in crop plants against various pathogens needs further experimentation and evaluation.

Biological Control

Biocontrol involves suppression of pests with other organisms. Biocontrol of plant pathogen has attracted great interest because of increasing regulation and restriction of pesticides or unsuccessful control attempts by other means. It is especially attracted for the soil borne diseases because these pathogens are difficult to control with specific pesticide and also such a venture may be highly expensive. Further, excessive use of broad spectrum or persistent chemicals may result soil/water contamination, development of resistance, deleterious effects on non-target organisms, and other harmful effects. However, despite much early optimism, as yet there are a very few established practical application of biocontrol.

Over the past one century research has repeatedly demonstrated that phylogenetically diverse microorganism can act as natural antagonist of various plant pathogens (Cook, 2000). Biocontrol action can be due to symbiosis, competition, induction of host resistance and predation. Of several bacterial and fungal antagonists of plant pathogens that have been screened on the basis of *in vitro* growth inhibition, only a few can suppress the plant disease under diversified growing conditions. Nonetheless, intensive screening has yielded numerous candidate organism of commercial development (McSpadden Gardener and Fravel, 2002). Some of the microbial taxa that have been successfully commercialized and are currently marketed as EPA–registered biopesticides in USA includes bacteria belonging to the genera of *Agrobacterium*, *Bacillus*, *Pseudomonas* and *Streptomyces* and fungi belonging to the genera of *Ampelomyces*, *Candida*, *Coniothyrum* and *Trichoderma*. Thus careful screening is a critical step in the development of a bioagent. Success of all subsequent stages depends on the ability of screening procedure to identify an appropriate candidate. Many useful bacterial biocontrol agents have been selected by observing zones of inhibition in petriplates. However, this method does not identify biocontrol agents with other modes of action such as parasitism, induced plant resistance or some other forms of competition.

Several commercially available biocontrol products have been released and used in USA to control plant diseases and pathogens of crop plants (Table 5.2). Biocontrol mechanism assumes specific significance in cotton, which is affected by a large number of soil borne fungi, causing root rots, wilts and seedling diseases. Seed treatment of cotton with *Gliocladium virens* and *B. subtilis* resulted in suppression of colonization of wilt pathogen *F. o.* fsp. *vasinfectum* (Zhang *et al.*, 1996). The result also indicated that such a treatment resulted in more vigorous and high yielding plants. Biological seed treatment for control of the seedling disease complex of cotton has been used successfully by Batson *et al.* (1999, 2000). A strain of *T. viride* successfully protected several crop plants against root rot caused by *R. solani*, *Sclerotium rolfsi* (Mathivanan, 2000). Seed treatment with talc-formulations of these bio-fungicides increased germination of seeds and seedling root/shoot length, besides protecting them from *R. solani* strain infection.

Use of endophytic bacteria for biological control of *Fusarium* wilt of cotton is another novel approach (Chen *et al.*, 1995). Six strains of endophytes isolated from internal tissue of cotton reduced the severity of vascular wilt of cotton caused by *F. o.* fsp. *vasinfectum* in cultivar Rowden. These strains were

Table 5.2: Commercial Biocontrol Formulations Used in USA for Control of Plant Diseases

Product	Biocontrol Agent	Disease/Target Pathogens	Crop	Formulation	Application Method	Manufacturer Distributor
Actigard	Acibenzolar-S-Methyl	Many	Tobacco, tomato, lettuce, spinach	Water dispersible granules	Drench	Syngenta Crop Protection, Greensboro, NC, USA
Actinovate	*Streptomyces lydicus*	Soilborne diseases	Greenhouse, nursery crops, turf grass	Water dispersible granules	Drench	Natural Industries Inc., Houston, TX, USA
AQ10	*Ampelomyces quisqualis*	Powdery mildew	Apple, cucurbits, grapes, strawberries, tomatoes	Water dispersible granules	Spray	Ecogen Inc., Langhorne, PA, USA
Aspire	*Candida oleophila*	*Botrytis* spp., *Penicillium* spp.	Citrus, pome fruits	Wettable powder	Fruit drench, drip or spray	Ecogen Inc., Langhorne, PA, USA.
Bioject Spot-less	*Pseudomonas aureofaciens*	*Dollar spot, anthracnose, Pythium aphanidematum etc.*	Turf and others	Liquid	Overhead irrigation	Eco Soil Systems, San Diego, CA, USA
Bio-save 10LP, 110	*Pseudomonas syringae*	*Botrytis* spp., *Penicillium* spp., *Mucor pyroformis, Geotrichum candidum*	Citrus, pome fruits, cherries, potatoes	Lyophilized product, Frozen cell concentrated pellets	Fruit drench, drip or spray	Village Farms LLC, Longwood. FL, USA
BlightBan A506	*Pseudomonas fluorescens* A506	*Erwinia amylovora* and russet-inducing bacteria	Almond, apple, apricot, blueberry, cherry, peach, pear, potato, strawberry, tomato	Wettable powder	Bloom time spray of flowers and fruits	NuFarm Inc, USA
Cedomon	*Pseudomonas chlororaphis*	Leaf stripe, net blotch, *Fusarium* sp., spot blotch, leaf spot etc.	Barley and oats	Seed treatment	Seed dressing	BioAgri AM, Uppsala, Sweden
Companion	*Bacillus subtilis GB03, other B. subtilis, B. lichenformis, B. megaterium*	*Rhizoctonia, Pythium, Fusarium* and *Phytophthora*	Green house and nursery	Liquid	Drench or spray	Growth Products, White Plains, NY, USA
Contans WG	*Coniothyrium minitans*	*Sclerotinia sclerotiorum* and *S. minor*	All crops	Water dispersible granules	Spray	Prophyta Biologischer Pflanzenschutz GmbH, Germany

Contd...

Table 5.2–Contd...

Product	Biocontrol Agent	Disease/Target Pathogens	Crop	Formulation	Application Method	Manufacturer Distributor
Deny	*Burkholderia cepacia*	*Rhizoctonia, Pythium, Fusarium*	Alfalfa, barley, beans, clover, cotton, peas, grains, sorghum, vegetable and wheat	Aqueous suspension	Seed dresser, drip Irrigation, Seedling drenches	Stine Microbial Products, Memphis, TN., USA
DiTera Biocontrol	*Fermentation product of killed Myrothecium verrucaria*	*Parasitic nematodes*	Cole crops, grapes, ornamentals, turf, trees	Powder and liquid suspension	Soil application	Valent Biosciences Inc., Walnut Creek, CA, USA
Galltrol	*Agrobacterium radiobacter strain 84*	*Agrobacterium tumefaciens*	Fruit, nut ornamental nursery	Agar grown culture	Bacterial suspension applied to seeds, seedlings, cuttings, roots, stems and soil drench	AgBioChem Inc., Orinda, CA, USA
HiStick N/T	*Bacillus subtilis* MB1600	*Fusarium, Rhizoctonia, Aspergillus*	Soybean, alfalfa, beans, peanuts	Slurry, damp and dry formulation	Seed inoculation	Becker Underwood Inc., Hemel Hempstead, UK
Intercept	*Burkholderia cepacia*	*Rhizoctonia, Pythium, Fusarium*	Maize, vegetables, cotton	Aqueous suspension	Seed dresser, drip Irrigation, Seedling drenches	Soil Tech. Corp., Fairfield, IA, USA
Kodiak	*Bacillus subtilis,* GB03	*Rhizoctonia solani, Fusarium* spp., *Alternaria* spp, *Aspergillus* spp.	Cotton, legumes	Dry powder with compatible fungicides	Slurry mix for seed treatment	Gustafson Inc., Plano, TX, USA
Messenger	*Erwinia amylovora* harpin protein	Many	Field, ornamental and vegetable crops	Powder	Drench or spray	EDEN Bioscience Corporation, Bothell, WA, USA
Mycostop	*Streptomyces griseoviridis* Strain K61	*Fusarium* spp., *Alternaria brassicola, Phomopsis* spp., *Botrytis* spp., *Pythium* spp., *Phytophthora* spp.	Field, ornamental and vegetable crops	Powder	Drench, spray or through irrigation system	Kemira Agro Oy, Helsinki, Finland

Contd...

Table 5.2–Contd...

Product	Biocontrol Agent	Disease/Target Pathogens	Crop	Formulation	Application Method	Manufacturer Distributor
Nogall	*Agobacterium radiobacter* K1026	Crown gall caused by *Agrobacterium tumefaciens*	Fruit, nut and ornamental nursery plants	Agar grown pure culture	Bacterial suspension applied to seeds, seedlings, cuttings, roots, stems or soil drench	Bio-Care Technology, Australia
Primastop	*Gliocladium catenulatum*	Soil borne pathogens causing rots and wilts	Ornamental, vegetable and tree crops	Powder	Drench, spray or through irrigation system	AgBio Development Inc., Westminster, CO, USA
RootShield	*Trichoderma harzianum* strain KRL-AG2	*Pythium* spp., *Rhizoctonia solani, Fusarium* spp.	Trees, shrubs, ornamentals and vegetables	Granules or wettable powder	Soil application as granules or drench	Bioworks Inc., Geneva, NY, USA
Serenade	*Bacillus subtilis Qst716*	Powdery mildew, downy mildew, Cercospora leaf spot, eearly & late blight, fire blight, brown rot	Cucurbits, fruits, vegetables	Wettable powder	Spray	AgraQuest Inc., Davis, CA, USA
SoilGard	*Gliocladium virens* GL-21	Damping off and root rot pathogens like *Rhizoctonia solani* and *Pythium* spp.	Ornamental and food crops	Granules	Soil application	Certis Inc., Columbia, MD, USA
YieldShield	*Bacillus pumilis* GB34	Soilborne fungi causing root rot	Soybean	Dry powder	Seed slurry treatment	Gustafson Inc., Plano, TX, USA

identified as members of *A. saperdae*, *B. pumilus*, *Phytobacterium rubiacearum*, *P. putida* and *Burkholderia solanacearum*. These bacteria can survive within cotton stems up to 28 days after inoculation.

Soil application of *T. harzianum* controlled the damping off of cotton caused by *R. solani* and gave crop stands comparable to or better than those in non-infested soils (Lewis *et al.*, 1996). Flores *et al.* (1997) improved the biological efficiency of *T. harzianum* against *R. solani* by increasing the copy number of the proteinase gene *prb1* through molecular manipulations. The transformants were stable and carried from 2-10 copies of *prb1* gene. High level of proteinase expression was detected when the improved biocontrol agent and *R. solani* were allowed to interact *in vitro*. Incorporation of *T. harzianum* transformants into pathogen infested soil sigficantly reduced the disease caused by *R. solani* in cotton.

In addition to direct application of biocontrol agents, the antimicrobial products or synthesised products of these bioagents can also be used for control of diseases:

1. Research on mechanism of biocontrol employed by effective bacterial strains revealed a variety of natural products that can be exploited for the development of chemical control measures. Pyrrolnitrin, a natural inhibition product of *Pseudomonas* spp has resulted in the synthesis and development of fludioxonil, a broad-spectrum fungicide used for seed treatment, foliar spraying or soil drenching (Ligon *et al.*, 2000). 2, 4 diacetyl phloroglucinol produced by strain of *P. fluorescens* was found to inhibit Xam *in vitro*. The natural compound also protected cotton from the infection when applied prior to inoculation (Mondal *et al.*, 2001)
2. Studies on biocontrol induced resistance in plant against bacterial pathogens has also resulted in the discovery of harpin, an effective protein that is now being used to activate crop defense mechanism prior to pathogen attack (Hutchinson, 1998).
3. A variety of pathogenic and nonpathogenic microorganisms can induce plant defence and may be useful as biocontrol agents (Van Loom *et al.*, 1998).

Research on the effects of organic amendments suggested that both chemical and biological component of compost amended soils can contribute to disease suppression (Abbasi *et al.*, 2002; Bulluck and Restaino, 2002). Investigations to identify the cause of such suppressive effects of organic amendments and their further exploitation may lead to improved biocontrol strategies. Further, advanced molecular techniques can be used to characterize the diversity, abundance and activities of microbes that inhabit in the rhizosphere/phyllosphere of plant, including those that significantly impact plant health. Recombinant DNA technology can be exploited to engineer biocontrol agents with improved and broad-spectrum efficacy.

Newly acquired expertise in plant transformation, continued discovery of plant *R* genes and novel genes as well as improvement of gene isolation techniques have intensified genetic engineering of plants for resistance against devastating diseases. Engineering both *R* genes and downstream defence responses will further enhance resistance. In addition to the progress being made on the plant components contributing to resistance, a thorough understanding of the pathogen side of the equation that contributes to its pathogenicity as well as the critical genes involved in pathogenicity are expected to open new horizon in plant protection. In future an integrated approach based on combined knowledge of the defence systems of the plant and assault systems of the pathogen should lead to development of designer genes for effective resistance against diseases. Plant biotechnology has already made significant strides towards development of disease resistant transgenic plants with several varieties being used commercially.

REFERENCES

Abbasi, P.A., Al-Dahmani, J., Sahin, F., Hoitink, H.A. J., and Miller, S. A. (2002). Effect of compost amendments on disease severity and yield of tomato in conventional and organic production system. *Plant Disease*, **86**: 156-161.

Anazi, H., Yoneyama, K. and Yamaguchi, I. (1989). *Mol. Gen. Genet.*, **219**: 492-494.

Anonymous (2000). Agriculture and Technology: Growing pains". The Economist, March 25, 2000, pages 1-16, available at: (www.economist.com).

Arce, P., Moreno, M., Gutierrez, M., Gebauer, M., Dell'Orto, P., Torres, H., Acuna, I., Oliger, P., Venegas, A., Jordana, X., Kalazich, J., Holuigue, L. (1999). Enhanced resistance to bacterial infection by *Erwinia carotovora* subsp. *atroseptica* in transgenic potato plants expressing the attacin or the cecropin SB-37 genes. *American Journal of Potato Research*, **76**: 169-177.

Asai, T, Tena, G., Plotnikova, J., Willmann, M.R., Chiu, W.L., Gomez L.G., Boller, T., Ausubel, F. M. and Sheen, J. *et al.*, (2002). MAP kinase signaling cascade in *Arabidopsis* innate immunity. *Nature*, **425**: 977-983.

Baker, B., Zambryski, B., Staskawicz, R., and Dinesh-Kumar, S.P. (1997). Signaling in plant-microbe interactions. *Science*, **276**: 726-733.

Bannerjee, A.K., Agrawal, D.C., Nalawade, S.M. and Krishnamurthy, K.V. (2002). Transient expression of beta-glucuronidase in embryo axes of cotton by Agrobacterium and particle bombardment methods. *Biologia Plantarum*, **45**: 359.

Batson, W., Cacers, J., Benson, M., Cubeta, M., Brannen, P., Kenny, D., Elliot, M., Huber, D., Hickman, M., Keinath, A., Dubose, V., Ownley, B., Newman, M., Rothrock, C., Schneider, R., and Summer, D. (2000). Evaluation of biological seed treatments for control of the seedling complex of cotton. *Biol. And Cul. Test*, **15**: 31-32.

Batson, W., Cacers, J., Benson, M., Cubeta, M., Brannen, P., Kenny, D., Elliot, M., Huber, D., Hickman, M., Keinath, A., Dubose, V., Ownley, B., Newman, M., Rothrock, C., Schneider, R., and Summer, D. (1999). Biological seed treatments for control of seedling disease complex of cotton. *Biol. And Cul. Test*, **14**: 24-26.

Baulcombe, D. C. (1996a). Mechanisms of pathogen-derived resistance to viruses in transgenic plants. *Plant Cell*, **8**: 1833-1844.

Baulcombe, D. C. (1996b). RNA as a target and an initiator of post-transcriptional gene silencing in transgenic plants. *Plant Mol. Biol.*, **32**: 79-88.

Baum, T.J., Hiatt, A, Parrott, W.A., Pratt, L.H., Hussey R.S. (1996). Expression in tobacco of a functional monoclonal antibody specific to stylet secretions of the root-knot nematode. *Mol. Plant Microbe. Interact.*, **9**: 382-387.

Bender, C.L., Alarcon-Chaidez, F., Gross, D.C. (1999). *Pseudomonas syringae* phytotoxins: Mode of action, regulation and biosynthesis by peptide and polyketide synthetases. *Microbiol. Mol. Biol. Rev.*, **63**: 266-292.

Birch, R.G. (1997). Plant transformation: Problems and strategies for practical application. *Annu. Rev. Plant Physiol. Plant Mol.*, **48**: 297-326.

Blasingame, D. and Patel, M.V. (2000). Cotton diseases loss estimate committee report. Pages 132-133. *In*: Proc. Beltwide Cotton Conf. P. Dugger and D. Richter (eds.). National Cotton Council in America, Memphis, Tn.

Bohlmann, H. (1994). The role of thionins in plant protection. *Crit. Rev. Plant Sci.*, **13**: 1-16.

Bolar, J.B. (1999). Genetic engineering of 'McIntosh' apple for resistance to scab (*Venturia inaequalis*). Ph.D Dissertation. Cornell University, Ithaca, NY. 131 pp.

Bolar, J.B., Norelli, J.L., Wong, K.W., Hayes, C.K., Harman, G.E., and Aldwinckle, H.S. (1999). Expression of endochitinase from *Trichoderma harzianum* in transgenic apple increases resistance to apple scab and stunts growth. *Phytopathology*, **90**: 72-77.

Boman, H.G. (1991). Antibacterial peptides: Key components needed in immunity. *Cell*, **65**: 205-207.

Brinkerhoff, L.A. 1970). Variation in *Xanthomonas malvacearum* and its relation to control. *Annu. Rev. Phytopathol.*, **8**: 85-110.

Brinkerhoff, L.A., Verhalen, L.M., Johnson, W.M., Essenberg, M. and Richardson, P.E. (1984). Development of immunity to bacterial blight of cotton and its implications for other diseases. *Plant Dis.*, **68**: 168-73.

Broekaert, W.F., Cammue, B.P.A., De Bolle, M.F.C., Thevissen, K., De Samblanx, G.W., Osborn, R.W. (1997). Antimicrobial peptides from plants. *Crit. Rev. Plant Sci.*, **16**: 297-323.

Broekaert. W.F. *et al.*, (1995). Plant defensins: novel antimicrobial peptides as components of the host defense system. *Plant Physiol.*, **108**: 1353-1358.

Broglie, K, Chet, I, Holliday, M, Cressman, R, Biddle, P, Knowlton, S, Mauvais, C, and Broglie, R. (1991). Transgenic Plants with Enhanced Resistance to the Fungal Pathogen Rhizoctonia solani. *Science*, **254**: 1194-1197

Bulluck, L. R. III, and Ristaino, J. B. (2002). Effect of synthetic and organic fertility amendments on southern blight, soil microbial communities, and yield of processing tomatoes. *Phytopathology*, **92**: 181-189.

Cao, H., Bowling, S.A., Gordan, A.S., and Dong, X. (1994). Characterization of an Arabidopsis mutant that is nonresponsive to inducers of systemic acquired resistance. *Pl. Cell*, **6**: 1583-1592.

Cao, H., Glazebrook, J., Clarke, J., Volko, S., and Dong, X. (1997). The Arabidopsis *npr1* gene that controls systemic acquired resistance encodes a novel protein containing ankyrin repeats. *Cell*: 57-63.

Cao, H., Li, X., and Dong, X. (1998). Generation of broad-spectrum disease resistance by overexpression of an essential regulatory gene in systemic acquired resistance. Proc. Natl. Acad. Sci., USA, **95**: 6531-6536.

Carlsson A., Nyström, T., de Cock, H., and Bennich, H. (1998). Attacin-an insect immune protein-binds LPS and triggers the specific inhibition of bacterial outer-membrane protein synthesis. *Microbiology*, **144**: 2179-2189.

Carmona, M.J., Molina, A., Fernandez, J.A., Lopez-Fando, J.J. and Garcia-Olmedo, F. (1993). Expression of the alpha-thionin gene from barley in tobacco confers enhanced resistance to bacterial pathogens. *Plant J.*, **3**: 457-462. *Cell*, **80**: 363-366.

Chakrabarty, P.K. (2003a). Development of sensitive molecular diagnostic tools for rapid detection and differentiation of races of *Xanthomonas campestris* pv. *malvacearum*. Final Report submitted to the DBT, New Delhi, 59 pp.

Chakrabarty, P.K. (2003b). Development of DNA based diagnostic tools for detection of strains of *Xanthomonas axonopodis* pv. *malvacearum*. *CICR Newsletter*, **19(1)**: 8-9.

Chakrabarty, P.K. (2003c). PCR based detection of *Xanthomonas axonopodis* pv. *malvacearum* and cotton leaf curl virus, two major pathogens of cotton. *CICR Newsletter*, **19(2)**: 4-5.

Chakrabarty, P.K. (1999). Pathogenicity genes cloned from *Xanthomonas campesrtis* pv. *malvacearum* as diagnostic tools. *ICAR News*, **5**: 1-3.

Chakrabarty, P.K. and Gabriel, D.W. (1995). Plasmid-borne pathogenicity genes of *Xanthomonas campestris* pv. *malvacearum*. *Phytopathology*, **85**: 1156.

Chakrabarty, P.K. and Gabriel, D.W. (1996). Sequence analysis of *Xanthomonas avr/pth* gene family member that defeats all known cotton *R* genes. Plant Molecular and Cellular Biology Workshop, March 30-31, 1996, Whitney Lab., Univ. of Florida, Marineland, USA.

Chakrabarty, P.K., and Raj, S. (2002). Molecular analysis of avirulence gene with special reference to Plant Pathogenic Xanthomonads. *In* Molecular Approaches in Plant Disease Management (Eds. G.S. Shekhawat, B.P. Singh and R.A. Singh). Malhotra Publishing House, New Delhi, pp. 63-87.

Chakrabarty, P.K., Chatterjee, A, Gassmann, W. and Chatterjee, A.K. (2003). Bacterial products important for plant disease- Toxins and Growth factors. *In* Encyclopedia of Plant and Crop Science (R.M. Goodman, Ed.). Marcel Dekker Inc., New York (in press).

Chakrabarty, P.K., Duan, Y.P. and Gabriel, D.W. (1997a). Cloning and characterization of a member of the *Xanthomonas avr/pth* gene family member that defeats all commercially available *R* genes in cotton in USA. *Phytopathology*, **87**: 1160-1167.

Chakrabarty, P.K., Gabriel, D.W. and Raj, S. (1996). Cloning and characterization of a pathogenicity gene from a highly virulent strain of *Xanthomonas campestris* pv. *malvacearum*. *In*: Plant pathogenic bacteria, (ed., A. Mahadevan). IX International Conference on Plant Pathogenic Bacteria, Aug. 26-29, 1996, University of Madras, Madras, India, Udayan offset, Chennai, pp. 236-246.

Chakrabarty, P.K., Gabriel, D.W. and Raj, S. (1997b). Pathogenicity gene *pthN* in *Xanthomonas axonopodis* pv. *malvacearum* pleiotropically encodes hyper-sensitive response in bacterial blight resistant cotton (*G.hirsutum*). Golden Jubilee International Conference on Integrated Plant Disease Management for Sustainable Agriculture, Nov. 10-15, 1997, IARI, New Delhi, p. 135.

Chakrabarty, P.K., Gabriel, D.W., Sravan Kumar, V. and Raj, S. (2001). Molecular basis of pathogenicity of *Xanthomonas axonopodis* pv. *malvacearum* and use of cloned pathogenicity genes as diagnostic tools. *In*: Role of Resistance in Intensive Agriculture (eds: S. Nagarajan and D.P. Singh), Kalyani Publishers, New Delhi, India, pp.115-135.

Chakrabarty, P.K., Mahadevan, A., Raj, S., Meshram, M.K. and Gabriel, D.W. (1995). Plasmid borne determinants of pigmentation, exopolysaccharide production, and virulence in *Xanthomonas campestris* pv. *malvacearum*. *Can. J. Microbiol*. **41**: 740-745.

Chawla, H.S. (2000). Biotechnology in the production of disease resistance plants. *In* Plant Biotechnology-Recent Advances (P.C. Trivedi, Ed.). Panima Publishing Corporation, New Delhi, pp. 407-434.

Chen, C., Bauske, E.M., Musson, G., Rodriguez K. R., Kloepper, J.W. (1995). Biological control of Fusarium wilt on cotton by use of endophytic bacteria. *Biological Control*, **5**: 83-91.

Chisholm, S.T., Mahajan, S.K., Whitham, S.A., Yamamoto, M.L., Carrington, J.C. (2000). Cloning of the Arabidopsis *RTM1* gene, which controls restriction of long-distance movement of tobacco etch virus. Proc. Natl. Acad. Sci., USA. **97**: 489-94.

Collins, N.C., Webb, C.A., Seah, S., Ellis, J.G., Hulbert, S.H. and Pryor, A. (1998). The isolation and mapping of disease resistance analogs in maize. *Mol. Plant-Microbe Interact.*, **11**: 968-978.

Cook, R. J. (2000). Advances in Plant Health Management in the twentieth Century. *Annu. Rev. Phytopathol.*, **38**: 95-116.

Creusot, A., *et al.* (1999). *Genome*, **42**: 254-264.

Dawson, W.D. (1996.) Gene silencing and virus resistance: A common mechanism. *Trends Pl. Sci.*, **1**: 107-108.

De Feyter, R. and Gabriel, D.W. (1991). At least six avirulence genes are clustered on a 90- kilobase plasmid in *Xanthomonas campestris* pv. *malvacearum*. *Mol. Plant-Microbe Interact.*, **4**: 423-432.

De Feyter, R., Yang, Y. and Gabriel, D.W. (1993). Gene-for-genes interaction between cotton *R* genes and *Xanthomonas campestris pv. malvacearum avr* genes. *Mol. Plant-Microbe Interact.*, **6**: 225-237.

Delaney, T.P., Friedrich, L., Ryals, J.A. (1995). Arabidopiss signal transduction mutant defective in chemically and biologically induced disease resistance. Proc. Natl. Acad. Sci. USA, **92**: 6602-6606.

Deng, Z., Huang, S., Ling, P., Chen, C., Yu, C., Weber, C.A., Moore, G.A. and Gmitter, F.G. (2000). Cloning and characterization of NBS-LRR class resistance gene candidate sequences in citrus. *Theo. Appl. genet.*, **101**: 814-822.

Dodds, P.N. Lawrence, G.J. and Ellis, J.G. (2001). Contrasting modes of evolution acting on the complex N locus for rust resistance in flax. *The Plant J.*, **27**: 439-453.

Düring, K., Porsch, P., Fladung, M. and Lörz, H. (1993). Transgenic potato plants resistant to the phytopathogenic bacterium *Erwinia carotovora*. *Plant J.*, **3**: 587-598.

Essenberg, M., Bayles, M.B., Samad, R.A. and Hall, J.A. (2002). Four near-isogenic lines with different genes for bacterial blight resistance. *Phytopathology* (In Press).

Flor, H.H. (1971). Current status of the gene for gene concept. *Annu. Rev. Phytopath.*, **9**: 275-296.

Flores, A., Chet, I. And Herrera Estrella, A. (1997). Improved biocontrol activity of Trichoderma harzianum by over-expression of the proteinase-encoding gene *prb1*. *Currrent Genetics*, **31**: 30-37.

Franconi, A., Roggero, P., Pirazzi, P., Arias, F.J., Desiderio, A., Bitti, O., Pashkoulov, D., Mattei, B., Bracci, L., Masenga, V., Milne, R.G., Benvenuto, E. (1999). Functional expression in bacteria and plants of an scFv antibody fragment against tospoviruses. *Immunotechnology*, **4**: 189-201.

Fuchs, M, Provvidenti, R., Slightom, J. L., and Gonsalves, D. (1996). Evaluation of transgenic tomato plants expressing the coat protein gene of cucumber mosaic virus WL under field conditions. Plant Dis., **80**: 270-275.

Fuchs, M. and Gonsalves, D. (1995). Resistance of transgenic hybrid squash ZW-20 expressing the coat protein genes of zucchini yellow mosaic virus and watermelon mosaic virus 2 to mixed infections by both potyviruses. Bio/Technol., **13**: 1466-1473.

Fuente-Martinez, J., Mosqueda-Cano, G., Alvarez-Morales, A. and Herrera-Estrella. (1992). Expression of a bacterial phaseolotoxin-resistant ornithil transcarbamylase in transgenic tobacco confers resistance to Psedumonas syringae pv. phaseolicola. *Bio/Technol.*, **10**: 905-907.

Gabriel, D.W. (1999). The *Xanthomonas avr/pth* gene family. *In*: Plant-Microbe Interactions (G. Stacey and N.T. Keen, Eds.), Vol. 4, APS Press, St. Paul, pp. 39-55.

Gabriel, D.W., Yuan, Q., Yang, Y. and P.K. Chakrabarty. (1997). Role of nuclear localizing signal sequences in three disease phenotypes determined by the *Xanthomonas avr/pth* gene family: Bacterial Determinants for Pathogenicity and Avirulence. In: Biology of Plant-Microbe Interactions, Volume 1 (Eds: Gary Stacey, Beth Mullin, and Peter M. Gresshoff. International Society for Molecular Plant-Microbe Interactions.

Gibbs, W.W. (1997). Biotechnology–Plantibodies. *Sci. Am.*, **277**: 44.

Glazebrook, J., Rogers, E.E., and Ausubel, F.M. (1996). Isolation of Arabidopsis mutants with enhanced disease susceptibility by direct screening. *Genetics*, **143**: 973-982.

Gonsalves, D. (1998a). Transgenic virus resistant papaya: new hope for controlling papaya ringspot virus in Hawaii. APSnet Feature article, September 1 through September 30, 1998.

Gonsalves, D. (1998b). Resistance to papaya ringspot virus. *Annu. Rev. Phytopath.*, **36**: 415-437.

Gonsalves, D. and Slightom, J. L. (1993). Coat-protein mediated protection: analysis of transgenic plants resistance in a variety of crops. *Sem. Virol.* **4**: 397-406.

Gosal, S.S. and Gosal, S.K. (2000). Genetic transformation and production of transgenic plants. *In* Plant Biotechnology-Recent Advances (P.C. Trivedi, Ed.). Panima Publishing Corporation, New Delhi, pp. 29-40.

Graham, M.A., Marek, L.F., Lohnes, D., Cregan, P. and Shoemaker, R.C. (2000). Expression and genome organization of resistance gene analogs in soybean. *Genome*, **43**: 86-93.

Gubba, A. (2000). Transgenic and natural resistance: Towards development of tomato and pepper plants with broad resistance to virus infection. PhD Dissertation, Cornell University, 196 pp.

Hannon G. J. (2002). RNA interference. *Nature*, **418**: 244-251.

Hayward, A.C. (1993). The hosts of Xanthomonas. *In*: *Xanthomonas* (J.G.Swing and E.L.Civerolo Ed.). Chapman & Hall, London, p. 1.

Hemenway, C., Fang, R. X., Kaniewski, W., Chua, N. H. and Tumer, N. (1988). Analysis of the mechanism of protection in transgenic plants expressing the potato virus X coat protein or its antisense RNA. *EMBO J.*, **7**: 1273-1280.

Hirt, H. (2002). A new blue print for plant pathogen resistance. *Nature Biotech.*, **20**: 450-451.

Huang, W., Cui, X., Tian, Y., Lin, M., Peng, X. (1994). Cloning of T7 lysozyme gene and construction of the vector for transgenic plants resistant to bacterial infection. *Weishengwu Xuebao*, **34**: 261-265.

Huang, Y., and McBeath, J.H. (1997). Plant promoters controlled expression of cecropin B gene in transgenic potatoes renders disease resistance to soft rot bacteria. *Phytopathology*, **87**: S45. (Abstr.)

Hudspeth, R.L., Hobbs, S.L., Anderson, D.M. and Grula, J.W. (1996). Characterization and expression of chitinase and 1, 3-beta-glucanase genes in cotton. *Plant Molecular Biology*, **31**: 911-916.

Hulbert, H. S; Craig, A. W; Shavannor, M. S and Sun, Q. (2001). Resistance gene complexes: Evolution and utilization. *Annu. Rev. Phytopathol.*, **39**: 285-312.

Hultmark, D., Engstrom, A., Andersson, K., Steiner, H., Bennich, H., and Boman, H.G. (1983). Insect immunity. Attacins, a family of antibacterial proteins from *Hyalophora cecropia*. *EMBO J.*, **2**: 571-576.

Hunter, R.E. and Brinkerhoff, L.A. (1961). Blight differentials. Pages 7-8. *In*: Proc. Cotton Dis. Council, Cotton Disease Council and National Cotton Council, Memphis, Tn.

Hutchinson, S.W. (1998). Current concepts of active defense in plants. *Annu. Rev. Phytopathol.*, **36**: 59-90.

Jach, G., Longemann, S., Wolf, G., Oppenheim, A., Chet, I., Schell, J., Longemann, J. (1992). Expression of a bacterial chitinase leads to improved resistance of transgenic tomato against fungal infection. *J. Biopractice*, **1**: 33-40.

Jaynes, J. M., Xanthopoulos, K.G., Destefano-Beltrán, L., and Dodds, J.H. (1987). Increasing bacterial disease resistance in plants utilizing antibacterial genes from insects. *Bioassays*, **6**: 263-270.

Johal, G.S. and Briggs, S.P. (1992). Reductase activity encoded by the Hm1 disease resistance gene in maize. *Science*, **258**: 985-987.

Jones, J.D.G. (1996). Plant disease resistance genes: structure, function and evolution. *Curr. Opin. Biotech.*, **7**: 155-160.

Jorgensen, J.H. (1992). Discovery, characterization and exploitation of *Mlo* powdery mildew resistance in barley. *Euphytica*, **63**, 141-152.

Kaniewski, W., Lawson, C. and Thomas, P. (1993). Agronomically useful resistance in Russet Burbank potato containing a PLRV CP gene. Abstract. IX International Congress of Virology, Glasgow, Scotland.

Kato, A., Nakamura, S., Ibrahim, H., Matsumi, T., Tsumiyama, C., and Kato, M. (1998). Production of genetically modified lysozymes having extreme heat stability and antimicrobial activity against Gram-negative bacteria in yeast and in plants. *Nahrung*, **42**: 128-130.

Kawata, N., Nakajima, Y., Yamamoto, T., Mori, K., Oikawa, T., Fukumoto, F., and Kuroda, S. (2003). Genetic Engineering for disease resistance in rice (*Oryza sativa*) using antimicrobial peptides. *JARQ*, **37**: 71-76.

Ko, K. (1999). Attacin and T4 lysozyme transgenic 'Galaxy' apple: Regulation of transgene expression and plant resistance to fire blight (*Erwinia amylovora*). Ph.D dissertation, Cornell University, Ithaca, NY. 194 pp.

Ko, K., Norelli, J.L., Reynoird, J.-P., Boresjza-Wysocka, E., Brown, S.K., Aldwinckle, H.S. (2000). Effect of untranslated leader sequence of AMV RNA 4 and signal peptide of pathogenesis-related protein 1b on attacin gene expression, and resistance to fire blight in transgenic apple. *Biotech. Lett.*, **22**: 373-381.

Krishnamurthy, K., Balconi, C., Sherwood J. E. and Giroux, M. J. (2001). Wheat puroindolines enhance fungal disease resistance in transgenic rice. *Mol. Plant Microbe Interact.*, **14**: 1255-1260.

Lewis, J.A., Lumsden, R.D. and Locke, J.C. (1996). Biocontrol of damping-off diseases caused by Rhizoctonia solani and Pythium ultimum with alginate prills of Gliocladium virens, Trichoderma hamatum and various food bases. *Biocontrol Science and Technology*, **6**: 163-173.

Ligon, J.M., Hill, D.S., Hammer, P.E., Torkewitz, N.R., Hofmann, D., Kempf, H. J., van Pee, K.H. (2000). Natural products with antifungal activity from Pseudomonas biocontrol bacteria. *Pest Man. Sci.*, **56**: 688-695.

Lin, W., Anuratha, C.S., Datta, K., Potrykus, I., Muthukrishnan, S., Datta, S.K. (1995). Genetic engineering of rice for resistance to sheath blight. *Bio/Technol.*, **13**: 686-691.

Linos, A. A., Bebeli, P. J. and Kaltsikes, P. J. (2002). Cultivar identification in upland cotton using RAPD markers. *Australian J. Agric. Res.*, **52(6)**: 637-642.

Liu, D., Gaghothama, K.G., Hasegawa, P.M. and Bressan, R.A. (1994). Proc. Natl. Acad. Sci., USA, **91**: 1888-1892.

Lius, S., Manshardt, R. M., Fitch, M. M. M., Slightom, J. L., Sanford, J. C. and Gonsalves, D. (1997). Pathogen-derived resistance provides papaya with effective protection against papaya ringspot virus. *Mol. Breed.*, **3**: 161-168.

Loesch-Fries, L. S., Merlo, D., Sinnen, T., Burhop, L., Hill, K., Krahn, K., Jarvis, N., Nelson, S. and Halk, E. (1987). Expression of alfalfa mosaic virus RNA 4 in transgenic plants confers virus resistance. *EMBO J.*, **6**: 1845-1852.

Lomonossoff, G. P. (1995). Pathogen-derived resistance to plant viruses. *Annual Review of Phytopathology*, **33**: 323-343.

Longemann, J., Jach, G., Tommerup, M., Mundy, J., Schell, J *et al.* (1992). Expression of a barley ribosome inactivating proteinleads to increased fungal protection in transgenic tobacco plants. *Bio/Technol.*, **10**: 305-308.

Lorito M., Woo, S.L., Fernandez, I.G., Colucci, G., Harman, G.E., Pintor-Toro, J.A., Filipponem, E., Muccifora, S., Lawrence, C.B., Zoina, A., Tuzun, S., and Scala, F. (1998). Genes from mycoparasitic fungi as a source for improving plant resistance to fungal pathogens. Proc. Natl. Acad. Sci., USA, **95**: 7860-7865.

Malehorn, D. E. *et al.* (1994)) Characterization and expression of an antifungal Zeamatin like protein (Zlp) gene from Zea mays. *Plant Physiol.*, **106**: 1471-1481.

Martin, G.B., Brommonschenkel, S.H., Chuwongse, J., Frary, A., Ganal, M.W., Spivey, R., Wu, T., Earle, E.D. and tanksley, S.D. (1993). Map based cloning of a protein kinase gene conferring disease resistance in tomato. *Science*, **262**: 1432-1436.

Mathivanan, N., Srinivasan, K. and Chelliah, S. (2000). Biological control of soil-borne diseases of cotton, egg plant, okra and sunflower by *Trichoderma viride*. *Zeitschrift fur Pflanzenkrankheitan and Pflanzenschutz*, **107**: 235-244.

McSpadden Gardener, B.B. and Fravel, D.R. (2002). Biological control of plant pathogens: Research, Commercialization, and Application in the USA. Online. Plant Health Progress doi: 10.1094/ PHP-2002-0510-01-RV.

Meyers, B.C., Dickerman, A.W., Michelmore, R.W., Sivaramakrishnan, S., Sobral, B.W. and Young, N.D. (1999). Plant disease resistance genes encode members of an ancient and diverse protein family within the nucleotide-binding superfamily. *Plant J.*, **20**: 317-332.

Michelmore, R. (1995). Molecular approaches to manipulation of disease resistance genes. *Annu. Rev. Phytopathol.*, **33**: 393-427.

Mitra, A. (2001). Genetic Engineering. *In*: Encyclopedia of Plant Pathology (O.C. Maloy and Murray, D. Eds.), VolI, pp. 507-512, John Wiley and Sons Inc., New York.

Mitra, A. and Zhang, Z. (1994). Expression of a human lactoferrin cDNA in tobacco cells produces antibacterial protein(s). Plant Physiol., **106**: 977-981.

Mondal, K. K., Gureja, P., Verma J.P. (2001). Management of *Xanthomonas campestris pv malvacearum*-induced blight of cotton Through phenolics of cotton Rhizobacterium. *Current Microbiology*, **43**: 336-339.

Montville, T.J. and Bruno, M.E.C. (1994). Evidence that dissipation of proton motive force is a common mechanism of action for bacteriocins and other antimicrobial proteins. *Int. J. Food Microbiol.*, **24**: 53-74.

Nakajima H., Muranaka, T., Ishige, F., Akutsu, K., Oeda, K. (1997). Fungal and bacterial disease resistance in transgenic plants expressing human lysozyme. *Plant Cell Rep.*, **16**: 674-679.

Namba, S., Ling, K., Gonsalves, C., Gonslaves, D. and Slightom, J. L. (1991). Expression of the gene encoding the coat protein of cucumber mosaic virus (CMV) strain WL appears to provide protection to tobacco plants against infection by several different CMV strains. *Gene*, **107**: 181-188.

Nelson, R. S., McCormick, S. M., Delanney, X., Dube, P., Layton, J., Anderson, E. J., Kaniewska, M., Prosksch, R. K., Horsch, R. B., Rogers, S. G., Fraley, R. T. and Beachy, R. N. (1988). Virus tolerance, plant growth, and field performance of transgenic tomato plants expressing coat protein from tobacco mosaic virus. *Bio/Technol.*, **6**: 403-409.

Norelli, J.L., Aldwinckle, H.S., Destéfano-Beltrán, L., and Jaynes, J.M. (1994). Transgenic 'Malling 26' apple expressing the attacin E gene has increased resistance to *Erwinia amylovora*. *Euphytica*, **77**: 123-128.

Norelli, J.L., Borejsza-Wysocka, E., Momol, M.T., Mills, J.Z., Grethel, A., Alkwinckle, H.S., Ko, K., Brown, S.K., Bauer, D.W., Beer, S.V., Abdul-Kader, A.M., Hanke, V. (1999). Genetic transformation for fire blight resistance in apple. Acta *Horticulturae*, **489**: 295-296.

Norelli, J.L., Mills, J.Z., Momol, M.T., Aldwinkle, H.S. (1998). Effect of cercropin-type transgenes on fire blight resistanc of apple. *Acta Hort.*, **489**: 273-278.

Oerke, E. -C., Dehne, H.–W., Schonbeck, F., and Weber, A. (1994). "Crop production and crop protection: estimated losses in major food and cash crops." Elsevier, Amsterdam.

Pang, S-Z., Slightom, J. L., and Gonsalves, D. (1993). Different mechanisms protect transgenic tobacco against tomato spotted wilt and Impatiens necrotic spot tospoviruses. *Bio/Technol.*, **11**: 819-824.

Peneula, S., Danesh, D. and Young, N.D. (2002). Targeted isolation, sequence analysis and physical mapping of non TIR NBS_LRR genes in soybean. *Theo. Appl. Genet.*, **104**: 261-272.

Perlak, F., Kaniewski, W., Lawson, C., Vincent, M. and Feldman, J. (1994). Genetically improved potatoes: their potential role in integrated pest management. *In*: Proceed. 3rd EFPP Conference (M. Manka, Ed). *J. Phytopath.*, pp 451-454.

Pink, D and Puddephat, I. (1999). Deployment of disease resistance genes by plant transformation-a "mix and match" approach. *Tr. Plant. Sci.*, **48**: 71-75.

Pooggin, M., Shivprasad, P.V. and Veluthambi, K. (2003). RNAi targeting of DNA virus in plants. *Nature Biotech.*, **21**: 1312-132.

Quemada, H. D., Gonsalves, D. and Slightom, J. L. (1991). Expression of coat protein gene from cucumber mosaic virus strain C in tobacco: Protection against infections by CMV strains transmitted mechanically or by aphids. *Phytopathology*, **81**: 794-802.

Reimann-Philipp, U., Behnke, S., Batschauer, A., Schafer, E and Apel, K. (1989). The effect of light on the biosynthesis of leaf-specific thionins in barley, *Hordeum vulgare*. *Eur. J. Biochem.*, **182**: 283-289.

Reynoird, J.P., Mourgues, F., Norelli, J.L., Aldwinckle, H.S., Brisset, M.N., and Chevreau, E. (1999). First evidence for differences in fire blight resistance among transgenic pear clones expressing attacin gene. *Plant Science*, **149**: 23-31.

Rommens, C. M. and Kishore, M. G. (2000). Exploiting the full potential of disease-resistance genes for agriculture use. *Curr. Opin. Biotech.*, **11**: 120-125.

Rommens, C.M., Salmeron, J.M., Oldroyd, G.E., Staskawicz, B.J. (1995). Intergenic transfer and functional expression of the tomato disease resistance gene *Pto*. *Pl. Cell.*, **7**: 1537-1544.

Ruess, W., Mueller, K., Knauf-Beiter, K.G. and Staub, T. (1996). Plant activator CGA-245704: An innovative approach for disease control in cereals and tobacco. *In*: *Proceedings of the 1998 Brighton Conference: Pests and diseases* (British Crop Protection Council Ed.), Brighton, UK, pp. 53-60.

Rungis, D., Llewellyn, D., Dennis, E. S. and Lyon, B, R. (2002). Investigation of the chromosomal location of bacterial blight resistance gene prersent in Australian cotton (*Gossypium hirsutum* L). *Australian J. Agric. Res.*, **53(5)**: 551-560.

Ryals, J., Weymann, K., Lawaton, K., Friedrich, L., Ellis, D., Steiner, H.Y., Johnson, J., Delaney, T.P., Jesse, T., Vos, P., and Uknes, S. (1997). The Arabidopsis NIM1 protein shows homology to the mammalian transcription factor inhibitor IkB. *Pl. Cell*, **9**: 425-439.

Sanders, P. R., Sammons, B., Kaniewski, W., Haley, L., Layton, J., Lavallee, B. J., Delannay, X. and Tumer, N. E. (1992). Field resistance of transgenic tomatoes expressing the tobacco mosaic virus or tomato mosaic virus coat protein genes. *Phytopathology*, **82**: 683-690.

Schots, A., deBoer, J., Schouten, A., Roosien, J., Zilverentant, J.F., Pomp, H., Bouwmansmits, L., Overmars, H., Gommers, F.J., Visser, B., Stiekema, W.J., Bakker, J. (1992). Plantibodies- a flexible approach to design resistance against pathogens. *Neth. J. Plant Pathol.*, **9**: 39-46.

Shah, J., Tsui., F., and Klessig, D.F. (1997). Characterization of a salicyclic acid insensitive mutant (*sai1*) of *Arabidopsis thaliana*, identified in a selective screen utilizing the SA-inducible expression of the *tms2* gene. Mol. *Plant-Microbe Interact.*, **10**: 69-78.

Shen, K.A., Meyers, B.C., Islam-Faridi, M.N., Chin, D.B., Stelly, D.M. and Michelmore, R.W. (1998). Resistance gene candidates identified by PCR with degenerate oligonucleotide primers map to clusters of resistance genes in Lettuce. *Mol. Plant-Microbe Interact.*, **11**: 815-823.

Smith, M.D. (1996). Antibody production in plants. *Biotechnol. Adv.*, **14**: 267-281.

Song, W.Y., Wang, G., Chen, L., Kim, H., Pi, L., Gardner, J., Wang, B., Holsten, T., Zhai, W., Zhu, L., Fuaguent, C., and Ronlad, P.C. (1995). A Receptor Kinase like protein encoded by the rice disease resistance gene *Xa21*. *Science*, **270**: 661-667.

Starr, M.P. (1981). The genus Xanthomonas. *In*: Phytopathogenic bacteria (M.P. Starr, Ed.), Springer-Verlag, NewYork, pp. 742-746.

Staub, T. (2001). Induced disease resistance in crop health management. Online. *Plant health progress*, doi: 10.1094/PHP-2001-0913-01-PS.

Staub, T., Kunz, W. and Oostendorp, M. (2001). Chemical activators of disease resistance in crop Protection. *In*: Encyclopedia of agrochemicals, John Wiley & Sons, New York.

Tacket C.O., and Mason H.S. (1999). A review of oral vaccination with transgenic vegetables; Microbes and Infection. **1**: 777-83.

Tai, T.H., Dahlbeck, D., Clark, E.T., Gajiwala, P., Pasion, R., *et al*. (1999). Expression of the Bs2 pepper gene confers resistance to bacterial spot disease in tomato. Proc. Natl. Acad. Sci., USA, **96**: 14153-58.

Tally, A., Oostendorp, M., Lawton K., Staub, T. and Bassi, B. (1999). Commercial development of elicitors of induced resistance to pathogens. Pages 357-369. In: *Induced plant defenses against pathogens and herbivores*. Agrawal, A.A. Tuzun, S. and Bent, E. (eds.). Ameriacn Phytopathological Society, St. Paul, Mn.

Tam, J. P. *et al*. (1999) An unusual structure motif of antimicrobial peptides containing end to end macrocycle and cystine knot disulfides. Proc. Natl. Acad. Sci., USA, **96**: 8913-8918.

Tenllado, F. and Diaz-Ruiz, J.R. (2001). Double-stranded RNA mediated interference with plant virus infection. *J. Virol.*, **75**: 12288-12297.

Thomma, B.P., Cammue, B.P. and Thevissen, K. (2002). Plant defensins. *Planta*, **216**: 193-202.

Tolin S.A. (2001). Disease Resistance Genes. *In*: Encyclopedia of Plant Pathology (O.C. Maloy and Murray, D. Eds.), John Wiley and Sons Inc., New York, **1**: 364-366.

Trudel. J., Potvin, C., Asselin, A. (1995). Secreted hen lysozyme in transgenic tobacco: Recovery of bound enzyme and in vitro growth inhibition of plant pathogens. *Plant Sci.*, **106**: 55-62.

Tu, J., Ona, I., Zhang, Q., Zhang, G., Khush, G.S. and Datta, S.K. (1998). Transgenic Rice variety "IR72" with *Xa21* is resistant to bacterial blight. *Theor. Appl. Genet.*, **97**: 31-36.

Tumer, N. E., Hemenway, C., O 'Connell, K., Cuozzo, M., Fang, R-X, Kaniewski, W. and Chua, N-H. (1987). Expression of coat protein genes in transgenic plants confers protection against alfalfa mosaic virus, cucumber mosaic virus and potato virus X. *In*: Plant Molecular Biology (D. von Wettstein and N-H. Chua, Eds). Plenum Publishing Corporation, pp. 351-356.

Van Loon, L.C. Bakker, P.A.H.M. and Pieterse, C.M. (1998). Systemic resistance induced by rhizosphere bacteria. *Annu. Rev. Phytopathol.*, **36**: 453-83.

Vrain, T.C. (1999). Engineering natural and synthetic resistance for nematode management. *J. Nematol.*, **31**: 424-436.

Vunnam, S., Juvvadi, P., and Merrifield, R.B. (1997). Synthesis and antibacterial action of cecropin and proline-arginine-rich peptides from pig intestine. *J. Peptide Res.*, **49**: 59-66.

Wang, C., Chin, C.K. and Chen, A. (1998). Expression of the yeast D-9 desaturase gene in tomato enhances its resistance to powdery mildew. *Physiol. Mol. Plant Pathol.*, **52**: 371-383.

Wang, G., Song, W.Y., Ruan, D., Sideris, S., and Ronald, P.C. (1996). The cloned gene *Xa21*, confers resistance to multiple *Xanthomonas oryzae* pv. *oryzae* isolates in transgenic plants. *Mol. Plant-Microbe Interact.*, **9**: 850-855.

Whitham, S., Dinesh-Kumar, S.P., Choi, D., Hehl, R., Corr, C. and Baker, B. (1994). The product of the tobacco mosaic virus resistance gene *N*: similarity to Toll and the interlukin-1 receptor. *Cell*, **78**: 1011-1115

Whitham, S., McCormick, S., Baker, B. (1996). The N gene of tobacco confers resistance to tobacco mosaic virus in transgenic tomato. Proc. Natl. Acad. Sci., USA, **93**: 8776-81.

Wilson, C.R. and Conner, A.J. (1995). Activity of antimicrobial peptides against the causal agents of common scab, black leg and tuber soft rot diseases of potato. *New Zealand Natural Sciences*, **22**: 43-50.

Wilson, T.M.A. (1993). Strategies to protect crop plants against viruses: pathogen-derived resistance blossoms. Proc. Nat. Acad. Sci., USA, **90**: 3134-3141.

Wu, G., Shortt, B.J, Lawrence, E.B., Levine, E.B., Fitzsimmons, K.C., and Shah, D.M. (1995). Disease resistance conferred by expression of a gene encoding H_2O_2-generating glucose oxidase in transgenic potato plants. *Plant Cell*, **7**: 1357-1368.

Wu, G.S., Shortt, B.J, Lawrence, E.B., Leon J., Fitzsimmons, K.C., Levine, E.B., Raskin, I., Shah, D.M. (1997). Activation of host defense mechanisms by elevated production of H_2O_2 in transgenic plants. *Plant Physiol.*, **115**: 427-435.

Yoshimura, S., Yamanouchi, U., Katayose, Y., Toki, S., Wang, Z. X., Kono, I., Kurata, N., Yano, M., Iwata, N., and Sasaki, T. (1998). Expression of *Xa1*, a bacterial blight-resistance gene in rice, is induced by bacterial inoculation. Proc. Natl. Acad. Sci., USA, **95**: 1663-1668.

Zhang, J.X., Howell, C.R. and Starr, J.L. (1996). Suppression of *Fusarium* colonization of cotton roots and fusarium wilt by seed treatments with *Gliocladium virens* and *Bacillus subtilis*. *Biocontrol Science and Technology*, **6**: 175-187.

Zhang, S. P., Song, W. Y., Chen, L.L., Ruan, D.L., Ronald, P. C., Beachy, R., and Fauquet, C. (1998). Genetic engineering of agronomical important Indica rice varieties for resistance to bacterial blight disease. *Mol. Breed.*, **4**: 551-558.

Zhang, X.D., Callahan, F.E., Jenkins, J.N., Ma, D.P., Karaca, M., Saha, S. and Creech, R.G. (2002). A novel root-specific gene, MIC-3, with increased expression in nematode-resistant cotton (*Gossypium hirsutum* L.) after root-knot nematode infection. *Biochmica et Biophysica acta-gene structure and expression*, **1576**: 214-218.

Zhang, Z., Coyne, D., Vidaver A. and Mitra, A. (1998). Expression of human lactoferrin cDNA confers resistance to *Ralstonia solanacearum* in transgenic tobacco plants. *Phytopathology*, **88**: 730-734

Zhang, Z.H., Wu, L.P. (1998). Research and development of expressing antibodies in plants. *Prog. Biochem. Biophys.*, **25**: 136-139.

Zhou, X.J., Lu, S., Xu, Y.H., Wang, J.W. and Chen, X.Y. (2002). A cotton cDNA (GaPR-10) encoding a pathogenesis-related 10 protein with *in vitro* ribonuclease activity. *Plant Science*, **162**: 629-636.

Zhu, Q., Maher, E.A., Masoud, S., Dixon, R.A., Lamb, C.J. (1994). Enhanced protection against fungal attack by co-expression of chitinases and glucanase genes in transgenic tobacco, 1994. *Bio/Technol.*, **12**: 807-812.

6

Integrated Plant Disease Management: Recent Approaches

MYANK U. CHARAYA[1] & R.S. MEHROTRA[2]

[1] *Department of Botany, C.C.S. University, Meerut (U.P.), India*
[2] *Retired Professor, Department of Botany, Kurukshetra University, Kurukshetra (Haryana), India*

Inspite of unprecedented human achievements in science and technology; including the "Green revolution" in the 20th century, our human family still faces the formidable challenges of widespread poverty, chronic hunger, disease and malnutrition. Nearly 800 million people still go hungry to bed every day; and about 17 million children under five die every year, with malnutrition contributing to at least one-third of these deaths (Hussain, 2002). As per projections by the United Nations, world population will increase to about 7.5 billion in 2020 (Anon, 1999). In order to close the large gap between food production and projected demands after 30-40 years, the world's farmers would have to produce 40 per cent more rice, wheat and other grains. The trends in the recent past, however, are not encouraging. Rate of increase in food production has come down from 3 per cent per annum in 1970s to about 1 per cent in 1990s (Sharma *et al.*, 2001). In view of the decreasing availability of land and water resources for food production, as well as environmental degradation that has already adversely affected our cropland, there is undoubtedly no place for complacency on the food production front; and "there is an urgent need to strengthen the ongoing process of farming systems, intensification, diversification and value addition" (Swaminathan, 2003).

Plant diseases constitute one of the major causes of crop losses all over the world. Oerke *et al.* (1994) estimated that about 14 per cent of the total agricultural production is lost due to pests. Crop losses in India have been estimated to vary between 10 to 30 per cent depending upon the crop, region and the severity of pest infestation–amounting to an annual loss of about Rs 60,000 crores (Agnihotri *et al.*, 1999). The need for minimizing the losses due to plant diseases is, thus, as essential in the present context as ever before.

Disease Control *vs.* Disease Management

The phrase "Control of Plant Disease" has been in use for a long time; and includes a variety of measures including cultural methods, chemical "control", biological "control" etc. However, at present, the term "Disease management" is preferred over control. There are a number of reasons for this:

1. The term 'Control' implies an absolute connotation-a situation which is almost impossible (at least very difficult) to obtain in a biological system.
2. The term 'control' conveys the impression that corrective remedies were implemented 'after' the disease had occurred.

On the other hand, 'Management' of disease conveys the idea of a 'continuous' process, which is more preventive–the loss from diseases being kept below a certain economic level.

Extensive Reliance Upon Pesticides: Adverse Effects

In the past many decades, chemicals have been relied upon extensively for preventing and controlling–rather protecting–the plants from the ravages of various pests. A tremendous variety of chemicals were identified and developed for use against the pests. These chemicals–called 'pesticides'–have made unequivocal contributions to human efforts for increasing crop yields. Dramatic success obtained in crop production through the use of the chemical pesticides has led to their overuse, and even abuse, over the past five decades. The consumption of pesticides in India has increased from 2350 metric tonnes in 1955-56 to about 85,000 metric tonnes in 1997 (Agnihotri *et al.*, 1999). At present, more than 150 pesticides are registered for use; out of these, about 62 are manufactured in India.

Inspite of a number of cautions including that by Carson (1962) regarding the adverse effects of pesticides on the environment, the consumption of pesticides and other chemicals continued to rise–rather the consumption of pesticides came to be regarded as an important factor in augmenting productivity. The indiscriminate overuse and abuse of chemical pesticides led to a number of undesirable side effects. These include:

1. Environmental pollution.
2. High pesticides residues in food and other products.
3. Development of resistance to pesticides.
4. Pest resurgence; and
5. Appearance of new pests.

Since pesticides are designed to 'kill' living organisms; and a vast majority of these are "non-specific' poisons, a chemical pesticide is almost sure to kill some 'non-target' species. All these and related aspects of pesticides have been discussed by Charaya (1993) and Sankaram (1999).

Emergence of the Concept of Integrated Pest Management

The rising concern about the use of chemicals led to the re-examination of prevailing methods of managing plant pests. It necessitated the seeking of new, ecologically sound, long-lasting methods of pest control, which would cause minimum disturbance to the ecosystem. Naturally, this means a combination of chemical pesticides with other methods of pest management. Consequently, emerged the concept of "Integrated Pest Management".

Integrated Pest Management (IPM), is the "Selection, integration and implementation of pest management based on predicted economic, ecological and sociological consequences". The basic assumption of IPM is that no single safe pest management method will be successful. IPM, therefore, seeks to use a variety of biological, physical and chemical methods integrated into a cohesive scheme designed to provide long-term protection. The development of a cohesive system for the management of pests is based upon the following principles:

1. Potentially harmful species (pathogens) will "continue to exist" at tolerable level of abundance. Complete eradication of any species from a given ecosystem is not advisable– only application of minimum level of control that will maintain the pest below the economic damage threshold, is sufficient.
2. The ecosystem is the unit of management. The ecosystem itself might be manipulated in such a way that the pests are held at tolerable levels while, at the same time, avoiding the disruption of the system itself;
3. Any control measure (not only chemical, but any other also) may produce unexpected and undesirable side effects; therefore, a judicious well-planned mix of best management strategies (multiple techniques) must be adopted.
4. The levels of pesticides use must be reduced as far as possible.
5. There must be least or no damage to the non-target organisms.
6. The use of natural control agents must be maximized. It involves the altering of pest environment so that the action of 'anti-pest' natural forces is enhanced. This is achievable by regulating the factors in the ecosystem which, in turn, regulate the normal growth of the pest. Such factors include food, weather, competition and natural enemies.
7. An interdisciplinary approach is essential; the system must be developed only after careful consideration of all the aspects including agronomical, economical, meteorological, engineering and sociological.

A variety of guidelines have been laid down for developing IPM programmes by different workers, but basic approach in all the guidelines are almost similar (Charaya, 1993):

1. Analyze the 'pest' status of each of the repeatedly injurious organism and establish economic threshold of 'real' pest. This may involve a serious consideration of economic importance of the pest in terms of lost vs. profit analysis of the management. A regular and serious pest deserves more consideration than a non-serious and regular ones, though some irregular but potentially serious pests also deserve to be included among key pests.
2. Devise schemes for lowering equilibrium position of key pests. This can be, achieved by resorting to measures like cultivating resistant varieties, introducing natural enemies, modification in the environment *etc.*
3. However, during emergency 'situation', remedial measures may be resorted to (including use of chemical pesticides). Efforts should be made to use these measures only to the extent absolutely necessary so that minimum ecological disruption is caused.

4. Devise techniques for continuous monitoring of weather, climate and other factors affecting pests as well as their natural enemies.

Integrated Disease Management

A perusal of plant pathological literature in the past would reveal that major thrust of concerned scientists was on "somehow controlling" the pest–chemically, physically or biologically. However, there is now increasing emphasis on managing different factors which affect disease development:

1. The plant and its relative health (healthy plants being less subject to many diseases).
2. The pathogen (and any vector involved).

According to Cook *et al.* (1993), "our crop plants and trees need only be healthy enough to yield to their full genetic potential within the physico-chemical limits imposed by the uncontrolled variables of climate, weather and soils". Thus, the focus has now shifted from 'Pathogen-centred-approach' (killing the pathogen) to 'Plant-centred approach (planning to save plants from getting/being diseased). In this connection, it must be remembered that there are many diseases of plants in which pathogens are not involved. Thus, the term "Integrated Diseases Management" (IDM) is gaining wider acceptance. A number of books, reviews and articles have appeared on the subject including that by Nagarajan *et al.* (1996), Rechcigl and Rechcigl (1997), Cook (2000), Paroda (2000), Srivastava (2000) and Prasad and Puri (2002).

Components of Integrated Plant Disease Management

Any successful Integrated Diseases Management Programme must ensure that the:

1. Diseases and pest are managed effectively so that the presence of pathogens/pests remains below the economic thresholds for the crop and the market;
2. A variety of pests (ranging from microorganisms to insects and large weeds) are managed.
3. The produce must meet the prescribed quality standards with respect to the extent of damage caused by the pathogen/pest, and pesticide residues (lowest possible or nil pesticide residues).

The costs of IDM (in terms of money involved) may be sometimes more than the conventional pests and disease control methods. However, it must be remembered that IDM offers immense long-term advantages through the use of fewer pesticides, and better maintenance of environment leading to sustainability.

Therefore, any ideal IDM programme must include one or more of the following management methods:

1. Proper identification of the pathogen.
2. Using the knowledge of environment/manipulating the environment to reduce disease incidence.
3. Use of natural enemies of pests (biological control) and selective pesticides (preferably biopesticides).
4. Selective minimum properly planned spray of broad–spectrum pesticides (if absolutely necessary).
5. Chosing cultivars resistant to or tolerant to pests/pathogens.

6. Appropriate cultural techniques (like time of planting, crop rotation, sanitation, adequate plant nutrition).

Though the use of these methods have been advocated for sufficiently long time, and many of these have been under practice, the rise of technology (especially biotechnology) has added new dimensions to their application. The following account aims at discussing some of these dimensions.

Detection and Correct Identification of Plant Pathogens

The first and foremost requirement of any IDM strategy is that the pest (pathogen) be identified properly so that a judicious use of the available literature may be made for deciding management strategies as quickly as possible. Of course, plant pathologists are able to identify many of these by morphological characteristics, cultural characteristics and pathogenicity tests; but it is quite often very time-consuming involving a lot of labour. With some genera, such as *Fusarium*, there is more of problem because such genera include a range of plant pathogenic and saprobic species. Sometimes, one may feel frustrated on being unable to detect or identify viable but not culturable pathogen. Technologies that would enable plant pathologists in identifying pathogens from plants, traps, baits or soil samples rapidly (but accurately) would be highly welcome for use not only in epidemiological and ecological studies, but for detecting initial inoculm in disease forecasting systems also.

The widespread application of recombinant DNA technology in recent years has dramatically improved the speed, sensitivity and specificity of assays–the direct diagnostic assays based on amplification or blotting methods allow the detection of even those bacteria which are viable but not culturable (Roszak and Colwell, 1987); and were previously very difficult to detect. Though several approaches to the identification of DNA sequences and development of DNA probes for plant pathogens have been pursued, most of the attempts in this direction are based upon the concept that "the ideal DNA probe would be based on known nucleic acid sequences that taxonomically define the target and are present in high copy number" (De Parrasis and Roth, 1990). As a result, many of the published reports are based upon species-specific application of PCR. If the DNA of the pathogen is present, species-specific primers will bind to the DNA to be amplified during the reaction. One may use multiple pairs of primers in a single reaction; and use the size differences of the PCR products for each different pair to test for several pathogens in a single reaction. An interesting variation of this technique–TaqMan™–uses different fluorescent labels attached to nested primers; this technique can be used to detect as well as quantify, in a single reaction, upto three pathogens at a time without the need to run an agarose gel (Anon., 1995). Such techniques for multi-pathogens assay would be immensely helpful in disease diagnostics especially for the measures like quarantine-testing.

Application of PCR to rDNA regions has provided specific-specific or subspecies-specific results especially for MLOs. Primers specific for apple proliferation and clover phyllody MLOs were obtained by Firrao *et al.* (1993). Sequential amplification of universal 16S rDNA primers specific to various rDNA clusters, was used for the first time to detect mixed infections of MLOs in plants (Lee *et al.*, 1995). Interestingly, sequence differences within the 16S rDNA locus need not be large for using them in specific detection assays–even a difference at the position of a single one of these nucleotides may be exploited for distinguishing between two species using ligase chain reaction (Wilson *et al.*, 1994). Highly variable internal transcribed sequences (ITS) of the rDNA clusters of fungi have been exploited for the detection of *Colletotrichum gloesporioides* (Mills *et al.*, 1992); *Cylindrocarpon heteronema* (Brown *et al.*, 1993); *Verticillium albo-atrum*, *V. dahliae* and *V. tricorpus* (Nazar *et al.*, 1991; Moukhamedov *et al.*, 1994); and *Pythium ultimum* (Levesque *et al.*, 1994).

Three non-homologous cryptic repetitive elements were identified in *Phytophthora parasitica* (not present in other species of *Phytophthora or Pythium*); probes so developed were useful for direct detection of *P. parasitica* in soil and plant host tissues (Goodwin *et al.*, 1989, 1990). Tripathi *et al.* (2003), conducted Southern hybridisation and RFLP analysis of PCR products of *Phytophthora* sp. infecting betelvine, and confirmed it as *P. nicotianae*. Similarly, Chowdappa *et al.* (2003) used ITS-RFLP and AFLP (amplified fragment length polymorphism) fingerprints to conclude that *Phytophthora meadii* (not *P. arecae*) was associated with fruit rot of areca nut in India since 1918–and later on, it moved to rubber and cardamom.

As with any other technique, the diagnostic assays (PCR assays in particular), have a risk of erroneous results. Therefore, complementary assays are always recommended before arriving at definite conclusions. The usefulness and limitations of molecular techniques for identifying plant pathogens have been nicely reviewed by Levesque (1997) and Hartung (1998).

Environment and Disease Management

Disease development is the result of the complex interaction between the pathogen(s), the host, and its environment. Environmental factors may directly affect the pathogen, or may have indirect effect on the pathogen through their influence on the functioning of plant itself. The physical environment influences all stages in the life cycles of plant pathogens, as well as different phases during pathogenesis. All these aspects and their applications in disease management have been extensively reviewed by McCartney (1997).

A knowledge of the environmental conditions not conducive for the development of a given disease (like unsuitable temperature, lack of enough moisture) can be helpful in devising 'disease avoidance' strategies. It is possible to identify the areas where the climate is suitable for a given disease; and growth of susceptible crops in such areas may be avoided. A knowledge of the geographical and seasonal distribution of microclimatic factors may be useful in identifying the periods when the microclimate favours the disease so that the particular period may be avoided by adjusting the sowing dates. Manipulation of microclimate may be restored to for discouraging disease. For example, in case of chickpea the pest damage was lower as a result of (a) sowing in the month of October; (b) wider spacing (Lal *et al.*, 1988); (c) intercropping of mustard, wheat, barley, linseed, and sorghum (Lal *et al.*, 1988 Das *et al.*, 1997). A knowledge of the relationship between environment, host and pathogen is also useful for building disease–forecasting system for better management. A disease forecaster is the predictor of "the outbreak or increase in intensity of a disease based on information on the weather, crop or pathogen." (Campbell and Madden, 1990). The system predicts one or more critical phases (initial infection, secondary infection, or disease development). More than 50 disease-forecasting systems have already been developed including BLIGHTCAST (for late blight of potato), TOMCAST (for early blight of tomato) and PLASMO (for downy mildew on grapevine). For details, readers are referred to the review by McCartney (1997). A computer-based disease- and pest-management system EPIPRE (Epidemics Prevention) was developed in the Netherlands for the; management of pest and disease in winter wheat (Zadoks *et al.*, 1984).

Of late, the use of Global Information System (GIS) and Pest Distribution Maps is gaining momentum all over the world. GIS is, infact, an organized collection of computer hardware, software, geographical data and personnel designed to efficiently capture, store, update, manipulate, analyse, and display all forms of geographically referenced information. Pest distribution maps are important tools for having an idea of the geographical distribution of pests under different agroclimatic conditions of the given region (over space and time). In India also, National Centre for Integrated Pest Management

(NCIPM), New Delhi has started using GIS for preparing distribution maps. Methods of utilization of GIS for disease management have been succinctly summarized by Kanojia *et al.* (2002).

Utilizing Resistance/Tolerance to Pathogens

The Use of Resistant Varieties

The use of resistant varieties has been one of the major approaches for containing disease development in several crops. Disease–resistant cultivars are:

1. Not more expensive than susceptible cultivars;
2. Do not require any special management practices; hence, these offer an attractive alternative not only to the use of pesticides, but to less economically feasible crop rotation as well.

The subject has been discussed at length by Carson (1997), Paroda (2000), Mehrotra and Aggarwal (2003). Paroda (2000) has emphasized the need for the replacement of varieties with vertical resistance every few years; and has provided a list of important resistant varieties of crops at present under cultivation.

Pathogen–resistant Transgenic Plants

Quite often, agronomically important crops lack effective resistance genes to control major diseases (Shah *et al.*, 1995). Therefore, new sources of resistance genes are required to develop disease–resistant crops. The tools of molecular biology and genetic engineering have provided us with unprecedented power to manipulate and develop novel crop genotypes thereby opening new rvenues for the development of disease–resistant crop plants. Fuchs and Gonslaves (1997) as well as James (2003) have published excellent reviews covering the background, and current status of development of transgenic plant resistant to insects and pathogens in general as also the associated risks. Sharma *et al.* (2003), have provided a list of major Indian developments in transgenic research and application in public sectors as well as a list of transgenic lines in advanced stage of development for field trials in private sectors. Grover and Pental (2003) have tried to sort out the breeding objectives and requirements for major fields crops of India (including the requirements for developing pathogen-resistance). Ranjekar *et al.* (2003); Grover and Gowthaman (2003); and Dasgupta et al. (2003) have reviewed the state of affairs in genetic engineering of crop plants for resistance to insects, to fungal pathogens, and to viral pathogens respectively. A glimpse of some of the important recordings made by them are as under:

1. There is unprecedented rapid adoption of transgenic crops–area under these increasing from 1.7 million hectares in 1996 to 44.2 million hectares in 2000. In the year 2000-2001, more growth in transgenic crop adoption was seen in developing countries than in industrial countries.
2. Use of transgenics for developing pathogen-resistant varieties is required for all crops. However, in case of wheat, development of varieties for resistance to Karnal bunt using transgenics emerged to be an important goal. For *Zea mays*, resistance to stem borer (*Chilo partillus*) appeared to be an important goal. In case of Sorghum, major objectives for transgenic research was to develop lines with resistance to grain mould (caused by a number of fungi), stem borer (*Chilo zonellus*) and shootfly (*Antherigona soccata*). For pearl millet, development of transgenics for resistance to downy mildew was given priority. In case of Chickpea, development of transgenics for resistance to *Helicoverpa armigera* was advocated; for cotton and potato, resistance to insects and late blight demands immediate attention. Development

of aphid–resistant cultivars in mustard; and YMV (Yellow mosaic virus)–resistant and fungal disease resistant varieties in Soyabean are the major requirements. It does not mean that other pathogens of these crops are not important; rather, these particular diseases were advised to be tackled through transgenics due to unsatisfactory results in these cases obtained through conventional plant breeding.

3. *Bacillus thuringiensis* (Bt) strains and their insecticidal crystal proteins (ICPs) have gained wide acceptance all over the world as ecofriendly biopesticides for extensive use in agriculture. The delivery of Bt ICPs through spray formulations suffers from many limitations like short half life, physical removal by wind and rain, as well as inability to reach burrowing insects. In order to overcome these disadvantages, attempts were made to express these crystals (Cry) proteins in plant cells at levels sufficient to kill the larvae. First transgenic plant using *cry* genes was developed in 1987. Till now, more than 30 plant species have been transformed with *Bt cry* genes–plants exhibiting insecticidal properties. Bt crops of cotton, potato and maize have been commercialized. In India also, attempts have been made to introduced *cry* genes in crop plants like rice, potato, cotton, sorghum and vegetables. Some success has already been obtained at TERI (Tata Energy Research Institute). Transgenic plants of tobacco carrying modified *cry tab* and *cry Tc* have been obtained- the plants have been found to be protected against tobacco caterpillar. Similar attempts for rice, brinjal and tomato are underway. Commercial release of Bt cotton has been approved by Govt. of India in March 2002. Interestingly, some novel insecticidal proteins during vegetative cell growth are produced by certain isolates of *B. thuringiensis*. These have been termed vegetative insecticidal protein (Vip). Another novel insecticidial protein has been discovered in the bacterium *Photorhabdus luminescens*. These proteins might provide further inputs for the development of transgenic crops with durable insect resistance.

4. A lot of basic work in the area of host- pathogen interaction has already been carried out to identify important genes which need to be introduced in the plants to improve their resistance to fungal pathogens. A number of the resistance genes have been identified and cloned. Many of the anti-fungal compounds which are synthesised by the plants to combat fungal infection have already been identified. Genes encoding many anti-fungal proteins have been used for raising fungus–resistant transgenic plants. Such proteins include pathogenesis related (PR) proteins, Ribosome inactivating proteins (RIPs), small cystein–rich proteins, lipid transfer proteins, storage albumins, Polygalacturonase inhibition proteins (PGIPs), antiviral proteins and non–plant anti-fungal proteins. The first report on developing fungus–resistant transgenic plant came from Broglie *et al.* (1991) who constitutively expressed chitinase gene in tobacco and *Brassica napus*, increasing the resistance of the plants to *Rhizoctonia solani*. Since then, a number of reports have appeared (see table 24: Grover and Gowthaman, 2003).

   Tobacco plants with a RIP encoding DNA sequence of barley showed better resistance to *Rhizoctonia solani*, while transenic tomato fruits constitutively expressing pear PGIPs showed reduced colonization by *Botrytis cinerea*.

   However, the introduction of a desired gene in the host plant under constitutively high–expressing promoter may cause silencing of the transgene and its endogenous homologues– as a result, a high proportion of progeny loses its enhanced resistance. Studies on gene silencing, therefore, would be required for further success in this direction.

Phytoalexins, the antimicrobial low molecular weight secondary metabolites produced in plants following pathogen attack, are also believed to play an important role in plant defence. Hain *et al.* (1993) introduced the gene encoding stilbane synthase from grapevine into tobacco plants (in which the enzyme is absent though the substrate is present). The result was the expression of stilbene synthase gene and production of resveratrol (a stilbene–type phytoalexin); and the transgenic tobacco showed greater resistance to *B. cinerea.* Tobacco plants have been engineered with multiple gene constructs in order to pyramid the effect of several antifungal proteins and to study the existence of synergism, if any. Interestingly, a combination of two genes (Chitinase and ß 1-1,3 glucanase; ribosome inactivating protein and Chitinase) provided substantially greater protection against *Cercospora nicotianae*; and against *Rhizoctonia soliani, Alternaria alternata* and *Botrytis cinerea* respectively (Zhu *et al.*, 1994; Jach *et al.,1995*).

5. Due to tremendous advances in our knowledge regarding plant–virus interaction, and in the techniques of genetic transformation, it has been possible to develop virus–resistant varieties utilising genes from pathogen itself–(PDR, pathogen derived resistance); and or from amongst host resistance genes. The use of viral coat protein (CP) as a transgene holds a great promise as numerous crops which have been transformed to express viral CP exhibit high levels of resistance (Tables 2 and 3: Dasgupta *et al.*, 2003). In India, success has been achieved in controlling physalis mottle virus using PDR in tobacco and tomato at Indian Institute of Science, Bangalore; and potato virus Y in tobacoo under collaborative programs between CPRI (Shimla) and BARC (Mumbai).

6. Phytotoxins are known to play a major role in pathogenesis. Naturally, inactivation of these toxins would relieve at least part of the symptom development. Hence, one possible strategy for plant disease management might involve the use of genes from phytotoxin–degrading microbes to engineer disease-resistant plants. Tomato leaf cuttings and intact plants treated with fusaric acid–detoxifying avirulent strain of *Pseudomonas solanacearun* were protected from wilt disease caused by *Fusarium oxysporum* f. sp. *Iyopersici* race 1 (Toyoda *et al.*, 1988; Ouchi *et al.*, 1989). A fusaric acid–resistant (fus) gene has been cloned from *Pseudomonas cepacia.* This gene or some other similar gene might be useful in obtaining transgenic plants resistant to fusaric acid. Further impetus to this strategy has been provided by the transgenic tobacco resistant to tabtoxin or infection by *Pseudomonas syringae* pv. *tabaci* (Anzai *et al.*, 1989). A *ttr* (tabtoxin resistance gene) isolated from *Ps. syringae pv. tabaci* was fused to the 355 promoter of cauliflower mosaic virus and the chimeric gene was introduced into tobacco *via Agrobacterium*–mediated transformation. The transgenic tobacco plants so produced exihibit high specific expression of the ttr gene, and do not develop any chlorotic symptoms either due to tabtoxin treatment or infection by *P. syringae pv. tabaci.*

Induction of Resistance

Enhancement in the resistance of a susceptible plant against pathogens in response to an external stimulus without a known alteration in the genome is referred to as "induced resistance". It is based on the activation or the enhancement of plant's own defence mechanism. The agents may induce a host response which protects the treated part of the plant (localized acquired resistance LAR); or part(s) of the plant distinct from treated area (systemic acquired resistance–SAR). LAR is restricted to the site of prior inoculation and, so far, only avirulent races of a pathogen and non–host pathogen are known to act as inducers (Ouchi *et al.*, 1976). SAR, on the other hand, is the result of necrosis brought about by inducing inoculation, and is usually detected by challenge inoculations at a different time and location on the plant–it is also not specific with respect to inducing agent. Inspite of a number of limitations like losses in yield, accumulation of undesirable secondary metabolites, non-availability of low molecular weight elicitors that can be easily absorbed by plants (Sequeria, 1983), extensive studies have established the efficacy of induced resistance in protecting crop plants in field conditions also; some of the examples are:

1. Systemic protection of tomato against *Verticillium* wilt (Matta and Garibaldi, 1977);
2. Protection of tobacco against TMV, *Pseudomonas tabaci* and *Peronospora tabacina* (Mclntrye and Dodds, 1979; Nesmith *et al.*, 1982; Tazun *et al.*, 1986);
3. Protection of watermelon against *Fusarium* wilt (Martyn *et al.*, 1991).

All the pros and cons of induced resistance in plants have been discussed at length by Kuc (1995); and Shekar Shetty & Kumar (2000).

Biological Control and Biopesticides

Biological control of plant pathogens is a natural phenomenon. But its possible exploitation for disease management came to light in 1926 when Sanford suggested that the control of potato scab by green manure was a biological control due to the increase in the population of certain saprophytic bacteria that multiplied on the organic material. Since then, a lot of activity has been undertaken in this field; and enormous amount of literature is available concerning this mode of managing plant diseases. For a concise but uptodate and balanced account, the readers are referred to the reviews by Backman *et al.* (1997); Mehrotra *et al.* (1997); Rush and Sherwood (1997).

More than 16 bacterial, fungal and viral products (called biopesticides) are now available in the market for pest control throughout the world. The biopesticides are microbial formulations that are mass produced, registered, marketed and applied inundatively like chemical pesticides. We have already discussed some of the transgenic aspects of biopesticides earlier in this very chapter. For a more detailed discussion, the readers may refer to the articles by Charaya (1993) and Aneja (2000).

Conclusions

This chapter has highlighted the basic concept of Integrated Plant Disease Management; as well as some facets of the recent developments in the technology and our knowledge regarding identification of the pathogens, their interactions with host and environment conferring resistance to diseases pathogens in the plants with special emphasis on their utility in disease management. We had to omit the details of many aspects. It is hoped that in near future, our scientists would be able to develop, coordinate and integrate all the findings from the "labs" across the world for the effective implementation of IPDM strategies on the "land".

REFERENCES

Agnihotri, N.P., Walia, S.C. and Gajbhiye, V.T. (1999). Towards Green Pesticides. In: *Green Pesticides, Crop Protection and Safety Evaluation*, (Eds. Agnihotri, N.P., Walia, S.C. and Gajbhiye, V. T.) pp. 1-20. Society of Pesticide Science, IARI, New Delhi, India.

Anonymous (1995). *TaqMan™ LS–50 B PCR Detection system*. Perkin–Elmer, Forest City, CA, USA.

Anonymous (1999). *The state of food insecurity in the world*. Food and Agriculture Organisation (FAO), Rome, Italy.

Anzai, H., Yonegama, K. and Yagamuchi, I. (1989) Transgenic tobacco resistant to a bacterial disease by the detoxification of a pathogenic toxin. *Mol. Gen. Genet.*, **219**: 492-494.

Backman, P.A. (1997). Bacteria for Biological Control of Plant Diseases. In: *Environmentally safe approaches to crop disease control*, (Eds. Rechcigl, N. A. and Rechcigl, J.E.). pp. 95-109. CRC Lewis Publishers, New York.

Broglie, K., Chet, I., Holliday, M., Cressman, R., Biddle, P., Knowlton, S., Mauvais, C.J. and Broglie, R. (1991). Transgenic plants with enhanced resistance to fungal pathogen *Rhizoctonia solani*. *Science*, **254**: 1194-1197.

Brown A.E., Muthumeenakshi, S., Sreenivasaprasad, S. and Mills, P.R. (1993). A PCR primer specific to *Cylindrocarpon heteronema* for the detection of the pathogen in apple wood. *FEMS Microbiol. Letters*, **108**: 117-120.

Campbell, C.L. and Madden, L.V. (1990). *Introduction to Plant Disease Epidemiology*. John Wiley and Sons, New York.

Carson, M.L. (1997). Plant breeding. In: *Environmentally safe approaches to crop disease control*, (Eds. Rechcigl, N.A. and Rechcigl, J.E.), pp. 201-220. CRC Lewis Publishers, New York.

Carson, R. (1962). *Silent Spring*. Fawcett Publications, Connecticut.

Charaya, M.U. (1993). From DDT to Microbial pesticides. In: *Pesticides Pollution*, (Eds. Kudesia, V.P. and Charaya, M.U.), pp. 207-220. Pragati Prakashan, Meerut, India.

Chowdappa, P., Brayford, D., Smith, J., and Flood, J. (2003). Identity of *Phytophthora* associated with arecanut and its relationship with rubber and cardamom isolates based on RFLP of PCR amplified ITS region of rDNA and AFLP fingerprints. *Current Science*, **85**: 585-587.

Cook, R.J. (1993). Making greater use of introduced microorganisms for biological control of plant pathogens. *Annu. Rev. Phytopathol.*, **31**: 53-80.

Cook, R.J. (2000). Ecologically sustainable plant disease management. In: *Proceedings International Conference on Integrated Plant Disease Management for Sustainable Agriculture*. Vol 1., pp. 15-23. Indian Phytopathological Society, IARI, New Delhi, India.

Das, S.B., Kandalkar, V.S. and Verma, O.P. (1997). Impact of intercropping on *Helicoverpa armigera* infestation on chick pea. In *Proceedings of National Seminar on Plant Protection towards sustainability*. Dec 22-24, 1997. Plant Protection Association of India, Hyderabad, pp. 33.

Dasgupta, I., Malathi, V.G. and Mukherjee, S.K. (2003). Genetic engineering for virus resistance. *Current Science*, **84**: 341-354.

De Parrasis, J. and Roth, D.A. (1990). Nucleic acid probes for the identification of phytobacteria: Identification of the genus-specific 16 S RNA sequences. *Phytopathology*, **80**: 618-621.

Firrao, G., Gobhi, E. and Locci, R. (1993). Use of polymerase chain reaction to produce oilgonucleotide probes for mycoplasma-like organisms. *Phytopathology* **83**: 602–607.

Fuchs, M. and Gonslaves, D. (1997). Genetic engineering. In *Environmentally Safe Approaches to Crop Disease Control*, (Eds. Rechcigl, N.A. and Rechcigl, J.E.), pp. 333- 368. CRC Lewis Publishers, New York.

Goodwin, P.H., Kirkpatrick, B.C. and Duniway, J.M. (1989). Cloned DNA probes for the identification of *Phytophthora parasitica*. *Phytopathology*, **79**: 716-721.

Goodwin, P.H., Kirkpatrick, B.C. and Duniway, J.M. (1990). Identification of *Phytophthora citrophthora* with cloned DNA probes. *Appl. Environ. Microbiol.,* **56**: 669-673.

Grover, A. and Gowthaman, R. (2003). Strategies for development of fungus-resistant transgenic plants. *Current Science,* **84**: 330-340.

Grover, A. and Pental, D. (2003). Breeding objectives and requirements for producing transgenics for major field crops of India. *Current Science,* **84**: 310–320.

Hain, R., Reif, H.J., Krause, E., Langerbartels, R., Kindl, H., Vorman, B., Wiese, W., Schmelzer, E., Schreir, P.H., Stocker, R.H. and Stentzel, K. (1993). Disease resistance results from foreign phytoalexin expression in a novel plant. *Nature,* **361**: 153-156.

Hartung, J.S. (1998). Molecular probes and assays useful to identify plant pathogenic fungi, bacteria and marked biocontrol agents. In: *Plant-Microbe Interactions and Biological Control,* (Eds. Boland, G. J. and Kuykendall, L.D.), pp. 393-413. Marcel Dekker Inc., New York.

Hussain, S. (2002). Challenges of reducing hunger and poverty. *The Hindu Survey of Indian Agriculture,* **2002**: 27-31.

Jach, G., Gornhardt, B., Mundy, J., Logemann, J., Pinsdorf, E., Leah, R., Schell, J. and Mass, C. (1995). Enhanced quantitative resistance against fungal disease by combinatorial expression of different barley antifungal proteins in transgenic tobacco. *Plant J.,* **8**: 97-109.

James, C. (2003). Global review of commercialised transgenic crops. *Current Science,* **84**: 303-309.

Kanojia, A.K., Singh, R.V., Kumar, S. and Trivedi, T.P. (2002). Geographic information system in IPM. In: *Crop Pest and Disease Management,* (Eds. Prasad, D. and Puri, S.N.). pp. 435- 442. Jyoti Publishers, New Delhi.

Kuc, J. (1995). Induced systemic resistance: An overview. In: *Induced resistance to disease in plants,* (Eds. Hammerschmidt, R. and Kuc, J.). Kluwer Academic Publisher, Netherland, pp. 169-175.

Lal, S.S., Yadav, C.P. and Das, C.A.R. (1988). Insect pests of pulse crops and their management. *Pesticides,* 66-77.

Lee, I.M., Bertaccini, A., Vibio, M. and Gunderson, D.E. (1995). Detection of multiple phytoplasmas in perennial fruit trees with decline symptoms in Italy. *Phytopathology,* **83**: 834-842.

Levesque, C.A. (1997). Molecular detection tools in integrated disease management: overcoming current limitations. *Phytoparasitica,* **25**: 1-7.

Levesque, C.A., Vrain, T.C. and De Boer, S.H. (1994). Development of a species–specific probe for *Pythium ultimum* using amplified ribosomal DNA. *Phytopathology,* **84**: 474-478.

Martyn, R.D., Biles, C.L. and Dillard, E.A. (1991). Induced resistance to *Fusarium* wilt of watermelon under simulated field conditions. *Plant Disease,* **75**: 874-877.

Matta, A. and Garibalidi, A. (1977). Control of *Verticillium* wilt of tomato by preinoculation with avirulent fungi. *Netherlands Journal* of *Plant Pathology,* **83**: 457-462.

McCartney, H.A. (1997). The influence of environment on the development and control of disease. In: *Environmentally safe approaches to crop disease control,* (Eds. Rechcigl, N.A. and Rechcigl, J.E.), pp. 3-31. CRC Lewis Publishers, New York.

McIntyre, J.L. and Dodds, J.A. (1979). Induction of localized and systemic protection against *Phytophthora parasitica* var. *nicotianae* by Tobacco Mosaic Virus infection of tobacco hypersensitive to the virus. *Physiological Plant Pathology,* **15**: 321-330.

Mehrotra, R.S., and Aggarwal A. (2003). *Plant Pathology*, 2nd Edn. Tata McGraw Hill Publishing Co. Ltd., New Delhi.

Mehrotra, R.S., Aneja, K.R. and Aggarwal, A. (1997). Fungal control agents. In: *Environmentally safe approaches to crop disease control*, (Eds. Rechcigl, N.A. and Rechcigl, J.E.), pp. 111–127. CRC Lewis Publishers, New York.

Mills, P.R., Sreenivasaprasad, S., and Brown, A.E. (1992). Detection and differentiation of *Colletotrichum gloesporioides* isolates using PCR. *FEMS Microbiol Lett.*, **98**: 137-141.

Moukhamedov, R., Hu, X., Nazar, R.N. and Robb, J. (1994). Use of polymerase chain reaction–amplified ribosomal intergenic sequences for the diagnosis of *Verlicillium tricorpus*. *Phytopathology*, **84**: 256-259.

Nagarajan, S., Sharma, A.K. and Kumar, J. (1996). Integrated Disease Management. In: *50 Years of Crop Science Research in India*, (Eds. Paroda, R.S. and Chadha, K.L.), pp. 72-87. ICAR, New Delhi.

Nazar, R.N., Hu, X., Schmidt, J., Culham, D. and Robb, J. (1991). Potential use of PGR–amplified detection and differentiation of *Verticillium* wilt pathogens. *Physiol. Mol. Plant Pathol.*, **39**: 1-11.

Nesmith, W.C., Cohen, Y., Kuc, J. and Spur, H. (1982). Field evaluation of systemic resistance to blue mold in tobacco induced by inoculation with the blue mold *pathogen. Phytopathology*, **72**: 360.

Oerke, E.C., Dehne, H.W., Schonbeck, F. and Weber, A. (1994). *Crop production and crop protection: Estimated losses in major food and cash crops.* Elsevier Publishing Co., Amsterdam, The Netherlands.

Ouchi, S., Oku, H. and Hibino, C. (1976). Localization of induced resistance and susceptibility in barley leaves inoculated with powdery mildew fungus. *Phytopathology*, **66**: 901.

Ouchi, S., Toyoda, H., Utsumi, R., Hashimoto, H. and Hadama T. (1989). A promising strategy for the control of fungal diseases by the use of toxin degrading microbes. In *Phytotoxins and plant pathogenesis*, (Eds. Granti, A., Durbin, R. D. and Ballio, A.). pp. 301-317, Springer, Berlin.

Paroda, R.S. (2000). Integrated management of plant diseases. In: *Proceedings International Conference on Integrated Plant Disease Management for Sustainable Agriculture*, Vol 1., pp. 1-12. Indian Phytopathological Society, IARI, New Delhi, India.

Prasad, D. and Puri, S.N. (2002). *Crop Pest and Disease Management.* Jyoti Publishers, New Delhi.

Ranjekar, P.K., Patankar, A., Gupta, V., Bhatnagar, R., Bentur, J. and Kumar, A.P. (2003). Genetic engineering of crop plants for insect resistance. *Current Science*, **84**: 321-329.

Rechcigl, N. and Rechcigl, J.E. (1997). *Environmentally Safe Approaches to Crop Disease Control.* CRC Lewis Publishers, New York.

Roszak, D.B. and Colwell, R.R. (1987). Survival strategies of bacteria in the natural environment. *Microb. Rev.*, **51**: 365-379.

Rush, C.M. and Sherwood, J.L. (1997). Viiral control agents. In: *Environmentally safe approaches to crop disease control*, (Eds. Rechcigl, N.A. and Rechcigl, J.E.), pp. 139-159. CRC Lewis Publishers, New York.

Sankaram, A. (1999). Integrated pest management: Looking back and forward. *Current Science*, **77**: 26-32.

Sequeria, L. (1983). Mechanism of induced resistance in plants. *Annual Review* of *Microbiology*, **37**: 51-79.

Shah, D.M., Rommens, C.M.T. and Beachy, R.N. (1995). Resistance to diseases and insects in transgenic plants progress and application to agriculture. *Trends Biotechnol.,* **13**: 362.

Sharma, H.C., Sharma, K.K., Seetharama, N. and Ortiz, R. (2001). Genetic transformation of crop plants: Risks and opportunities for rural poor. *Current Science,* **80**: 1495-1508.

Sharma, M., Charak, K.S. and Ramanaiah, T.V. (2003). Agricultural biotechnology research in India: Status and politics. *Current Science,* **84**: 297-302.

Shekar Shetty, H. and Kumar, V.U. (2000). Systemic Acquired Resistance: an ecofriendly method for managing Pearlmillet downy mildew disease. In: *Glimpses in Plant Sciences,* (Eds. Aneja, K.R., Charaya, M.U., Aggarwal, A., Hans, D.K. and Khan, S.A.), pp. 99-109, Pragati Prakashan, Meerut.

Srivastava, M.P. (2000). Integrated plant disease management for sustainable agriculture. In: *Glimpses in Plant Sciences,* (Eds. Aneja, K.R., Charaya, M.U., Aggarwal, A., Hans, D.K. and Khan, S.A.), pp. 77-81, Pragati Prakashan, Meerut.

Swaminathan, M.S. (2003). Enhancing our agricultural competitiveness. *Current Science,* **85**: 886-895.

Toyoda, H., Hashimoto, H., Utsumi, R., Kobayashi, H. and Ouchi, S. (1988). Detoxification of fusaric acid by a fusaric acid–resistant mutant of *Pseudomonas solanacearum* and its application to biological control of Fusarium wilt of tomato. *Phytopathology,* **78**:1307-1311.

Tripathi, A., Singh, R., Raj, S.K., Singh, A.P., and Johri, J.K. (2003). Molecular identification of *Phytophthora nicotianae* isolates causing leaf rot of betel vine (*Piperbetle L.*), *Current Science,* **84**: 22-24.

Tuzun, S., Nesmith, W., Ferrises, R. and Kuc, J. (1996). Effects of stem infections with *Peronospora tabacina* on growth of tobacco and protection against blue mold in the field. *Phytopathology,* **76**: 938-941.

Wilson, W.J., Wiedmann, M., Dillard, H.R. and Batt, C.A. (1994). Identification of *Erwinia stewartii* by ligase chain reaction assay. *Appl. Environ. Microbiol.,* **60**: 278-284.

Zadoks, J.C. and Schein, R.D. (1979). *Epidemiology and Plant Disease Management.* Oxford University Press, Oxford.

Zhu, Q., Maher, E.A., Masoud, S., Dixon, R.A. and Lamb, C.J. (1994). Enhanced protection against fungal attack by constitutive co-expression of chitinase and glucanase genes in transgenic tobacco. *Bio/Technology,* **12**: 807-812.

7

Biocontrol Potential of Microorganisms–An Overview: Focus on *Trichoderma* as Biofungicide for the Management of Plant Diseases

N. MATHIVANAN, V.R. PRABAVATHY & K. MURUGESAN

Centre for Advanced Studies in Botany, University of Madras, Guindy Campus, Chennai – 600 025, India

ABSTRACT

Biological control of various plant pathogens using different biocontrol agents has been attempted and effective biocontrol agents belonging to the group bacteria and fungi were identified besides, different endophytes and actinomycetes. The biocontrol potential of *Fusarium solani* and *Fusarium chlamydosporum* against groundnut rust pathogen, *Puccinia arachidis* has been studied. Both the species of *Fusarium* significantly reduced the rust development on groundnut. Several strains of *Trichoderma* were isolated from soil samples and screened for their biocontrol potential against different fungal pathogens. An effective *Trichoderma* strain was identified, multiplied using molasses yeast medium and formulated with talc powder as biofungicide. This formulation was distributed to the farmers in Andhra Pradesh and Karnataka states for the management of root diseases in different crops. Impact assessment study on the distribution of biofungicide (*Trichoderma*) has revealed that the application of *Trichoderma* effectively controlled the soil-borne root diseases in cotton, chilli, sugarcane and acid lime. Distribution of *Trichoderma* in large scale made tremendous impact on farmers in Andhra Pradesh, who are at present coming forward to buy the commercial products of biocontrol agents from the local market. The mechanisms of biological control exerted by the biocontrol agents were studied and these organisms could be commercially exploited as biofungicides for the management of various soil and foliar plant pathogens. The protoplast fusion technique has been exploited for the successful development of fusant progenies in *Trichoderma* with increased biocontrol potential than their parental strains.

Key Words: *Biological control, Endophytes, Actinomycetes, Plant pathogens, Fusarium, Trichoderma, Plant diseases, Chitinase, Protoplast fusion.*

In agriculture, about 30000 plant diseases were recorded all over the world affecting the crop plants and in India alone 5000 plant diseases are known to be prevalent (Rangaswamy, 1996). Indian agriculture contributes about 40 per cent of the nation's economy with more than 65 per cent population directly engaged in agriculture. The crop losses due to various factors are the serious concern and it is estimated from 24 to 40 per cent in various crops.(Anonymous, 2000). Due to biotic stresses like pathogens and insect pests, crop losses are to the tune of Rs. 29240 crores per annum in India (Saksena, 2001). Hence, there is a need for better understanding of the pathogens, diseases caused by them and the host pathogen relationship. So that suitable control measures can be developed for the management of crop diseases and minimize the yield losses.

Protection of agricultural crops against diseases and insect pests is managed by different methods *viz.* cultural, chemical and biological. The cultural method of crop protection includes crop rotation, alteration of sowing time, sanitation, etc. By this method, there is reduction in the proliferation and attack of pathogens, but it cannot give guarantee of complete protection.

Chemical pesticides and diverse antibiotics were widely used to achieve high levels of crop protection. However, the indiscriminate use of chemicals create serious problems such as development of resistance to pesticides, resurgence of target pests, secondary pest outbreak, killing of non-target organisms, residual toxicity and environmental pollution (Brent, 1995; Sharma *et al.*, 1998; Urech *et al.*, 2000). In agriculture, the crop protection chemicals are sprayed again and again in the form of insecticides and fungicides, so the crop become slowly toxic and the land become sterile due to residual toxicity. The constant and injudicious use of chemicals had to be curtailed owing to their hazardous effects on non-target organisms and because of the undesirable changes they inflict on the environment. With the increasing concern over the use of chemical pesticides in the environment, there is an urgent need for alternative disease control strategies. Biological control, therefore, assumes special significance in being eco-friendly cost effective strategy, which can be used in integration with other disease management systems to afford greater levels of protection and sustain crop yield.

Biological control of plant diseases is gaining momentum in recent years since use of chemical pesticides lead to various health hazards to human beings (Brent, 1995; Sharma *et al.*, 1998). Biological control by antagonistic organisms is a potential non-chemical tool for crop protection against phytopathogens (Papavizas, 1985). In the last two decades, numerous biological disease control agents are being examined and the use of microorganism based biopesticides is being increased for the management of various crop diseases. Biological control is centered more towards the basic competition among microorganisms for substrate in the specific form and specific condition called antagonism. Antagonists include bacteria, fungi, viruses and predatory microfauna such as protozoa, nematodes, rotifers and mites. There is an increasing commercial and environmental interest in the use of microbial agents as an alternative or in combination with chemicals for controlling the spread and severity of a wide range of crop diseases. Among microorganisms, fungi are the potential candidates for biological control of plant pathogens and several antagonistic fungi *viz.*, *Trichoderma*, *Gliocladium*, *Talaromyces*, *Penicillium*, *Aspergillus*, *Coniothyrum* have shown enormous potential as biocontrol agents against different root and foliar diseases.

Microorganisms Used in Biological Control

Bacteria, fungi and actinomycetes play a major role in biological control of plant diseases. Various microbial antagonists including the strains of bacterial genera *Bacillus*, *Pseudomonas* and *Enterobacter* and fungal genera *Trichoderma* and *Gliocladium* have shown biocontrol activity against damping-off disease in several crops (Hader *et al.*, 1983; Anderson and Guerrra, 1985; Weller, 1988; Adams, 1990; Handelsman *et al.*, 1990). All these microbes are reported to secrete some secondary metabolites (antibiotics) and extracellular cell wall degrading enzymes (lytic enzymes). One of the most common mechanisms of interaction between populations of microorganisms in natural environment is enzyme induced lysis, one species eliminating another by digesting the cells or hyphae of the second species (Bruce *et al.*, 1995; Inbar *et al.*, 1996). A directed biological control of pathogenic fungi in nature may be achieved by modifying the indigenous microbial community in such a way as to favor destruction of the pathogen, probably through a preferential stimulation of microorganisms capable of degrading fungal cell walls (Mitchell and Alexander, 1962).

Endophytes as Biocontrol Agents

Endophytic bacteria have recently been a focus of interest as biocontrol agents. Such bacteria are indigenous to most plant species colonizing the tissue locally or systemically and both intercellularly and intracellularly (Hallmann *et al.*, 1997; Quadt-Hallmann and Kloepper, 1996). Endophytic bacterial biocontrol agents can be divided into two groups. One is, strains that extensively colonize the internal plant tissues and suppress invading pathogens by niche occupation and antibiosis or both. The other is, strains that primarily colonize the root cortex where they stimulate general plant defense/resistance mechanisms. More extensive and continuous colonization of plants might be required for the first type because coincidence with pathogen propagules would be necessary for antagonism (Hallmann *et al.*, 2001).

Endophytic bacteria are predominantly observed in the rhizosphere (Hallmann *et al.*, 1999; Mahaffee and Kloepper, 1997) where they presumably use wounds and natural openings to enter the plant. Lytic enzymes produced by these bacteria might also contribute to more efficient penetration and colonization (Quadt-Hallmann *et al.*, 1996). Compared with rhizosphere colonizers, internal colonizers will provide additional benefits as biocontrol agents because the plant provides shelter and nutrients, the bacteria can develop under less competitive conditions and shield the plant interior against pathogens. Through bacteria mediated induced resistance, the close association formed between endophytic bacteria and the plant might result in a stronger and more persistent plant protection.

The non-pathogenic endophytic fungi exhibit intercellular association with plants and complete their entire life cycle with the plants aerial portion. Many grass species harbor endophytic fungi (Clay, 1990), which are the most widely used biological plant protection agents for forage and turf grasses (Clay, 1989; Kimmons *et al.*, 1990; Gwinn and Gavin, 1992; Funk *et al.*, 1993; Rowan and Latch, 1994). Endophyte infected grasses also showed other properties such as growth stimulation, improved survival and drought tolerance (Arechavaleta *et al.*, 1989). Endophytes play a symbiotic role in the tropical grasses (Kelemu and Takayama, 1998). An endophytic fungus, *Acremonium implicaum* associated with the tropical forage grass, *Brachiaria brizantha* significantly reduced the leaf infection caused by *Drechslera* sp. (Kelemu *et al.*, 2001). Some endophytes, which isolated from medicinal plants and healthy cotton plant proved to be best biocontrol agents for important phytopathogens. *Scopulariopsis bravicola* isolated from a healthy plant from a severely wilted field when tested using dual plate technique showed effective inhibition of important fungal pathogens. *Nodulosporium* sp. and *Aspergillus ochraceous* also inhibited many phytopathogens and proved to be the ideal biocontrol agents (Mathivanan *et al.*, 2003b).

Bacteria as Biocontrol Agents

Isolates of *Pseudomonas, Bacillus, Erwinia,* and *Serratia* are some of the bacteria used as important biocontrol agents. The biocontrol bacteria may also provide plant protection through induction of host plant defense mechanisms, elimination of plant signals that trigger pathogen development or competition for nutrition or by production of antagonistic compounds such as antibiotics, siderophores, cyanide and hydrolytic enzymes. Effective bacterial biocontrol agents with activity against phytopathogenic fungi often synthesize a variety of antifungal metabolites and enzymes (Gutterson *et al.*, 1990; Hass *et al.*, 1991; Lam *et al.*, 1993). In the biocontrol of *Pythium* damping-off by the strains of *Erwinia cloacae,* there was no production of antibiotics or enzymes but agglutination of cell wall fragments of *Pythium ultimum* and inhibition of mycelial growth was observed (Nelson *et al.*, 1986).

Bacteria have been reported to play an important role in biological control of plant pathogens (Cook and Baker, 1983). *Pseudomonas fluorescence* CHAO produces an antibiotic pyoluteorin and 2,4-diacetyl phloroglucinol, which protect various plants from disease caused by soil-borne pathogenic fungus (Maurhofer *et al.*, 1992). They also reported that the production of antibiotics in bacterial biocontrol agent increased several folds against cucumber disease caused by *P. ultimum.* The antibiotic over producing strains showed improved biocontrol abilities in several host pathogen systems. In addition, they observed enhanced antibiotic production and improved protection against damping-off of cucumber. Kamala Nalini (2000) has reported that *P. chlororaphis* was an effective biocontrol agent against *Curvularia lunata, Alternaria* sp., *Rhizoctonia solani* and *Pyricularia oryzae.*

Ganesan and Gnanamanickam (1987) reported the effective control of *Sclerotium rolfsii* in groundnut using antagonistic bacterium *P. fluorescens.* Beneficial effect of fluorescent and non-fluorescent bacteria isolated from South Indian rice fields showed antagonistic activity towards *R. solani* (Vasanthadevi *et al.*, 1989). Seed treatment or foliar application of *Pseudomonas aureofaciens* reduced the disease incidence and increased grain yield of rice at International Rice Research Institute (Mew and Rosales, 1992). Thara and Gnanamanickam (1994) demonstrated the production of chitinase and antibiotic substances by *P. fluorescens* and their role in suppression of sheath blight disease. Seed treatment with *P. fluorescens* and *P. chlororaphis* significantly reduced the blast disease caused by *Xanthomonas oryzae* pv. *oryzae* in rice (Vasandadevi *et al.*, 1989; Valasubramanian, 1994; Kamala Nalini, 2000).

Suppression of *Fusarium* wilt and take all disease caused by *Gaeumannomyces graminis* was observed in pseudomonad colonized plants (Kloepper *et al.*, 1980; Scher *et al.*, 1985). Two chitinolytic bacterial strains, *Paenibacillus* and *Streptomyces* sp. suppressed *Fusarium* wilt of cucumber caused by *Fusarium oxysporum.* Several mechanisms have been proposed to be involved in the suppression of *F. oxysporum* by these antagonistic microorganisms (Weller, 1988; Leeman *et al.*, 1995). Competition for iron between pathogenic *F. oxysporum* and fluorescent pseudomonads was proposed to contribute to the suppression of *Fusarium* wilt, although the bacteria also produce several antifungal metabolites. The mixture of two antibiotic producing bacteria or a non-pathogenic *F. oxysporum* and fluorescent pseudomonads enhanced the efficacy of biological control of *Fusarium* wilt as compared with individual strains (Weller, 1988).

Various bacterial species such as *Aeromonas, Enterobacter, Chromobacterium* and *Serratia* have been reported to be effective biological control agents against S. *rolfsi* and *R. solani* (Snech *et al.*, 1984). *Pseudomonas cepacia* from the surface of leaves, fruits of apple and lemon have been reported to be effective against *Botrytis cinerea* on apple fruits (Janisiewicz, 1988). Rhizosphere bacteria are excellent agents to control soil-borne plant pathogens. Bacterial species like *Bacillus, Pseudomonas, Serratia* and *Arthrobacter* have been proved in controlling the fungal diseases (Ordentlich *et al.*, 1987). Soil bacterium, *Bacillus* sp. isolated from the rhizosphere of sorghum, showed high antagonistic activity against *C. lunata* (Basha and Ulaganathan, 2002).

Actinomycetes as Biocontrol Agents

Actinomycetes are important in maintaining a satisfactory biological balance in soil, largely because of their ability to produce antibiotics. Many of them are effective against bacteria, fungi and actinomycetes as well. Many have the ectoenzyme system necessary to breakdown proteins, cellulose and chitin and are potential in biocontrol of microbes with chitinous walls (Mitchell and Alexander, 1962; Vruggnik, 1970). Biocontrol of fire blight of apple and pear have been successful by using streptomycin (Beer *et al.*, 1984; Vanneste and Yu, 1990). Actinomycetes and their products have been used in the control of *R. solani* (Martin, 1996). Actinomycetes are the most commonly isolated filamentous soil bacteria and many members of the group particularly *Streptomyces* are characterized by their ability to produce a diverse range of secondary metabolites in response to nutrient availability (Bibb, 1996).

The actinomycetes are known to produce various antifungal and antibacterial compounds, which are being exploited commercially for the control of several plant diseases (Lange, 1992; Lange *et al.*, 1993; Lange and Lopez, 1996). More recently novel therapeutic compounds such as megalomicin and calicheamycin have been isolated from *Micromonospora* sp. and have been shown to have antiviral, anticancer and antiparasitic activity (Lee *et al.*, 1987). Oleandomycin is a clinically important 14 membered macrolide antibiotic produced by *Streptomyces antibioticus* (Rodriguez *et al.*, 2001), and spiramycin produced by *Streptomyces ambofaciens* is a 16 membered antibiotic (Lounes *et al.*, 1995). *Streptomyces griseoviridis* has been commercially exploited as biofungicide to control various fungal pathogens (Copping, 1996). Actinomycetes activity was found against different sp. of *Fusarium* and *Rhizoctonia* but not towards *Pythium* sp. (Kerr and Bambieries, 1969; Okafor, 1970). *Streptomyces spiroverticillatus* uses a UV-irradiation regime, which is effective against the causal pathogens of the wilt of French bean and the bacterial wilt of banana. In China, *Streptomyces* strain 5406 has been used with success for several decades in the treatment of cotton seeds (Yin *et al.*, 1965).

In our laboratory research on actinomycetes has been initiated in order to study their biocontrol potential against plant pathogens. About 60 different strains of actinomycetes were isolated from soil samples collected from different ecological habitats. Some of the unidentified actinomycetes showed antifungal activity against fungal pathogens in *in vitro* condition. An unidentified actinomycete inhibited the sclerotial development in *Rhizoctonia solani* and *Sclerotium rolfsii* when the antagonist and pathogens grew in dual plates (Prabavathy, unpublished data). The culture filtrates and crude metabolites significantly inhibited the conidial and sclerotial germination of *F. oxysporum* and *R. solani*, respectively.

Fungi as Biocontrol Agents

Fungi are extremely diverse group of organisms with about 2,30,000 species widely distributed in essentially every ecosystem. Muller and Loeffler (1976) estimated that the weight of fungi on Earth exceeds that of humans. *Armellaria bulbosa,* a tree root pathogen is reported to be among the largest and oldest organisms on the planet (Smith *et al.*, 1992). Among the 2,30,000 species of fungi, only limited number of species are used in biological control. The antagonistic effect of different fungi is being exploited in the control of plant diseases. Some of the fungal biocontrol species used are, *Trichoderma, Ampelomyces, Coniothyrium, Trichothecium, Gliocladium, Talaromyces, Phlebiopsis, Tilletiopsis, Lactisaria, Penicillium, Aspergillus* and *Sporodesmia.*

Trichoderma and *Gliocladium* have been extensively exploited due to their high efficacy of broad spectrum of activity and amenable for large-scale multiplication and formulation. They are reported to control collar rot, root rot, wilt and damping-off caused by *Sclerotium, Rhizoctonia, Fusarium, Phytophthora* and *Phythium* that affect crops like cotton, sunflower, pulses, vegetables, sugarcane, turmeric, ginger and plantation crops like pepper, cardamom, banana, etc. Granular formulations developed with biomass of *Gliocladium* and *Trichoderma* were effectively reduced the damping-off caused by *Rhizoctonia solani* (Lewis and Larkin, 1997; Lewis *et al.*, 1998). The first report on biocontrol potential of *Trichoderma* suggested that this fungus could serve as an effective biocontrol agent against soil-borne fungal pathogens (Weindling, 1932).

Gliocladium virens was used to control damping-off diseases of vegetables and ornamental plant seedlings caused by *R. solani* and *Pythium* sp. (Lumsden *et al.*, 1996). *G. roseum* was active against *B. cinerea* on strawberry (Sutton, 1990). Biocontrol of *Botrytis* leaf blight of onion has been achieved using *Gliocladium* as spray (James and Sutton, 1996). *Ampelomyces quisqualis* was the first biocontrol fungus developed specifically for controlling powdery mildew. AQ10, the product of *A. quisqualis* is a water disposable formulation used against powdery mildew of strawberry, tomato, grape, fruit trees and ornamental plants (Dik *et al.*, 1998). *Coniothyrium minitans* was a mycoparasite to sclerotia of *Sclerotinia sclerotiorum* and *Sclerotinia minor.* A white yeast *Candida oleophila,* is a biocontrol agent that reduces the rot diseases on citrus fruit and apples. It also reduces green and blue mold caused by *Penicillium digitatum* and *Penicillium italicum,* respectively (Dropy *et al.*, 1998). The yeast was effective against sour rot caused by *Geotrichum candidum* (Wilson *et al.*, 1993). In this article, we have elaborately discussed our experience on two fungal genus namely *Fusarium* and *Trichoderma* as biocontrol agents against fungal phytopathogens.

Fusarium

Several species of *Fusarium viz., F. heterosporum, F. semitectum* var. *majus, F. decemcellulare, F. longipes, F. chlamydosporum* and non-pathogenic *F. oxysporum* have been reported as biocontrol agents against various phytopathogenic fungi (Kapooria and Sinha, 1969; Hornok and Walcz, 1983; Gill and Chahal, 1988; Rao and Thakur, 1988; Amorim *et al.*, 1993; Navi and Singh, 1993). Jayapal Gowdu (1986) isolated *F. solani* from the rust pustules on groundnut. Spraying the conidial suspension and culture filtrate of this antagonist reduced the pustule formation on groundnut leaves indicating the release of extracellular antifungal substances such as toxin and lytic enzymes. Later, *F. solani* was found to produce extracellular chitinases, glucanases and cellulases in the culture filtrates (Mathivanan and Balasubramanian, 1990) and the crude enzyme preparation inhibited the spore germination and germ tube growth of *P. arachidis* (Mathivanan, 2000).

Fusarium chlamydosporum was isolated from pustules of groundnut rust, *P. arachidis* and it was also found to be non-pathogenic to groundnut plants (Mathivanan, 1995). The biocontrol potential of

F. chlamydosporum against *P. arachidis* was studied in different laboratory and greenhouse experiments. The application of conidia of *F. chlamydosporum* reduced the pustule number in both detached and intact groundnut leaves. The antagonist rapidly colonizes the rust pustules and as a result, the uredopsores of *P. arachidis* greatly lost its ability to germinate. The colonization of *F. chlamydosporum* also caused damage by bursting and degradating the cell walls of uredospores (Mathivanan and Murugesan, 2000). *F. chlamydosporum* produces extracellular chitinase (Mathivanan *et al.*, 1997a) and the purified chitinase inhibited the uredospore germination and lysed the cell walls of uredospores and germ tube of *P. arachidis* (Mathivanan *el al.*, 1998). Further, *F. chlamydosporum* also produced an antifungal metabolite of p-disubstituted aromatic in nature having carbonyl and methylene groups with the molecular weight of 257. This metabolite completely inhibited the uredospore germination at 30 µg/ml concentration (Mathivanan and Murugesan, 1999). *F. chlamydosporum* has been successfully identified as antagonist to groundnut rust and the mechanisms of biological control were demonstrated. This organism can be commercially promoted as biocontrol agents; however, large-scale production, effective formulation and field experiments at different agroclimatic conditions are to be carried out before commercialization.

Trichoderma

Among biocontrol agents, fungi in the genus *Trichoderma* are considered the most important because they control various root diseases caused by a wide range of fungal pathogens (Alagarsamy *el al.*, 1987; Mukhopadhyay *et al.*, 1992; Latunde-Dada, 1993; Abada 1994; Arisan-Atac *et al.*, 1995; Lo *et al.*, 1996; Mathivanan *et al.*, 2000a; 2003a). The species of *Trichoderma* are considered as multi-potential organisms and their novel biological properties are being exploited as fabric detergent, animal feed, fuel production, alternative to conventional bleaching, effluent treatment, degradation of pesticides (Manoharachary, 2003). Weindling (1932) reported for the first time, the potential of *Trichoderma* as an effective biocontrol agent against soil-borne fungal pathogens. Later several researchers across the world have demonstrated the control of a wide range of plant pathogens using different species of *Trichoderma* (Elad *et al.*, 1981; Sivan and Chet, 1986; Baker, 1989; Adams, 1990; Devaki *et al.*, 1992; Haran *et al.*, 1996; Mathivanan *et al.*, 2000a). Brown and Bruce (1999) assessed the biocontrol potential of *T. viride* under field trials and it reported to protect the woods from soft rot and basidiomycete decay (Brown *et al.*, 1999). Seed treatment or soil amendments of conidia or other formulations of *Trichoderma* reduced root and foliar diseases in many agricultural and plantation crops (Sivan and Chet, 1986; Alagarsamy *et al.*, 1987; Thomas *et al.*, 1993; Cliquet and Scheffer, 1996; Latorre *et al.*, 1997; Prasad *et al.*, 2002; Batta, 2004). Further commercial formulations of *Trichoderma* are available globally for the farmers' use (Whipps 1997; Mukopadhyay and Mukherjee, 1996). A talc-based formulation of *Trichoderma* significantly controlled the root diseases in field grown cotton, egg plant, okra and sunflower crops (Mathivanan *et al.*, 2000a).

The biocontrol mechanisms exercised by *Trichoderma* could be attributed to competition for nutrients, release of toxic metabolites and extracellular hydrolytic enzymes (Elad, 2000). *Trichoderma* was reported to produce volatile and non-volatile antibiotic compounds, which inhibit the fungal growth at very low concentration (Weindling and Emerson, 1936; Weindling, 1941). According to Dennis and Webster (1971), the possible mechanisms involved in *Trichoderma* antagonism towards pathogenic fungi are due to production of antibiotics and cell wall degrading enzymes. A phenol like compound isolated from *Trichoderma harzianum* inhibited the uredospore germination of rust pathogen of groundnut, *Puccinia arachidis* (Govindasamy and Balasubramanian, 1989). Cotton seedling disease incited by *R. solani* has been suppressed by *Trichoderma viride* due to mycoparasitism and antibiotic production (Baker, 1988; Howell *et al.*, 2000).

Trichoderma antagonizes other fungi by coiling, hooks or appresorium like bodies and penetrates the hosts by secreting cell wall lytic enzymes (Weindling, 1932; Inbar and Chet, 1995). *Trichoderma* is also reported to coil tightly and penetrate the hyphae of *Rhizoctonia solani* (Manibhushanrao *et al.*, 1989) and to destroy the mycelium by secreting the extracellular hydrolytic enzymes (Srinivasaprasad and Manibhushanrao, 1990; Kumaresan and Manibhushanrao, 1991). Inbar *et al.*, (1996) studied the hypal interaction between *T. harzianum* and *Sclerotinia sclerotiorum* and established its role in biological control.

T. harzianum, which produces an endochitinase and N-acetyl- -glucosaminidase was reported to be a potent biocontrol agent against several phytopathogenic fungi like *R. solani*, *S. rolfsii* and *P. ultimum* (Cook and Baker, 1983; Ahmad and Baker, 1987a; 1987b). Lorito *et al.*, (1993) have reported that purified endochitinase and chitobiosidase from *T. harzianum* inhibits spore germination and germ tube elongation of *Botrytis cinerea*, *Fusarium solani*. *Trichoderma* produced the extracellular hydrolytic enzyme, chitinase and laminarinase in liquid culture and these enzymes effectively involved in biological control of wood decay fungi (Bruce *et al.*, 1995). Viterbo *et al.*, (2001) isolated a novel endochitinase gene (chit36) from *T. harzianum* and they demonstrated its antifungal activity against fungal pathogens. Further, Prabavathy *et al.*, (2003) investigated the production of extracellular chitinase, by *T. harzianum* and *T. viride* and its possible role of inhibiting conidial and sclerotial germination of *F. oxysporum* and *R. solani*.

Trichoderma as a Potent Biofungicide: A Success Story

Trichoderma is a saprophyte present abundantly in soils containing rich organic matters and hence soil is an excellent source for isolation of different species of *Trichoderma*. Screening of various soil samples from different ecological habitats would yield *Trichoderma* strains with diverse characteristics. Basic and applied researches have been carried out at Nagarjuna Agricultural Research and Development Institute (NARDI), Hyderabad to standardize mass multiplication and formulation technologies for *Trichoderma*. Several strains of *Trichoderma* were isolated from soil samples collected from differents parts of Southern India. All the isolated strains of *Trichoderma* were screened against fungal pathogens *viz.*, *Fusarium oxysporum*, *R. solani* and *S. rolfsii* in dual culture resulted in identifying five efficient antagonists, two of *T. viride* and *T. harzianum*, and a strain of *T. longibrachiatum*. One of these strains, *T. viride* (NCC 34), was found highly effective in inhibiting mycelial growth of fungal pathogens *viz.*, *F. oxysporum*, *R. solani* and S. *rolfsii*. Owing to its superior performance against these pathogens as compared to the other strains, it was selected for mass multiplication and formulation as biofungicide (Mathivanan *et al.*, 2000a; Chelliah and Mathivanan (2001).

Mass Multiplication and Formulation of *Trichoderma*

The species of *Trichoderma* grow in most of the artificial media containing different nutrients. Among different media, molasses yeast medium suggested by Papavizas *et al.*, (1984} is a simple, efficient and cost effective medium for mass cultivation of *Trichoderma*. The selected strain of *T. viride*, NCC 34 was grown in sterile molasses yeast medium and harvested between 7 and 10 days after incubation, homogenized and formulated as biofungicide. This biofungicide formulation contained conidia, chlamydospores and mycelial fragments of *Trichoderma* (Mathivanan *et al.*, 2000a).

Low Cost Production Technology of *Trichoderma*

Majority of the Indian farming community consists of small and marginal farmers and it is quite natural to have reservation over the cost of biological products, which are being used in crop protection, irrespective of their efficacy. From the farmers' point of view the cost should be cheaper than the existing crop protection chemicals. Considering the farmers's concern, new low cost technology has been developed for the mass cultivation of *Trichoderma,* which ultimately resulted in achieving very low cost of production per unit biofungicide product. Chelliah and Mathivanan (2001) developed a cost effective production technology for mass multiplication and formulation of *Trichoderma.* In this method, molasses yeast medium was sterilized separately using conical flasks/pressure cooker. Round plastic tubs measuring 40 cm diameter were sterilized by wiping out thoroughly by commercial disinfectant followed by spirit. One liter molasses yeast medium was poured into each tub and placed in a well-protected inoculation room for 45 min for UV sterilization. Then, each tub was seeded with 100 ml fresh mother culture of *T. viride,* prepared in 500 ml conical flasks and incubated at room temperature. After 7–10 days, the cultures were harvested and formulated with talc carrier as mentioned above. The mass multiplication of *Trichoderma* in this method is much easier and economical. *Trichoderma* grows profusely in tub culture than in flasks and the quality of final product was always better. Care must be taken to avoid other microbial contamination in tub culture however, we notice < 1 per cent contamination, which is negligible as compared to its beneficial aspects.

Quality Control of Biofungicide Product of *Trichoderma*

The viability of biocontrol agents is an important factor in deciding the quality of commercial biocontrol products. It should have a minimum 2×10^8 cfu/g of product in order to be an effective agent (Ragunathan, 1997). The biofungicide product developed by Mathivanan *et al.* (2000a) was subjected to quality control in terms of *Trichoderma* population. It was estimated as 22×10^{13} cfu/g, at the time of manufacture. It declined to 14×10^9 cfu due to storage for three months at 28 ± 2°C. After 4 months of storage, the number of cfu was reduced to 28×10^8/g (Mathivanan *et al.,* 2000a). The population of *T. viride* in biofungicide was > 10^9 cfu/g which is much higher than the recommended level of 2×10^8 cfu/g of product. Moreover, the biofungicide could be used even after four months of storage at 28°C, as 28×10^8 cfu/g remained viable.

Effects of Biofungicide on Seed Germination

Seed treatment is one of the most successful delivery systems for biocontrol agents. However, seed treatment may occasionally inhibit germination since fungal biocontrol agents often produce antibiotics. Although *Trichoderma* species are known to produce antibiotics, the biofungicide formulation did not inhibit seed germination even at higher doses. Significant increase was recorded due to coating with the biofungicide at 25 and 10 g/kg seeds compared to control in cotton and okra. In fact, treated seeds in general germinated better than control (Mathivanan *et al.,* 2000a). Hence, it appears to be a safe product for seed treatment. Enhanced germination, root and shoot lengths recorded due to seed coating with biofungicide might have been due to production of plant growth promoting substances by *T. viride.* These observations are in conformity with findings of Windham *et al.,* (1986), Sharma (1994) and Raguchander *et al.* (1997), who reported several species of *Trichoderma* to increase seed germination, plant growth and yield in different crops.

Effect of Biofungicide on Plant Diseases

In eggplant, seed treatment followed by soil application of the biofungicide significantly controlled wilt disease caused by *Fusarium solani* as compared to chemical fungicides and control. Highest fruit yield was recorded in eggplant due to application of the biofungicide. In sunflower, the above treatment (seed treatment followed by soil application) of biofungicide significantly reduced the plant mortality due to damping-off (*R. solani*) at seedling and collar rot (*S. rolfsii*) at later growth stages as compared to the rest of the treatments (Mathivanan *et al.,* 2000a).

Density of fungal antagonist propagules is an important factor in achieving effective control of plant pathogens in soil. Adams (1990) indicated that *Trichoderma* requires a minimum of 10^5 cfu/g of soil to achieve effective disease control. Because of the high density of propagules in seed treatments followed by soil application of biofungicide in this study, the establishment and multiplication of *T. viride* in soil could be rapid. This may have prevented the accrual of pathogen populations. Cano and Catedral (1994) also recorded a decrease in wilt disease of cotton caused by *Fusarium oxysporum* f. sp. *vasinfectum* following seed treatment and soil drenching with spores of *Trichoderma.* In both eggplant and sunflower there was no plant mortality recorded in plots with biofungicide application both through seed and soil treatment. This could be possible due to the high density of propagules of *T. viride* received by this treatment. These propagules might have proliferated faster, colonized the rhizosphere region and protected the plants till maturity. Chemical fungicides, however, were active only for a shorter period of time and therefore, may be ineffective in protecting the plants at later growth stages.

In *R. solani* infested soil, the biofungicide was found to give effective control of root diseases in cotton, okra and sunflower. Although biocontrol agents are effective for a narrow spectrum of diseases and specific crops (Jacobson and Backman, 1993), the biofungicide was found effective against a wide range of pathogens in several crops. Reduction in plant mortality was greatly influenced by the dose of the biofungicide. Application of biofungicide at 5.0 and 2.5 kg/ha, significantly reduced plant mortality compared to other treatments (Mathivanana *et al.*, 2000a). Increment in yield recorded by application of biofungicide could be due to lowering of plant mortality and enhancement of growth.

Spraying of biofungicide formulation of *Trichoderma* effectively controlled the early, late and rust diseases of groundnut caused by *Cercospora personata, Cercospora arachidicola* and *Puccinia arachidis,* respectively and leaf blight of sunflower caused by *Alternaria alternata.* Among the biocontrol agents evaluated, *T. viride* was the best and its efficacy in controlling major diseases of groundnut and leaf blight of sunflower was on par with many commonly used fungicides (Mathivanan *et al.*, 2000b).

Trichoderma is an excellent component for integrated disease management system because it reported to show tolerance to various fungicides. An integrated use of chemical fungicides with *Trichoderma* was reported to give excellent control of dry root rot in soybean (Vyas, 1994). The combined use of carbendazim and *T. viride* was tried in cotton to control soil-borne root diseases under field conditions. The seed treatment with *T. viride* reduced the plant mortality up to 56 per cent, whereas, integrated use of carbendazim and *T. viride* reduced plant mortality up to 72 per cent (Mathivanan *et al.*, 1997b).

Despite many successful applications of fungal biocontrol agents in both green house and field, only a few have been commercialized (Whipps, 1997). This is mainly because of short shelf life, a wide gap between results of controlled and field conditions, varying performance in different environments and inability to offer quick results compared to chemical fungicides (Burpee, 1990; Whipps, 1992). In light of the results obtained from this study, the *Trichoderma* strain NCC 34 of *Trichoderma viride* merits commercialization.

Distribution of Biofungicide

The biofungicide was distributed to the farmers in eight districts (Adilabad, Karimnagar, Warangal, Khammam, Krishna, Guntur, Prakasam, Nellore) in Andhra Pradesh (AP) state and Bellary district in Karnataka state for the control of soil-borne diseases in different crops (Chelliah and Mathivanan (2001; Mathivanan *et al.*, 2003a). The entire quantities of *T. viride* were distributed free to all the growers through different agencies *viz.*, Farm Management Services (FMS), Hyderabad, Gayathri

Sugar Complex Ltd (GSCL), Nellore, Centre for World Solidarity (CWS), Hyderabad, Navajyothi, Ramayampet, Medak district, AP.

About 10,783 ha area was covered with the biofungicide, spread over eight districts in AP and an adjacent Bellary district in Karnataka. The crop-wise details for the distribution along with total villages, districts and beneficiaries are presented in Table 7.1. The biofungicide has also been supplied to the acid lime growers in Nellore district in AP for the control of wilt disease.

Table 7.1: Distribution of Biofungicide (*T. viride*) in AP

Crop	*No. of Villages*	*No. of Districts*	*No. of Farmers*	*Area Covered (ha)*
Cotton-ST	218	7	4598	8898
Cotton-SD	8	1	50	20
Chilli-SA	83	4	1039	1169
Chilli-SD	2	1	4	4
Chickpea	NR	NR	NR	4
Sunflower	15	1	31	14
Turmeric	3	1	4	4
Sugarcane	55	1	274	670
Acid lime	24	1	50	1600 trees

ST: Seed treatment; SD: Soil drench; SA: Soil application; NR: List not received

Source: Mathivanan *et al.*, 2003a.

Andhra Pradesh has the dubious distinction of using maximum pesticides in India. Five years ago, the use of biocontrol agents was not very popular among farmers for the control of insect pests and diseases due to several reasons. Realizing the gaps and deficiencies in promoting these agents so far, the technological aspects particularly with reference to the mass production of bioagents, effective packaging, timely delivery and guidance in their effective use by the farmers have been improved. Under a scheme sponsored by the Department of Biotechnology, Government of India, sincere and systematic efforts were made to popularize the biological control in eight districts of AP. In addition to the distribution, a number of trials were conducted in the farmers' fields to demonstrate the efficacy of biocontrol agents. Several farmers' meets and field days were organized in different regions of AP to popularize the use of biocontrol agents at a large scale.

The impact assessment surveys conducted on the distributions of biofungicide (*T. viride*) have indicated that majority of the farmers are willing to use the biocontrol agents. They are also eager to adopt this technology and incorporate it in their regular package of practices, if suitable arrangements exist for their regular production and supply at economical costs. The awareness on biocontrol agents has increased several folds among farmers since 1998 due to constant efforts taken by us with the help of other distributing agencies. Distribution of *T. viride* in a wide coverage of area within short period has made tremendous impact on the farmers in AP, who are now coming forward to buy the biocontrol agents from local market at this regions for controlling insect pests and diseases (Chelliah and Mathivanan, 2001).

Strain Improvement in *Trichoderma* by Protoplast Fusion

The basic microbiological techniques such as isolation and screening are useful in identifying newer fungal strains or isolates, which can be used for industrial production of various metabolites and also for biological control in agriculture. However, the protoplast fusion is a novel approach for the strain improvement programme especially in filamentous fungi. Fungal protoplasts are an important tool in physiological and genetic research (Isaac *et al.*, 1986; Peberdy, 1979; 1989). Genetic manipulation in filamentous fungi can be successfully achieved through fusion of proptoplats and it is an effective tool in fungal biotechnology. Interspecific, intraspecfic and intergeneric hybridation could be done by this technique and hence strain improvement of fungal strains is possible (Ohnuki *et al.*, 1982; Quigley *et al.*, 1987; Lalithakumari, 2000).

Isolation, regeneration and fusion of protoplasts have been reported in different fungal species (Gallmetzer *et al.*, 1999; Tschen and Fang Li, 1994; Faweett *et al.*, 1973; Hamlyn *et al.*, 1981; Issac *et al.*, 1986; Stasz and Harman, 1988; Ishikawa and Oishi, 1985; Peberdy *et al.*, 1976; Roncal *et al.*, 1991; Lalithakumari, 1996; 2000; Balasubramanian *et al.*, 2003). *Trichoderma* is an ideal model for the protoplast fusion techniques and several researchers have demonstrated the isolation and regeneration of protoplasts from this genus and also interspecific, intraspecific and intergeneric fusion of protoplasts for enhancing the biocontrol potential for the control of pathogenic fungi (Sivan and Harman, 1989; 1991; Pe'er and Chet, 1990; Mrinalini and Lalithakumari, 1996; Balasubramanian, 2003). This technique has extensively been exploited in this centre to enhance the biocontrol potential of *Trichoderma* (Lalithakumari *et al.*, 1996; Mrinalini and Lalithakumari, 1996; Balasubramanian, 2003) and also to develop novel strains capable of producing multi enzyme complex (Chellappa, 1998; Mathivanan *et al.*, unpublished data).

Biocontrol of the plant pathogenic fungi is increasingly becoming an important alternative/ supplement for plant disease control. *Trichoderma* spp. are the important fungi, both as sources of useful enzymes and as antagonists against plant pathogens. Enhanced antagonistic potential of biocontrol agent is important for the success in control of plant diseases and their ultimate acceptance in integrated disease management. Genetic manipulation of biocontrol agents especially *Trichoderma* species is essential for strain improvement. Manipulation by sexual cycle and anastomosis is not so successful. Protoplast fusion is an effective alternate technique for studying the genetic and biochemistry of fungi leading to new approaches in genetic manipulation.

Isolation and self-fusion of protoplasts were achieved in *T. hazianum* and *T. viride*. The self-fusion progenies showed enhanced antagonistic activity against various fungal pathogens in dual culture (Mathivanan *et al.*, Unpublished data). Protoplast fusion between two species of *T. harzianum* and *T. longibrachiatum* bring together the characters of the parents in to the fusant progenies. The former is known for its antagonism against several plant pathogens and the latter for tolerance to $CuSO_4$ and Carbendazim. The fusant progenies possess antagonistic activity with tolerance to the heavy metal and fungicide (Lalithakumari *et al.*, 1996). Protoplast fusion increased the efficacy of the biocontrol strain though enhanced antagonistic potential integrated with acquired tolerance to fungicides against a broad spectrum of plant pathogens. The fusants exhibited similar morphological characters (growth pattern and pigmentation) of *T. harzianum* and tolerance $CuSO_4$ and Carbendazim as in *T. longibrachiatum*, but the spore size was intermediate to both parent strains.

The antagonistic ability of fusants was evaluated in pot culture experiment on paddy and they established their saprophytic competency. Paddy plants treated with the fusants showed an increased growth, tillering and increased height when compared to other treatments. The sheath blight disease caused by *R. solani* was significantly reduced due to treatment of fusants strains as compared to untreated control (Mrinalini, 1997). Balasubramanian (2003) has also reported the enhanced biocontrol potential in fusant strains obtained through protoplast fusion between *T. harzianum* and *T. viride*. These fusants effectively inhibited the mycelial growth of several fungal pathogens as compared to their parental strains. The intergeneric fusion between *T. viride* and G. *virens* yielded a diverse variation in fusant progenies. These fusants have also showed enhanced lytic enzymes production and antagonistic potential against fungal pathogens (Mathivanan *et al.*, Unpublished data).

Conclusions

Many biocontrol agents have been successfully identified for the control of various plant pathogens and the mechanisms of biological control were demonstrated in most cases. Commercial products of few biocontrol agents are available at present for the use in crop protection. However, it is necessary to identify newer biocontrol agents for the management of various crop diseases across the world and their biocontrol potential might be exploited, which could significantly bring down the indiscriminate use of chemical pesticides. *Trichoderma,* a potent biocontrol agent has gained greater attention and extensive research has been carried for developing this organism as biofungicide as alternative or supplement to chemical fungicides. The out come of biofungicide (*Trichoderma*) distribution under the lab to land programme has clearly broughtout the role of *Trichoderma* in the management of plant diseases in field conditions which merit commercialization. The protoplast fusion technique could be highly useful for the strain improvement programme in filamentous fungi through which more efficient strains of *Trichoderma* with excellent performance in different agroclimatic and adverse ecological conditions could be developed.

REFERENCES

Abada, K.A. (1994). Fungi causing damping-off and root rot on sugar beet and their biological control with *Trichoderma harzianum. Agri. Ecosys. Environ.*, **51**: 333-337.

Adams, P.B. (1990). The potential of mycoparasites for biological control of plant diseases. *Ann. Rev. Phytopathol.*, **28**: 59-72.

Ahmad, J.S. and Baker, R. (1987a). Rhizosphere competence of *Trichoderma harzianum* with altered antibiotic production on selected in vitro. *Can. J. Microbiol.,* **77**: 182-189.

Ahmad, J.S. and Baker, R. (1987b). Competitive saprophytic ability and cellulolytic activity of rhizosphere competent mutants of *Trichoderma harzianum. Phytopathology* **77**: 358-362.

Alagarsamy, G., Mohan, S. and Jeyarajan, R. (1987). Effect of seed pelleting with antagonists in the management of seedling disease of cotton. *J. Biol. Control,* **1**: 66-67.

Amorim, E.P.R., Pio-Ribeiro, G., Menezes, M. and Coelho, R.S.B. (1993). The pathogenicity and hyperparasitic action of *Fusarium decemcellulare* on *Puccinia psidii* in guava (*Psidium guajava*). *Fitopatol. Bras.*, **18**: 226-229.

Anderson, A.J. and Guerra, D. (1985). Responses to root colonization with *Pseudomonas putida* in a hydrophonic system. *Phytopathology,* **75**: 992-995.

Anonymous, (2000). Role of pesticides in Indian agriculture. *Pesticides World,* **5**: 29-31.

Arechavaleta, M., Bacon, C.W., Hoveland, C.S. and Radcliffe, D.E. (1989). Effect of the tall fesue endophyte on plant response to environmental stress. *Agron. J.*, **81**: 83-90.

Arisan-Atac, I., Heidenreich, E. and Kubicek, C.P. (1995). Randomly amplified polymorphic DNA fingerprinting identifies subgroups of *Trichoderma viride* and other *Trichoderma* sp. capable of chestnut blight biocontrol. *FEMS Microbiol. Lett.*, **126**: 249-256.

Baker, R. (1988). Mycoparasitism: Ecology and Physiology. *Can. J. Plant Pathol.*, **9**: 370-379.

Baker, R. (1989). Some perspectives on the application of molecular approaches to biocontrol problems. In: *Biotechnology of fungi for improving plant growth.* (Eds. Whipps, J.M. and Lumsden, R.D), Cambridge University Press, Cambridge. pp. 220-233.

Balasubramanian, N. (2003). *Strain improvement of Trichoderma spp. by protoplast fusion for enhanced lytic enzyme and biocontrol potential.* Ph. D., thesis, University of Madras, Chennai, India.

Balasubramanian, N., Annie Juliet, A., Srikalavani, P. and Lalithakumari, D. (2003). Release and regeneration of protoplasts from *Trichothecium roseum. Can. J. Microbiol.,* **49**: 263-268.

Basha, S. and Ulaganathan, K. (2002). Antagonism of *Bacillus* species (strain BC121) towards *Curvularia lunata. Curr. Sci.,* **82**: 1457-1463.

Batta, Y.A. (2004). Postharvest biological control of apple gray mold by *Trichoderma harzianum* Rifai formulated in an invert emulsion. *Crop Prot.,* **23**: 19-26.

Beer, S.V., Rundle, J.R. and Wodzinski, R.S. (1984). Recent progress in development of biological control for fire blight- a review. *Acta. Hortic.,* **151**: 195-201.

Bibb, M. (1996). The regulation of antibiotic production in *Streptomyces coelicolor* A3. *Microbiology,* **142**: 1335-1344.

Brent, K.J. (1995). *Fungicide resistance in crop pathogens: How can it be managed?* Groupement International des Associations Nationales de Fabricants de Produits Agrochimiques. Brussels, Belgium, 48p.

Brown, H.L. and Bruce, A. (1999). Assessment of the biocontrol potential of a *Trichoderma viride* isolate. Part I: Establishment of field and fungal cellar trials. *Internatl. Biodeter. Biodegrad.,* **44**: 219-224.

Brown, H.L., Bruce, A. and Staines, H.J. (1999). Assessment of the biocontrol potential of a *Trichoderma viride* isolate. Part II: Protection against soft rot and basidiomycete decay. *Internatl. Biodeter. Biodegrad.,* **44**: 225-231.

Bruce, A., Srinivasan, U., Staines, H.J. and Highley, T.L. (1995). Chitinase and laminarinase production in liquid culture by *Trichoderma* spp. and their role in biocontrol of wood decay fungi. *Internatl. Biodeter. Biodegrad.,* **40**: 337-353.

Burpee, L.L. (1990). The influence of abiotic factors on biological control of soilborne plant pathogenic fungi. *Can. J. Plant Pathol.,* **12**: 308-317.

Cano, L.C. and Catedral, I.G. (1994). Efficacy of *Trichoderma* sp. as biological control agent against *Fusarium oxysporum* f. sp. *vasinfectum. Cotton Res. J.,* **7**: 64-74.

Chellappa, G. (1998). *Strain improvement through protoplast fusion technology for enhanced lytic enzyme production.* Ph.D., thesis, University of Madras, Madras, India.

Chelliah, S. and Mathivanan, N. (2001). Final project report on *Development, production and demonstration of biocontrol agents under IPM.* Submitted to the Department of Biotechnology, Government of India, New Delhi.

Clay, K. (1989). Clavicipitaceous endophytes of grasses: Their potential as biocontrol agents. *Mycol. Res.,* **92**: 1-12.

Clay, K. (1990). Fungal endophytes of grasses. *Annu. Rev. Ecol. Syst.,* **21**: 275-297.

Cliquet, S. and Scheffer, R.J. (1996). Biological control of damping-off caused by *Pythium ultimum* and *Rhizoctonia solani* using *Trichoderma* spp. applied as industrial film coatings on seeds. *Euro. J. Plant Pathol.,* **102**: 247-255.

Cook, R.J. and Baker, K.F. (1983). *The Nature and Practice of Biological Control of Plant Pathogens.* Cst. Paul, Minn. American Phytopathological society.

Copping, L.G. (1996). *Crop Protection Agents from Nature: Natural Products and Analogues.* The Royal Society of Chemistry, Cambridge, UK. 501p.

Dennis, L. and Webster, J. (1971). Antagonistic properties of species group of *Trichoderma* III hyphal interactions. *Trans. Br. Mycol. Soc.,* **57**: 363-369.

Devaki, N., Shankara Bhat, S.S., Bhat, S.G. and Manjunatha, K.R. (1992). Antagonistic activities of *Trichoderma harzianum* against *Pythium aphanidermatum* and *Pythium myriotylum* on tobacco. *J. Phytopathol.,* **136**: 82-87.

Dik, A.J., Verhaur, M.A. and Belenger, R.R. (1998). Comparison of three biological control agents against cucumber powdery mildew (*Sphaerotheca fulginea*) in semi- commercial scale glass house trials. *Euro. J. Plantpathol.,* **104**: 413-423.

Droby, S.L., Cohen, A., Daus, B., Weiss, B., Horev, E., Chalutz, H., Katz, M., Kerren, T. and Shachnai, A. (1998). Commercial testing of Aspire: A yeast preparation for the biological control of post harwal decays of citrus. *Biol. Control,* **12**: 97-101.

Elad, Y. (2000). Biological control of foliar pathogens by means of *Trichoderma harzianum* and potential modes of action. *Crop Prot.,* **19**: 709-714.

Elad, Y., Chet, I. and Henis, V. (1981). Biological control of *Rhizoctonia solani* in strawberry fields by *Trichoderma harzianum. Plant Dis.,* **60**: 245-254.

Fawcett, P.A., Loder, P.B., Duncan, M.J., Beesley, T.J. and Abraham, E.P. (1973). Formation and properties of protoplasts from antibiotic producing strains of *Penicillium chrysogenum* and *Cephalosporium acremonium. J. Gen. Microbiol.,* **79**: 293-309.

Funk, C.R., White, R.H. and Breen, J.P. (1993). Importance of *Acremonium* endophytes in turfgrass breeding and management. *Agric. Ecosyst. Environ.,* **44**: 215-232.

Gallmetzer, M., Burgstaller, W. and Schinner, F. (1999). An optimized method for the isolation of protoplast from *Penicillium simplicissimum* to produce sealed plasma membrane vesicles. *Mycologia,* **91**: 206-212.

Ganesan, P. and Ganamanickam, S.S. (1987). Biological control of *Sclerotium rolfsii* sacc. in peanut by inoculation with *Pseudomonas fluorescence. Soil Biol. Biochem.,* **19**: 35-38.

Gill, K.S. and Chahal, S.S. (1988). Growth inhibition of *Claviceps fusiformis* with culture filtrate of *Fusarium chlamydosporum. Plant Dis. Res.,* **3**: 64-65.

Govindasamy, V. and Balasubramanian, R. (1989). Biological. control of groundnut rust, *Puccinia arachidis* by *Trichoderma harzianum. Z. PflKrankh. PflSchutz,* **96**: 337-345.

Gutterson, N., Howie, W. and Suslow, T. (1990). Enhancing efficacies of biocontrol agents by the use of biotechnology. In: *New Directions in Biocontrol* (Eds. Baker, R., Dunn, P. and Liss, A.R.) New York, pp. 749- 765.

Gwinn, K.D. and Gavin, A.M. (1992). Relationship between endophyte infestation level of tall fescue seed lots and *Rhizoctonia zeae* seedling disease. *Plant Dis.,* **76**: 911-914.

Hader, Y., Harman, G.E., Taylor, A.G. and Norton, J.M. (1983). Effects of pregerminating pea and cucumber seeds and of seed treatment with *Enterobacter cloacae* on rots caused by *Phytium* spp. *Phytopathology,* **73**: 1322-1325.

Hallmann, J., Quadt-Hallmann, A., Mahaffee, W.F. and Kloepper, J.W. (1997). Bacterial endophytes in agricultural. crops. *Can. J. Microbiol.*, **43**: 895-914.

Hallmann, J., Quadt-Hallmann, A., Miller, W.G., Sikora, R.A. and Lindow, S.E. (2001). Endophytic colonization of plants by the biocontrol agent *Rhizobium etli* G12 in relation to *Meloidogyne incognita* infection. *Phytopathology*, **91**: 415-422.

Hallmann, J., Rodriguez-kabana, R. and Kloepper, J.W. (1999). Chitin mediated changes in bacterial communities of the soil, rhizosphere and with in the roots and with in roots of cotton in relation to nematode control. *Soil Biol. Biochem.*, **31**: 551-560.

Hamlyn, P.F., Bradshaw, R.E., Mellon, F.M., Santiago, C.M. and Peberdy, J.F. (1981). Efficient protoplast isolation from the fungi using commercial enzymes. *Enzyme Microbiol. Technol.*, **3**: 321-325.

Handelsman, J.S., Raffel, E.H., Master, L., Wunderlich, C.R. and Grau, L. (1990). Biological control of damping-off alfalfa seedlings with *Bacillus cereus* UW85. *App. Environ. Microbiol.*, **56**: 713-718.

Haran, S., Schikler, H. and Chet, I. (1996). Molecular mechanisms of lytic enzymes involved in the biocontrol activity of *Trichoderma harzianum.*, **142**: 2321-2331.

Hass, D.C., Keel, J., Laville, M., Maurhofer, T., Oberhansli, U., Schinder, C., Voisard, B. Wathrich and Defago, G. (1991). Secondary metabolites of *Pseudomonas fluorescence* strain CHAO involved in the suppression of root disease. In: *Advanced Molecular Genetics of Plant Microbe Interactions.* (Eds. H. Hennecke and D.P.S. Venna). Kluwer Academic Publishers, Dor drechat, Netherlands. pp. 450-456.

Hornok, L. and Walcz, I. (1983). *Fusarium heterosporum*, a highly specialized hyperparasite of *Claviceps purpurea. Trans. Br. Mycol. Soc.*, **80**: 377-380.

Howell, C.R., Hanson, L.E., Stipanovic, R.D. and Puckhaber, L.E. (2000). Induction of terpenoid synthesis in cotton roots and control of *Rhizoctonia solani* by seed treatment with *Trichoderma virens. Phytopathology*, **90**: 248-252.

Inbar, J. and Chet, I. (1995). The role of recognition in the induction of specific chitinase during mycoparasitism by *Trichoderma harzianum. Microbiology*, **141**: 2823.

Inbar, J., Menendez, A. and Chet, I. (1996). Hyphal interaction between *Trichoderma harzianum* and *Sclerotinia sclerotiorum* and its role in biological. control. *Soil Biol. Biochem.*, **28**: 757-763.

Isaac, S., Gokhale, A.V. and Wyatt, A.M. (1986). Selectivity of K+ and Na+ of protoplast fractions isolated from different regions of *Aspergillus nidulans hyphae. J. Gen. Microbiol.*, **132**: 1173-1179.

Ishikawa, F. and Oishi, K. (1985). Formation and regeneration of protoplasts from *Conidiobolus lamprauges. J. Gen. Microbiol.*, **131**: 3311-3316.

Jacobson, B.J. and Backman, P.A. (1993). Biological and cultural plant disease controls: Alternatives and supplements to chemicals in IPM systems. *Plant Dis.*, **77**: 311-315.

James, T.D. W. and Sutton, J.C. (1996). Biological control of *Botrytis* leaf blight of onion by *Gliocladium roseum* applied as sprays and with fabric applicators. *Phytopathology*, **86**: 265.

Janisiewicz, W.J. (1988). Biological control of post harvest disease of temperate fruits; Microbe interaction and biological control (Eds. J. Boland and L.D. Kuykendall) Marcel Dekker, Inc, New York, pp. 171-198.

Jayabal Gowdu, B. (1986). *Studies on some biological aspects of fungal associates on groundnut rust, Puccinia arachidis Speg.* Ph.D., thesis, University of Madras, Chennai, India.

Kamala Nalini, S.P. (2000). *Strain improvement and characterization of Pseudomonas chlororaphis: A broad spectrum biocontrol agent of phytopathogens.* Ph. D., thesis, University of Madras, Chennai, India.

Kapooria, R.G. and Sinha, S. (1969). Phylloplane mycoflora of pearl millet and its influence on the development of *Puccinia pennisetti. Trans. Br. Mycol. Soc.,* **53**: 153-155.

Kelemu, S. and Takayama, Y. (1998). An endophytic fungus in the tropical forage grass *Brachiaria brizantha*: Effect on a leaf spot disease. *Phytopathology,* **88**: Abstr. S46.

Kelemu, S., White Jr., J.F., Munoz, F. and Takayama, K. (2001). An endophyte of the tropical forage grass *Brachiaria brizantha*: Isolating, identifying, and characterizing the fungus, and determining its antimycotic properties. *Can. J. Microbiol.,* **47**: 55-62.

Kerr, A. and Bumbieris, M. (1969). Effect of amendments on numbers of soil microorganisms and on the root-rot *Fusarium* wilt complex of peas. *Annu. Rep.,* **69**: 77-78.

Kimmons, C.A., Gwinn, K.D. and Bernard, E.C. (1990). Nematode reproduction on endophyte-infected and endophyte-free tall fescue. *Plant Dis.,* **74**: 757-761.

Kloepper, J.W., Schorth, M. V. and Miller, T.D. (1980). Effects of rhizosphere colonization by the plant growth promoting *Rhizobacteria* on potato plant development and yield. *Phytopathology,* **70**: 1078-1082.

Kumaresan, S. and Manibhushanrao, K. (1991). Studies on the biological control of rice sheath blight disease. *Indian J. Pl. Pathol.,* **9**: 64-70.

Lalithakumari, D. (1996). Protoplasts-A biotechnological tool for plant pathological studies. *Indian Phytopath.,* **49**: 199-212.

Lalithakumari, D. (2000). *Fungal protoplast: A biotechnological tool.* Oxford & IBH Publishing Co., Pvt., Ltd., New Delhi, 184p.

Lalithakumari, D., Mrinalini, C., Chandra, A.B. and Annamalai, P. (1996). Strain improvement by protoplast fusion for enhancement of biocontrol potential integrated with fungicide tolerance in *Trichoderma* spp. *J. Plant Dis. Prot.,* **103**: 206-212.

Lam, S.T., Gaffney, T.D., Frazelle, R.A., Gates, K., Dimaio, J., Torkewitz, N., Ligon, J., Hills, S., Goodwin, S. and Kempf, H.J. (1993). Two genes, which regulates the coordinated expression of antifungal activities in *Pseudomonas fluorescence.* In: *Abstracts of the Fourth International Symposium on Pseudomonas: Biotechnology and Molecular Biology,* (Eds. R.W. Hancock and B.C. Vancouver), Canada.

Lange, L. (1992). Microbes and microbial products in plant protection. *Prog. Bot.,* **53**: 252.

Lange, L. and Lopez, S.C. (1996). Microorganisms as source of biologically active secondary metabolites. In: *Crop Protection Agents from Nature: Natural Products and Analogues,* (Ed. Copping, L.G.) The Royal Society of Chemistry, Cambridge, UK. pp. 1-26.

Lange, L., Breinholdt, J., Rasmussen, F.W. and Nielsen, R.I. (1993). Microbial fungicides-the natural choice. *Pestic. Sci.,* **39**: 155.

Latorre, B.A., Agosin, E., San Martin, R. and Vasquez, G.S. (1997). Effectiveness of conidia of *Trichoderma harzianum* produced by liquid fermentation against *Botrytis* bunch rot of table grape in Chile. *Crop Prot.,* **16**: 209-214.

Latunde-Dada, A.O. (1993). Biological control of southern blight disease of tomato caused by *Sclerotium rolfsii* with simplified mycelial formulations of *Trichoderma koningii. Plant Pathol.,* **42**: 522-529.

Lee, M.D., Dunne, T.S., Siegel, M.M., Chang, C.C., Morton, G.O. and Borders, D.B.X. (1987). Calichemicin. A novel family of antitumour antibiotic. Chemistry and partial structure of calichemicin. *J. Am. Chem. Soc.*, **109**: 3664-3466.

Leeman, M.J.A., Van Pelt, F.M., Denouden, M., Heinsbroek, P.A., Baker, H.M. and Schippers. (1995). Induction of systemic resistance against *Fusarium* wilt of radish by lipopolysaccharides of *Pseudomonas fluorescens*. *Phytopathology*, **85**: 1021-1027.

Lewis, J.A. and Larkin, R.P. (1997). Extruded granular formulation with biomass of biocontrol *Gliocladium virens* and *Trichoderma* spp. to reduce damping-off of eggplant caused by *Rhizoctonia solani* and saprophytic growth of the pathogen in soil-less mix. *Biocontrol Sci. Technol.*, **7**: 49-60.

Lewis, J.A. and Larkin, R.P. and Rogers, D.L. (1998). A formulation of *Trichoderma* and *Gliocladium* to reduce damping-off caused by *Rhizoctonia solani* and saprophytic growth of the pathogen in soilless mix. *Plant Dis.*, **82**: 501-506.

Lo, C.T., Nelson, E.B. and Harman, G.E. (1996). Biological control of turfgrass diseases with a rhizosphere competent strain of *Trichoderma harzianum*. *Plant Dis.*, **80**: 736-741.

Lorito, M., Harman, G.E., Hayerss, C.K., Broadway, R.M., Tronsmo, A., Woo, S.L. and Pietro, A. (1993). Chitinolytic enzymes produced by *Trichoderma harzianum*: Antifungal activity of purified endochitinase and chitobiosidase. *Phytopathology*, **83**: 302-307.

Lounes, A., Lebrihi, A., Benslimane, C., Lefebure, G. and Germain, P. (1995). Regulation of valine catabolism by ammonium in *Streptomyces ambfaciens*, producer of spiramycin. *Can. J. Microbial.*, **41**: 800-808.

Lumsden, R.D., Walter, J.F. and Baker, C.P. (1996). Development of *Gliocladium virens* for damping-off disease control. *Can. J. Plant Pathol.*, **8**: 463-468.

Mahaffee, W.F. and Kloepper, J.W. (1997). Temporal changes in the bacterial communities in the soil, rhizosphere and endorhiza. *Microbial. Ecol.*, **34**: 210-233.

Manibhushanrao, K., Srinivasaprasad, S., Baby, U.I. and Joe, Y. (1989). Susceptibility of rice sheath blight pathogen to mycoparasites. *Curr. Sci.*, **58**: 515-518.

Manoharachary, C. 2003. Role of *Trichoderma* Pers. Ex. Fr. as a biocontrol agent. *National Conference on Biological Control of Crop Diseases: Present Perspectives and Future Challenges.* March 5-6, 2002. Centre for Advanced Studies in Botany, University of Madras, Chennai – 600 025, pp. 2.

Martin, M.K. (1996). Potential use of actinomycetes and their products for control of *Rhizoctonia solani. Rhizoctonia News Lett.*, **1**: 1.

Mathivanan, N. (1995). *Studies on extracellular chitinase and secondary metabolites produced by Fusarium chlamydosporum, an antagonist to Puccinia arachidis, the rust pathogen of groundnut.* Ph. D., thesis, University of Madras, Madras, India.

Mathivanan, N. (2000). Chitinase and -1,3 glucanase of *Fusarium solani*: Effect of crude enzymes on *Puccinia arachidis. J. Mycol Plant Pathol.*, **30**: 327-330.

Mathivanan, N. and Balasubramanian, R. (1990). Extracellular and intracellular chitinases, glucanases and proteinases of *Fusarium solani* (Mart.) Sacc. *Indian J. Microbiol.*, **30**: 433-437.

Mathivanan, N., Bharati N. Bhat, Prabavathy, V.R., Srinivasan, K. and Chelliah, S. (2003a). *Trichiderma viride*: Lab to land for the management of root diseases in different crops. In: *Innovative Methods and Techniques for Integrated Pest and Disease Management* (Eds. Mathivanan, N., Prabavathy, V.R.

and Gomathinayagam, S.), Centre for Advanced Studies in Botany, University of Madras, Chennai, India. pp. 52-58.

Mathivanan, N., Kabilan, V. and Murugesan, K. (1997a). Production of chitinase by *Fusarium chlamydosporum*, a mycoparasite to groundnut rust, *Puccinia arachidis. Indian J. Exp. Biol.*, **35**: 890-893.

Mathivanan, N., Kabilan, V. and Murugesan, K. (1998). Purification, characterization and antifungal activity of chitinase from *Fusarium chlamydosporum*, a mycoparasite to groundnut rust, *Puccinia arachidis. Can. J. Microbiol.*, **44**: 646-651.

Mathivanan, N. and Murugesan, K. (1999). Isolation and purification of an antifungal, metabolite from *Fusarium chlamydosporum*, a mycoparasite to *Puccinia arachidis*, the rust pathogen of groudnut. *Indian J. Exp. Biol.*, **37**: 98-101.

Mathivanan, N. and Murugesan, K. (2000). *Fusarium chlamydosporum*, a potent biocontrol agent to groundnut rust, *Puccinia arachidis. Z. PflKrankh. PflSchutz.*, **107**: 225-234.

Mathivanan, N., Srikalaivani, P., Niveditha, S.V., Malarvizhi, K., Prabavathy, V.R., Murugesan, K. and Lalithakumari, D. (2003b). Exploitation of different biocontrol agents for the management of plant pathogens. In: *Innovative Methods and Techniques for Integrated Pest and Disease Management* (Eds. Mathivanan, N., Prabavathy, V.R. and Gomathinayagam, S.), Centre for Advanced Studies in Botany, University of Madras, Chennai, India, pp. 26-33.

Mathivanan, N., Srinivasan, K. and Chelliah, S. (1997b). Evaluation of *Trichoderma viride* and carbendazim for integrated management of root diseases in cotton. *Indian J. Microbiol.*, **37**: 107-108.

Mathivanan, N., Srinivasan, K. and Chelliah, S. (2000a). Biological control of soil- borne diseases of cotton, eggplant, okra and sunflower by *Trichoderma viride. J. Plant Dis. Prot.*, **107**: 235-244.

Mathivanan, N., Srinivasan, K. and Chelliah, S. (2000b). Field evaluation of biocontrol agents against foliar diseases of groundnut and sunflower. *J. Biol. Control*, **14**: 31-34.

Maurhofer, M.C., Keel, U., Schinder, C., Voisard, D. and Defago, G. (1992). Influence of enhanced antibiotic production in *Pseudomonas.fluorescens* strain CHAO on its disease suppressive capacity. *Phytopathology*, **82**: 190-195.

Mew, T.W. and Roseles, A.M. (1992). Control of *Rhizoctonia* sheath blight and other disease of rice by seed bacterization. In: *Progress in Rainfed Low Land Rice.* IRRI Manila, Philippines, pp. 339-348.

Mitchell, R. and Alexander, M. (1962). Microbiological processes associated with the use of chitin for biological control. *Proc. Soil. Sci. Soc. Am.*, **26**: 556-558.

Mrinalini, C. (1997). *Strain improvement of enhancement and integration of antagonistic and fungicide tolerance potential in Trichoderma* spp. Ph. D., thesis, University of Madras. Chennai, India.

Mrinalini, C and Lalithakumari, D. (1996). Protoplast fusion: A biotechnological tool for strain improvement of *Trichoderma* spp. In: *Recent Development in Biocontrol of Plant Pathogens.* (Eds. Manibhushanrao, K. and Mahadevan, A.), Today and Tomorrow's Printers and Publishers, New Delhi, pp. 133-146.

Mukhopadhyay, A.N. and Mukherjee, P.K. (1996). Fungi as fungicides. *Int. J. Trop. Plant Dis.*, **14**: 1-17.

Mukhopadhyay, A.N., Shrestha, S.M. and Mukherjee, P.K. (1992). Biological seed treatment for control of soil-borne plant pathogens. *FAO Plant Prot. Bull.*, **40**: 21-30.

Muller, E. and Loeffler, W. (1976). *Mycology: An Outline for Science and Medical Students.* Thieme Publishers, Stuttgart, Germany.

Navi, S.S. and Singh, S.D. (1993). *Fusarium longipes*–a mycoparasite of *Sclerospora graminicola* on pearl millet. *Indian Phytopath.*, **46**: 365-368.

Nelson, E.B., Chao, W.L., Norton, J.M., Nach, G.T. and Harman, G.E. (1986). Attachment of *Enterobacter cloacae* to hyphae of *Pythium ultimum*: possible role in the biological of *Pythium* pre-emergence damping-off. *Phytopathology*, **76**: 327-335.

Ohnuki, Y., Etoh, Y. and Beppu, T. (1982). Intraspecific and interspecific hybridization of *Mucor pusillus* and *M. miehei* by protoplast fusion. *Agric. Biol. Chem.*, **46**: 451-458.

Okafor, N. (1970). Influence of chitin on mycoflora and length of roots of wheat seedlings. *Trans. British. Mycol. Soc.*, **55**: 483-514.

Ordentlich, A., Elad, Y. and Chet, I. (1987). Rhizosphere colonization of *Serratia, marcescens* for the control of *Sclerotium rolfii. Soil. Biol. Biochem.*, **19**: 747-751.

Papavizas, G.C. (1985). *Trichoderma* and *Gliocladium*: biology, ecology and potential for biocontrol. *Annu. Rev. Phytopathol.*, **23**: 23-54.

Papavizas, G.C., Dunn, M.T., Lewis, J.A. and Beagle-Ristanio, J. (1984). Liquid fermentation technology for experimental production of biocontrol fungi. *Phytopathology*, **74**: 1171-1175.

Peberdy, J.F. (1979). Fungal protoplasts: Isolation, reversion and fusion. *Ann. Rev. Microbiol.*, **33**: 21-29.

Peberdy, J.F. (1989). Fungi without coats–Protoplasts as tools for mycological research. *Mycol. Res.*, **93**: 1-20.

Peberdy, J.F., Buckley, C.E., Daltrey, D.C. and Moore, P.M. (1976). Factors affecting protoplast release in some filamentous fungi. *Trans. Br. Mycol. Soc.*, **67**: 23-26.

Pe'er, S. and Chet, I. (1990). *Trichoderma* protoplast fusion; a tool for improving biocontrol agents. *Can. J. Microbiol.*, **36**: 6-9.

Prabavathy, V.R., Bharati N. Bhat and Mathivanan, N. (2003). Role of chitinase from *Trichoderma* in biolol control of fungal pathogens. *National Conference on Biological Control of Crop Diseases: Present Perspectives and Future Challenges.* March 5-6, 2003. Centre for Advanced Studies in Botany, University of Madras, Chennai, India, pp 33.

Prasad, R.D., Rangeswaran, R., Hegde, S. V. and Anuroop, C.P. (2002). Effect of soil and seed application of *Trichoderma harzianum* on pigeonpea wilt caused by *Fusarium udum* under field conditions. *Crop Prot.*, **21**: 293-297.

Quadt-Hallmann, A., Benhamou, N. and Kloepper, J.W. (1996). Bacterial endophytes in cotton: Mechanisms of entering the plants. *Can. J. Microbiol.*, **43**: 577-582.

Quadt-Hallmann, A. and Kloepper, J. W. (1996). Immunological detection and localization of the cotton endophyte *Enterobacter asburia* JM22 in different plant species. *Can. J. Microbiol.*, **42**: 1144-1154.

Quigley, D.R., Taft, C.S., Stark T. and Selitrennikoff, C.P. (1987). Optimal conditions for the release of protoplasts of *Neurospora* using Novozym 234. *Exp. Mycol.*, **11**: 236-240.

Raguchander, T., Rajappan, K. and Samiappan, R. (1997). Evaluating methods of application of biocontrol agent in the control of mung bean root rot. *Indian Phytopath.*, **50**: 229-234.

Ragunathan, V. (1997). Quality control of biopesticides. In: *Souvenir, National Conference on Biopesticides, Department of Agriculture and Cooperation.* Government of India and All India Biotech Association, New Delhi, April 10, 1997, pp. 18-32.

Rangaswamy, G. (1996). *Disease of crop plants in India.* Prentice-Hall of India Private Limited, New Delhi.

Rao, V.P. and Thakur, R.P. (1988). *Fusarium semitectum* var. *majus*–a potential biocontrol agent of ergot (*Claviceps fusiformis*) of pearlmillet. *Indian Phytopath.*, **41**: 567-574.

Rodriguez, L., Rodriguez, D., Olano, C., Brana, A.F., Mendez, C. and Salas, J.A. (2001). Functional analysis of ole-YL-oleandrosyl 3-0-methyltranstrase of the oleandomycin biosynthetic pathway in *Streptomyces antibioticus. J. Bacterial.*, **183**: 5358-5363.

Roncal, T., Ugalde, U.O., Barnes, J. and Pitt, D. (1991). Production of protoplasts of *Penicillium cyclopium* with improved viability and functional properties. *J. Gen. Microbiol.*, **137**: 1647-1651.

Rowan, D.D. and Latch, G.C.M. (1994). Utilization of endophyte-infected perennial rye grasses for increased insect resistance. In: *Biotechnology of endophytic fungi of grasses*, (Eds. Bacon, C.W. and White Jr, J.F.), CRC Press, Boca Raton, Fla., pp. 169-183.

Saksena, S. (2001). Pesticide Industry: In a climate of signal shifts. *Pesticides World*, **5**: 27-44.

Scher, F.M., Kloepper, J.W. and Singleton, C.A. (1985). Chemotaxis of fluorescent *Pseudomonas* spp. to soyabean seed exudates and in soil. *Can. J. Microbiol.*, **31**: 570-574.

Sharma, B.K. (1994). Efficacy of biocontrol agents for the control of chickpea stem rot. *J. Biol. Controls*, **8**: 115-117.

Sharma, A.K., Singh, D.P. and Singh, A.K. (1998). Epidemiological studies in biological control of plant pathogens. In: *Biological Suppression of Plant Diseases, Phytoparasitic Nematodes and weeds*, (Eds. Singh, S.P. and Hussaini, S.S.), Project Directorate of Biological Control, Bangalore, India, pp. 110-117.

Smith, M., Bruhn, J. and Anderson, J. (1992). The fungus *Armillaria bulbosa* is among the largest and oldest living organisms. *Nature*, **326**: 428-431.

Snech, B.M., Dupler, Elad, Y. and Baker, R. (1984). Clamydospore germination of *Fusarium oxysporum* f. sp. *cuemrinum* as affected by fluorecent and lytic bacteria from *Fusarium* suppressive soils. *Phytopathology*, **74**: 1115-1124.

Sivan, A. and Chet, I.. (1986). Biological control of *Fusarium* spp. in cotton, wheat and muskmelon by *Trichoderma harzianum. J. Phytopathol.*, **116**: 38-47.

Sivan, A. and Harman, G.E. (1989). Increase in rhizosphere competence of *Trichoderma harzianum* following protoplast fusion. *Phytopathology*, **79**: 1159.

Sivan, A. and Harman, G.E. (1991). Improved rhizosphere competence in a protoplast fusion progeny of *Trichoderma harzianum. J. Gen. Microbiol.*, **137**: 23.

Srinivasaprasad, S. and Manibhushanrao, K. (1990). Antagonistic potential of *Gliocladium virens* and *Trichoderma longibrachiatum* to phytopathogenic fungi. *Mycopathologia*, **109**: 19-26.

Stasz, T.E., Harman, G.E. and Weeden, N.F. (1988). Protoplast preparation and fusion in two biocontrol strains of *Trichoderma harzianum. Mycologia*, **80**: 141-150.

Sutton, J.C. (1990). Epidemiology and management of *Botrytis*, Leaf of onion and gray mold of Strawberry: a comparative analysis. *Can. J. Plant. Pathol.*, **12**: 100-110.

Thara, K.V. and Gnanamanickam, S.S. (1994). Biological control of rice sheath blight in India: Lake of correlation between chitinase and production by bacterial antagonistic and sheath blight suppression. *Plant Soil*, **160**: 227-280.

Thomas, J., Bhai, R.S., Vijayan, A.K. and Naidu R. (1993). Evaluation of antagonists and their efficacy in managing root diseases of cardamom. *J. Biol. Control*, **7**: 29-36.

Tschen, J.S.M. and Li, I.F. (1994). Optimization of formation and regeneration of protoplasts from biocontrol agents of *Trichoderma* sp. *Mycoscience*, **35**: 257-263.

Urech, P.A., Theo, S. and Gunther, V. (2000). Resistance: A limiting factor in crop protection. *Pesticides World*, **5**: 46-51.

Valasubramanian, R. (1994). *Biological control of Rice blast: Role of antifungal antibiotic in disease suppression.* Ph. D., thesis, University of Madras, Chennai, India.

Vanneste, J.L. and Yu, J. (1990). Control of fire flight of *Ervinia herbicola* Eh 252 in an experimental rehard in Dax (South west of French). *Acta. Hortic.*, **273**: 409-410.

Vasanthadevi, T., Malarvizhi, R., Sakthivel, R. and Gnanamanickam, S.S. (1989). Biological control of sheath blight of rice in India with antagonistic bacteria. *Plant Soil*, **119**: 325-330.

Viterbo, A., Haran, S., Friesem, D., Ramot, O. and Chet, I. (2001). Antifungal activity of a novel endochitinase gene (chit36) from *Trichoderma harzianum* Rifai TM. *FEMS Microbiol. Lett.*, **200**: 169-174.

Vruggnik, H. (1970). The effect of chitin amendment on actinomycetes in soil and on the infection of potato tubers by *Streptomyces scabies. Neth. J. Plant Pathol.*, **76**: 293-295.

Vyas, S.C. (1994). Integrated biological control and chemical control of dry root rot on soybean. *Indian J. Mycol. Pl. Pathol.*, **24**: 132-134.

Weindling, R. (1932). *Trichoderma lignorum* as a parasite of other fungi. *Phytopathology*, **22**: 837.

Weindling, R. (1941). Experimental consideration of the mould toxins of *Gliocladium and Trichoderma. Phytopathology*, **31**: 991-1003.

Weindling, R. and Emerson, O.H. (1936). The isolation of a toxic substance from the culture of a *Trichoderma. Phytopathology*, **26**: 1068-1070.

Weller, D.M. (1988). Biological control of soil borne plant pathogens in the rhizosphere with bacteria. *Annu. Rev. Phytopathol.*, **26**: 379-407.

Whipps, J.M. (1997). Developments in the biological control of soil-borne plant pathogens. *Adv. Bot. Res.*, **26**: 1-13.

Whipps, J.M. (1992). Status of biological disease control in horticulture. *Biocontrol Sci. Technol.*, **2**: 3-24.

Wilson, C.L., Wishiewski, M.E., Droby, S. and Chalutz, E. (1993). A selection strategy for microbial antagonists to control post harvest diseases of fruits and vegetables. *Scientific Hort.*, **50**: 105-112.

Windham, M.T., Elad, Y. and Baker, R. (1986). A mechanism for increased plant growth by *Trichoderma* spp. *Phytopathology*, **76**: 518-521.

Yin, S.Y., Chang, J.K. and Xun, P.C. (1965). Studies in the mechanisms of antagonistic fertilizer '5406' IV. The distribution of the antagonist in soil and its influence on the rhizosphere. *Acta Microbiol. Sinica*, **11**: 259.

8

Effect of *Sphacelia sorghi* Culture Filtrate on Callus and Cell Suspension Cultures of Sorghum

NICKY JOHNSON[1] & A.H. RAJASAB[2]

Plant Pathology and Mycology Lab, Gulbarga University, Gulbarga – 585 106, Karnataka, India

ABSTRACT

Sorghum bicolor (L.) Moench callus showed differential growth response to culture filtrate (CF) of *Sphacelia sorghi* anamorph of *Claviceps sorghi* at varying concentrations. Callus mortality increased with the increase in concentration. There was a leveling effect beyond 60 per cent concentration. CF treated callus showed decline in fresh weight as compared to controls. However, there was no difference in dry weight of treated and normal callus. Biochemical analysis revealed that, as the concentration of CF increased the proline and polyphenol accumulation in the CF treated callus was increased. While, the protein content in callus decreased with increase in CF concentration. Decrease in the cell size, increased plasmolysis was observed in cell suspension culture incorporated with CF. A small proliferation of callus mass was observed in 60 per cent CF concentration of in vitro screening technique and development of resistant cell lines to ergot pathogen.

Key Words: *Claviceps, Sorghum, Tissue culture.*

[1] nickyj@rediffmail.com and [2] rajasab55@rediffmail.com

Pathotoxins/crude culture filtrates/toxin analogs have been extensively used for selection of disease resistant cell line in cell and tissue culture system (Carlson, 1973). There are very few reports on comparative study of the culture filtrate (CF) and partially purified toxins preparations on different selective units (Chauhan and Singh, 1995). Bajaj and Saetler (1968) reported the work on the effect of pathotoxin at the cellular level and later to select toxin tolerant cell cultures. Bajaj *et al.* (1980) worked on the effect of *Claviceps fusiformis* sclerotial extract on the growth and development of embryos, excised segments of seedlings and callus culture of Pearl millet. The results showed differential tolerance of callus. The extract at higher concentration completely stopped callus growth. However certain cells survived, showing the possibility of selecting some of the resistant cell line, and later regenerating entire plant from them.

Most promising male sterile varieties are highly susceptible to ergot disease and cause serious yield decline. So there is need to utilize *in vitro* selection to evolve sorghum ergot resistant cultivar. Pathotoxins play an important role in pathogenesis and their lethal effect has been discussed in relation to the susceptibility of a host to a particular pathogen. *C. sorghi,* the causal organism of ergot of sorghum has been reported to produce toxic alkaloids. In this direction, attempts have been made to assess the effect of culture filtrate of *S. sorghi* on growth responses of susceptible sorghum callus and cell. Biochemical changes of treated callus during the CF effect was investigated.

Materials and Methods

Callus was initiated from immature inflorescence of M-35-1 and CSH–9 varieties of sorghum on Murashige and Skoog's medium (1962) supplemented with 2,4-D (2.5mg/l) and BAP (0.5 mg/l). The pH of the medium was adjusted to 5.6 prior to adding agar; the cultures were maintained at 25 ± 2°C under continuous light (2000lux). After 30 days the callus was sub-cultured twice on fresh medium at an interval of 10 days with same concentration of growth hormones. Such sub-cultured callus without any trace of the explant was used to investigate the effect of *S. sorghi* culture filtrate.

To prepare culture filtrate, the panicles of sorghum showing honeydew symptoms were collected from field. *S. sorghi* was isolated on Modified Kirchoff's medium and 5mm block of agar from pure culture was inoculated in 250ml conical flasks containing 100ml potato dextrose broth. Ten such conical flasks were prepared and they were kept in rotatory shaker at 120rpm. for 21 days, culture was filtered thrice through Whatman filter paper No. 1. The culture filtrate obtained although free from conidia, it was thoroughly shaken and subjected to centrifugation at 3,000rpm. for 10min. and the supernatant was used as the culture filtrate of 100 per cent concentration. This solution was diluted to 20 per cent, 40 per cent, 60 per cent and 80 per cent respectively by adding distilled water. The culture filtrate of *S. sorghi* was incorporated to MS medium to give a final concentration of 20 per cent, 40 per cent, 60 per cent 80 per cent and 100 per cent and used to observe the effect of CF on the callus growth. The growth of callus was measured by dry weight and fresh weight of callus after 15 and 30 days respectively. The affected CF treated callus was biochemically evaluated for the quantitative difference of Protein, Proline and Polyphenols with controlled callus. The Protein, Proline and Polyphenol were estimated by following the Lowry *et al.* (1951), Bates *et al.* (1973) and Follin and Denis (1915) methods respectively.

To study the effect of culture filtrate on cells under suspension culture the MS medium supplemented with 2mg/l 2,4-D and 0.5mg/l BAP maintained on rotatory shaker at 120 rpm. for a week and sub cultured at an interval of 8 days to fresh medium. The cell suspension culture was treated with 20 per cent, 40 per cent, 60 per cent 80 per cent and 100 per cent CF within MS liquid medium and after 4 days the cell size and shape were observed under microscope by using Evan's blue stain. The viability of the cells in treated and untreated CF were compared.

Results and Discussion

Callus grown on MS medium supplemented with *S. sorghi* culture filtrate showed progressive browning of the callus. The percentage of browning increased with increase in the concentration' of CF with duration of incubation period. The cell growth was not observed at higher concentrations (*i.e.* 80 per cent and 100 per cent), which caused the death of the callus. The results are in confirmation with those of Sharma and Chahal (1989).

Effect of CF on Callus Growth

Growth of the callus in terms of fresh and dry weight of the callus was measured following treatment. Most of the callus showed arrested growth at higher concentration of CF supplemented medium (Table 8.1). The inhibition of growth was more at higher concentration of CF with the increase in the duration of culture period. The fresh weight of the callus reduced by 19.7 per cent, 29.03 per cent, 39.36 per cent, 49.28 per cent and 60.7 per cent from normal callus at 20 per cent,40 per cent,60 per cent,80 per cent and 100 per cent concentration of CF respectively after 30 days. From results, it was clear that the CF at higher concentration caused excessive browning, growth retardation and ultimate death of callus. The browning of callus observed in the present study is probably due to excess of phenolic production in CF treated callus. The mortality occurred in most of the cultures at 60 per cent CF concentration. However in certain culture tubes, pockets of cells are seen growing. These surviving cells may be considered as tolerant to CF.

Table 8.1: Effect of CF on Growth of *Sorghum bicolor* (L.) Moench Callus

Medium	*Concentration of CF*	*15 Days*		*30 Days*	
		FW (mg)	*DW (mg)*	*FW (mg)*	*DW (mg)*
MS+2,4-D (2.5 mg/l) + BA (0.1mg/l)	Control	396 ± 12	39 ± 04	558 ± 15	55 ± 06
	20%	380 ± 11	38 ± 04	448 ± 13	45 ± 05
	40%	366 ± 10	36 ± 04	396 ± 11	39 ± 04
	60%	280 ± 09	28 ± 03	308 ± 09	31 ± 03
	80%	250 ± 08	25 ± 02	250 ± 08	25 ± 02
	100%	250 ± 08	25 ± 02	250 ± 08	25 ± 02

* Fresh weight of the inoculum = 150 ± 20mg

FW = Average fresh weight (mg) of callus.

DW = Average dry weight (mg) of callus.

Effect of CF on the Protein, Proline, Polyphenol Content of the Callus

The protein content of the treated callus decreased with the increase in the concentration of CF (Correlation coefficient = – 0.0004). The protein content decreased 5.45 per cent, 10.09 per cent, 15.45 per cent, 20.9 per cent and 27.27 per cent respectively at 20 per cent, 40 per cent, 60 per cent, 80 per cent and 100 per cent CF level after 30 days (Table 8.2). The rapid accumulation of proline was observed with the increase in the concentration of CF (Correlation coefficient = + 0.00067). The proline accumulation was high at 80 per cent CF concentration, about 5–7 times more than the normal level. At 100 per cent CF concentration the proline level is slightly less as compared to callus at 80 per cent CF concentration. In general, phenolic content of the treated callus increased with an increase in the concentration of CF (Correlation coefficient = + 0.0083). *S. sorghi* culture filtrate has shown toxic effect on sorghum callus and cell suspension cultures. Similar results have been reported where; extracts of ergot (c. *fusiformis*) affect the growth and development of pearl millet (Bajaj, 1981). Although at higher concentration, the CF stopped the growth of the callus, certain cells survived at 60 per cent CF concentration showing the possibility of *in vitro* screening technique and development of resistant cell line to ergot pathogen.

Table 8.2: Amount of Protein, Polyphenol and Proline in CF Treated Callus Cultures

Concentration of CF in MS Medium	*Protein (mg/100gm) ± SE*	*Polyphenol (mg/gm) ± SE*	*Proline (mg/gm) ± SE*
Control	111.33 ± 0.66	0.840 ± 0.005	0.083 ± 0.004
20%	103.33 ± 0.66	1.505 ± 0.000	0.135 ± 0.004
40%	98.33 ± 0.34	1.638 ± 0.009	0.171 ± 0.004
60%	92.66 ± 0.34	1.978 ± 0.004	0.233 ± 0.003
80%	85.66 ± 0.88	2.634 ± 0.004	0.266 ± 0.007
100%	79.33 ± 0.55	2.923 ± 0.004	0.225 ± 0.004
CD at 5% ($P<0.05$)	2.104	2.128	2.100
Correlation coefficient	– 0.00045	+ 0.00827	+ 0.00067

Effect of CF on Cell Size and Shape

The size of cell decreased with increase in CF treatment in cell suspension. The cell permeability also changed with increase in the concentration of CF. The plasmolysis of cell increased with increase in concentration of CF (Fig. 8.1 A-F). At 80 per cent and 100 per cent CF treatments the mortality of the cells was observed.

REFERENCES

Bajaj, Y.P.S. (1981). Production of Disease-Resistant plants through cell culture: A new approach. *Journal of Nuclear Agricultural Biology*, **10**: 1-5.

Bajaj, Y.P.S. and Saettler, A.W. (1968). *Phytopathology*, **58**: 1041

Bajaj, Y.P.S., Phul, P.S., and Sharma, S.K. (1980). Differential tolerance of tissue culture of Pearl millet to ergot extract. *Indian Journal of Experimental Biology*, **18**: 429-432.

Bates, L.S., Waldeen, R.P., and Teare, I.D. (1973). Rapid determination of proline for water stress studies. *Plant and Soil*, **39**: 205-207.

Carlson, P.S. (1973). Methionine sulfoximine–resistant mutant of tobacco, *Science*, **180**: 1366-1368.

Chauhan, R.S. and Singh, B.M. (1995). *Indian National Science Academy*, **61(B)**: 479.

Follin, O. and Denis, W. (1915). A calorimetric estimation of phenols (and Phenol derivatives) in Urine. *J. Biol. Chem.*, **23**: 305-308.

Lowry, O.H., Rosenbrough, N.J., Farr, A.L. and Randall, R.J. (1951). Protein measurement with the Follin Phenol Reagent. *J. Biol. Chem.*, **193**: 265-275.

Murashige, T. and Skoog, F. (1962). A revised medium for rapid growth and bioassay with tobacco tissue culture. *Physiologia Plantarum*, **15**: 473-497.

Sharma, S.B., and Chahal, S.S. (1989). Effect of culture filtrate of ergot pathogen on pearl millet seedlings and callus cultures.. *Indian Journal of Experimental Biology*, **27**: 187-188.

9

Molecular Basis of Plant Disease Resistance

S.M. PAUL KHURANA, SWARUP K. CHAKRABARTI & DEBASIS PATTANAYAK

Central Potato Research Institute, Shimla – 171 001, India

Introduction

Microbes are ubiquitous. They inhabited earth much before the emergence of terrestrial plants. During a long period of co-existence, plant-microbe interactions evolved in divergent ways. Endosymbiotic existence of prokaryotic chloroplasts in plant cells is an extreme example of this type of interaction. Development of root-nodules in legumes by *Rhizobium* is another example of symbiosis between plants and microbes. Many a times, however, interactions between plants and microbes result in some sort of physiological disorders in plants causing a disease. The importance of plant diseases was realized as soon as mankind shifted from nomadic to agrarian life style about 10, 000 years back. Ravages of plant disease are documented in almost all ancient literature (Ainsworth, 1981). In fact occurrence of potato late blight famine during 1840's triggered scientific study of plant diseases giving birth to the discipline of plant pathology. Though several remedial measures for controlling plant diseases were developed by ancient people, control measures based on scientific facts emerged only during 19th century. Initially, chemical control using sulfur, copper and mercury-based fungicides were the most popular choice (McNew, 1967). The first sulfur-based fungicide was used during 1802 to treat mildew in trees (Ainsworth, 1981). Similarly, physical methods like heat treatment, solarization etc. were found effective in reducing crop damage due to diseases. But breeding for improving natural resistance of crop plants started only during early 20th century. Shortly after scientists became aware of Mendel's law of inheritance, yellow rust resistance in wheat was demonstrated to be inherited in a Mendelian fashion (Biffen, 1904). This stimulated research to identify and incorporate disease resistance in crops through breeding. On the other hand variability in pathogenicity of causal organisms was also reported during this period. The concept of physiological races in rust fungi was forwarded by Stackman and Piemesel in 1917. Mendelian nature of virulence in pathogen was demonstrated by 1940 (Johnson and Newton, 1940). With this basic information on Mendelian inheritance of disease resistance as well as pathogen's virulence the stage was set for the discovery of gene-for-gene paradigm.

Gene for Gene Interaction

The gene-for-gene hypothesis was proposed by Prof. Harold H Flor, USDA, North Dakota Agril.Experimental Station, Fargo, USA in 1942 as the simplest explanation of the results of studies on the inheritance of pathogenicity in the flax rust fungi, *Melampsora lini*. On varieties of flax, *Linum usitatissimum* that have one gene for resistance to the avirulent parent race, F2 cultures of the fungus segregated into monofactorial ratios. On varieties having 2, 3, or 4 genes for resistance to the avirulent parent race, the F2 cultures segregated into bi-, tri, or tetrafactorial ratios respectively. This suggested that for each gene that conditioned reaction in the host there is a corresponding gene in the parasite that conditioned pathogenicity. Each gene in either member of a host-parasite system may be identified only by its counterpart in the other member of the system. In interactions that follow a gene-for-gene relationship, the expression of resistance or susceptibility of the host to a particular pathogen is conditional on the pathogen genotype, while the degree of pathogen virulence observed is conditional on the host genotype. Compatibility is the outcome of a host-pathogen combination unless an allele for resistance at a particular host locus is specifically matched by an allele for avirulence at a particular pathogen locus. Gene-for-gene hypothesis, therefore advocates that compatability (disease) is non-specific while resistance (healthy condition) is a specific outcome (Vanderplank, 1984). This is contrary to the popular belief in plant pathology that good health is the rule (non-specific) while disease is an exception (specific) (Table 9.1). Several other genetic systems were also reported which do not follow the classical *R-Avr* interaction.

Table 9.1: Diagonal Check for Specificity in Gene-for-Gene Relations

Pathogen Genotype	Plant Genotype	
Type I: Wheat Rust		
	R-	rr
A-	I	C
aa	C	C
Type II: cmsT maize vs *Biopolaris maydis* race T		
	S-	ss
V-	D	N
vv	N	N
Type III: HMI maize vs *Cochliobolus carbonum*		
	R-	rr
V-	N	D
vv	N	N

Note: R: dominant resistant gene; r: recessive allele of R; A: dominant avirulent gene; a: recessive allele f A; I: incompatibility or resistance reaction; C: compatibility or disease reaction; S: dominant susceptibility gene; s: recessive allele of S; V: dominant virulent gene; v: recessive allele of V; D: disease; N: no disease.

In potato-late blight pathosystem, an altogether different interpretation of gene-for gene relation is used for identifying pathogenic races. The system was devised by Black *et al.* (1953) for races of *Phytophthora infestans* and resistance genes in potato. As illustrated below (Table 9.2), for 4 resistance genes in potato (out of 11 or more known in *Solanum*), there will be $2^4 = 16$ different races of *P. infestans*. The genes are numbered R1, R2, R3 and R4 and are entered singly and in combination. A plant with the gene R1 is susceptible to race(1) or any other race with 1 in its designation, but resistant to races (2), (3), and (4) or races with 2, 3, or 4 in any combination in their designation. A plant with the two genes R1 and R2 is susceptible to races (1, 2) or any other race with both the numbers 1 and 2, but is resistant to all other races (Table 9.2).

The more the resistance genes the clearer it becomes that susceptibility is specific and resistance nonspecific. In a 4-gene system a plant with the genes R1, R2, R3, and R4 it is susceptible only specifically to race (1, 2, 3, 4) and resistant to any or all of the other 15 races. Similarly, in a 10 gene system (with 2^{10}=1024 races) a plant with the 10 genes, R1 to R10, is susceptible only to race (1, 2, 3, 4, 5, 6, 7, 8, 9, 10) but resistant to any of the other 1023 races. To explain all the above types of plant-parasite interaction which follow gene-for-gene pattern, a generalized definition was forwarded by Person and associates (1962) as follows: "A gene-for-gene relationship exists when the presence of a gene in one population is contingent on the continued presence of a gene in another population, and where the interaction between the two genes leads to a single phenotypic expression by which the presence or absence of the relevant gene in either organism may be recognized." Among the three diagonal checks for gene-for-gene interactions (Table 9.1), the first one illustrated by wheat rust is by far the most common and well-studied system. The *R-Avr* gene pairs in this interaction results in a single phenotypic expression, *i.e.* incompatibility which is epistatic (exhibit nonallelic dominance) over gene pairs that would otherwise result in compatibility. Gene pairs conditioning higher degrees of incompatibility are in general epistatic over gene pairs associated with lower degrees of incompatibility (Crute and Pink, 1996). Though genes encoding resistance are specific in a gene-for-gene interaction the plant defense responses triggered by these *R* genes are almost similar thus non-specific in nature (Hammond-Kosack and Jones, 1996). Several models have been developed to give a general explanation of all disease situations through gene-for-gene hypothesis.

The first model explaining molecular basis of gene-for-gene interaction was proposed by Albersheim and Anderson-Prouty in 1975 as a surface carbohydrate elicitor-receptor model. According to this model *R* gene(s) of hosts encode a surface receptor. The elicitor(s) of active defense are surface glycoprotein of pathogens, which are specifically modified by glycosyl transferase encoded by *avr* gene(s). Receptor-elicitor matching evokes a generalized host defense, *i.e.* hypersensitive response (HR). Later on, it has been demonstrated that elicitors can also be extracellular metabolites of the pathogen or even degraded plant components (endogenous elicitors). Mechanism of defense activation was supposed to be similar in homologous and heterologous interaction. In this model compatibility is achieved by suppressing, degrading, or avoiding an active defense response, which is its major weak point. If the pathogen can degrade, or is insensitive to, the effector molecules (phytoalexins, phenolics, lignins etc.) made by a plant as part of its defense, how could the same molecules be effective when triggered by gene-for-gene interactions? Secondly, *avr* gene products have rarely been demonstrated to be associated with elicitors. A third weakness of this model is the lack of testable mechanism explaining how R-gene receptors activate plant defense response genes. Even compatible pathogens kill plant cells and trigger some defense responses, which is called normo-sensitive response. What causes the difference between the normal and hypersensitive response?

Table 9.2: International System of Designating Races of *P. infestans* and Resistance Gene in Potato

Gene (s)	Race															
	(0)	(1)	(2)	(3)	(4)	(1, 2)	(1, 3)	(1, 4)	(2, 3)	(2, 4)	(3, 4)	(1, 2, 3)	(1, 2, 4)	(1, 3, 4)	(2, 3, 4)	(1, 2, 3, 4)
R	S	S	S	S	S	S	S	S	S	S	S	S	S	S	S	S
R1	R	S	R	R	R	S	S	S	R	R	R	S	S	S	R	S
R2	R	R	S	R	R	S	R	R	S	S	R	S	S	R	S	S
R3	R	R	R	S	R	R	S	R	S	R	S	S	R	S	S	S
R4	R	R	R	R	S	R	R	S	R	S	S	R	S	S	S	S
R1, R2	R	R	R	R	R	S	R	R	R	R	R	S	S	R	R	S
R1, R3	R	R	R	R	R	R	S	R	R	R	R	S	R	S	R	S
R1, R4	R	R	R	R	R	R	R	S	R	R	R	R	S	S	R	S
R2, R3	R	R	R	R	R	R	R	R	S	R	R	S	R	R	S	S
R2, R4	R	R	R	R	R	R	R	R	R	S	R	R	S	R	S	S
R3, R4	R	R	R	R	R	R	R	R	R	R	S	R	R	S	S	S
R1, R2, R3	R	R	R	R	R	R	R	R	R	R	R	S	R	R	R	S
R1, R2, R4	R	R	R	R	R	R	R	R	R	R	R	R	S	R	R	S
R1, R3, R4	R	R	R	R	R	R	R	R	R	R	R	R	R	S	R	S
R2, R3, R4	R	R	R	R	R	R	R	R	R	R	R	R	R	R	S	S
R1, R2, R3, R4	R	R	R	R	R	R	R	R	R	R	R	R	R	R	R	S

S: means the potato plant is susceptible; R: it is resistant.

Two other models have been proposed to address the drawbacks of the above model. The dimer model, proposed by Ellingboe in 1982, envisages that incompatibility is not effected by elicitation of phytoalexins or defense compounds, but instead involves a potentially wide array of developmental disruptions, involving the formation of dimers. The dimers are viewed as being formed by the primary protein products of the *R* genes and *avr* genes, directly binding one another in a protein-protein interaction. Alternatively, one of them might bind the gene or mRNA transcript of the other. The formation of these dimers was proposed as "the cause for incompatibility". According to this model, the hypersensitive response is viewed as a consequence, but not the cause of incompatibility. Heterologous incompatibility observed between most plants and most microbes is explained by the lack of specific virulence genes, and not necessarily by host defense. The major problems of this model are: (i) sequence data of *avr* gene and its product does not reflect its protein or nucleic acid binding property, (ii) direct evidence of *avr* gene encoding secondary metabolites without having dimer forming capacity, (iii) no direct evidence showing the formation of protein-protein or protein-nucleic acid dimer, and (iv) detail is lacking on how the dimers effect incompatibility.

The third model is called ion-channel defense model proposed by DW Gabriel in 1988. According to this model, the plant cell plasmalemma is equipped with a large number of transmembrane protein receptors capable of being opened into ion channels when bound by specific or semispecific elicitors. Incompatibility is the result of the opening of ion channels due to the binding of *avr* gene product. Due to ion channel formation, host cell death occurs or it may lead to elicitation of active defense. Like the elicitor-receptor model, it also recognizes the role of endogenous and exogenous elicitors as well as secondary metabolites of pathogens in elicitation of host's active defense. Like the elicitor-receptor model it also proposes that *R* genes encode transmembrane protein receptors in host plants. However, it agrees with the dimer model in emphasizing the role of specific virulence factors in conditioning basic compatibility. A distinctive feature of the model is that signal transduction is due to electrolyte fluxes and, in particular by the calcium ion, which is known to control a wide array of cellular responses. As a result of movement of this ion, membrane depolarization occurs leading to greater electrolyte fluxes and the collapse of the energized state of the plasma membrane, resulting the onset of host cell death.

Isolation of *R* Genes

Though genes specifying resistance have been identified during 1940's, isolation of *R* genes could be attempted only after standardization of molecular techniques for cloning plant genes of unknown structure or molecular function. Techniques like transposon tagging, T-DNA mutagenesis and positional cloning enabled isolation of several *R* genes during the last decade. The first cloning of a *R* gene was achieved through transposon tagging of *Hm1* from maize, which was distinct from other classical *R* genes. This gene controls resistance to race 1 isolates of *Cochlioboluos carbonum*. *Hm1* encodes a NADPH-dependent HC-toxin reductase (HRTC). HRTC inactivates HC-toxin, a pathogenicity factor produced by the fungus *C. carbonum* Nelson race 1, that permits the fungus to infect certain genotypes of maize. Unfortunately, studies of *Hm1* did not suggest a structure or function for classically defined *R* genes because the toxin degrading strategy of *Hm1* does not involve pathogen *Avr* genes, induction of hypersensitive plant cell death, or other hallmarks of gene-for-gene interactions. Perfection of positional cloning (chromosome walking) technique and heterologous transposon tagging provided feasible approaches to the isolation of *R* genes, and a number of successes have been reported in recent years (Table 9.3).

Table 9.3: A Summary of Cloning and Characterization of Plant *R* Genes

Class	*R Gene*	*Plant*	*Pathogen*	*Avr gene*	*Structure*
1	*RSP2*	*Arabidopsis*	*Pseudomonas syringae* pv. *tomato*	*AvrRpt2*	LZ-NBS-LRR
	RPM1	*Arabidopsis*	*P. syringae* pv. *maculicola*	*AvrRpm1*, *avrB*	LZ-NBS-LRR
	Prf	Tomato	*P. syringae* pv. tomato	*AvrPto*	LZ-NBS-LRR
	Mla1	Barley	*Blumeria graminis* f sp *hordei*	*AvrMla1*	LZ-NBS-LRR
	N	Tobacco	Tobacco mosaic virus	Unknown	TIR-NBS-LRR
	L^6	Flax	*Malampsora lini*	AL^6	TIR-NBS-LRR
	M	Flax	*M. lini*	*AM*	TIR-NBS-LRR
	RPP5	*Arabidopsis*	*Peronospora parasitica*	*AvrPp5*	TIR-NBS-LRR
	I2	Tomato	*Fusarium oxysporum*	Unknown	NBS-LRR
	Pi-ta	Rice	*Magnaporthe grisea*	*AvrPita*	NBS-LRR (?)
	RPW 8	*Arabidopsis*	*Erysiphe cruciferarum*, *E. cichoracearum*	Unknown	NBS-LRR (?)
	Ctv	*Poncirus trifoliata*	citrus tristeza virus	—-	NBS-LRR
2	*Pto*	Tomato	*P. syringae* pv. tomato	*AvrPto*	Protein kinase
3	*Cf-9*	Tomato	*Cladosporium fulvum*	*Avr9*	LRR-TM
	Cf-2	Tomato	*C. fulvum*	*Avr2*	LRR-TM
	Cf-4	Tomato	*C. fulvum*	*Avr4*	LRR-TM
4	*Xa21*	Rice	*Xanthomonas oryzae* pv. *oryzae*	Unknown	LRR, Protein kinase
5	*Hm1*	Maize	*Cochliobolus carbonum* (race 1)	None	Toxin reductase

Pto of tomato, which confers resistance against *Pseudomonas syringae* pv *tomato* bacteria expressing the *Avr* gene *avrPto*, was the first *Avr*-gene specific *R* gene to be isolated. A positional cloning strategy was used to isolate *Pto*. This gene was found to encode a protein with similarity to serine-threonine kinases. The next five *R* genes that were isolated bore a striking similarity to each other but had no resemblance to *Pto*. These genes contain a leucine-rich repeat (LRR) domain. Sequence analysis of this later class of *R* genes revealed a number of structural themes in addition to the LRR, and new aspects are added to these themes with the characterization of each additional cloning of *R* gene. Over 20 *R* genes with recognition-specificity for defined *avr* genes have been isolated from seven plant species, including both monocots and dicots (Table 9.3). These genes are effective against bacterial, viral, and fungal pathogens and in one interesting case, the *Mi* gene from tomato, against both nematodes and an aphid species. The most striking observation was that *R* genes from diverse plant species with specificity for a wide variety of viral, bacterial and fungal pathogens often encode structurally similar proteins. This similarity suggests a high degree of mechanistic conservation among the pathways that plant use to trigger defense response. There is great diversity in the lifestyles and pathogenic mechanisms of disease-causing organisms and it was, therefore, somewhat surprising that *R* genes were found to encode proteins with certain common motifs. Five classes of *R* genes are now recognized: intracellular protein kinases; receptor-like protein kinases with an extracellular leucine-rich repeat (LRR) domain; intracellular LRR proteins with a nucleotide binding site (NBS) and a leucine zipper (LZ) motif; intracellular NBS–LRR proteins with a region with similarity to the Toll and interleukin-1 receptor (TIR) proteins; and LRR proteins that encode membrane-bound extracellular proteins. The major structural motifs of *R* genes are briefly described below.

Structural Motifs of *R* Protein

Leucine-Rich Repeat (LRR)

LRRs are multiple, serial repeats of a motif about 24 amino acids in length. They contain leucine or other hydrophobic residues at regular intervals and can also contain regularly spaced prolines and asparagines. LRR domain usually takes a tertiary structure of -helical arrays. In terms of function, the LRR domains mediate protein-protein interactions like interactions between enzymes and enzyme inhibitor, binding of peptide hormones by transmembrane receptors etc. In gene-for-gene resistance the LRR motif of *R* genes probably serve as the binding domain for a ligand produced by *Avr* gene activity. Alternatively, LRRs may facilitate the interactions of *R* gene products with other proteins that participate in defense signal transduction.

Nucleotide Binding Sites (NBS)

Many resistance genes that encode LRRs also encode amino acid sequences with strong similarity to known nucleotide binding sites. These NBS domains occur in diverse proteins with ATP or GTP binding activity. The presence of the highly conserved NBS domain suggests that nucleotide triphophate binding is essential for the functioning of these proteins. Site-specific mutations that alter key residues within the proposed NBS have been found to eliminate the HR-inducing function of *RPS2* of *Arabidopsis* and *N* of tobacco. Though the mechanistic role of NBS domains in resistance activation remains unclear, it is proposed that nucleotide triphosphate binding may alter the interaction between *R* gene products and other members of the defense signal transduction cascade.

Leucine Zippers

Some *R* genes encode proteins that have probable leucine zipper sequences between the N terminus and the NBS and LRR domains. Proteins with this type of motif facilitate protein-protein interactions by promoting the formation of coiled-coil structure. LZs are well known for their role in homo- and heterodimerization of eukaryotic trascription factors. Work is underway to test whether *R* gene products can undergo homodimerization.

Toll/Interleukin-1 Receptor Similarity

Some *R* gene like *N*, *RPP5* and *L6* encode a large N-terminal domain with similarity to the cytoplasmic signaling domain of the *Drosophilla* Toll protein and mammalian interleukin receptors. Interleukin-1 receptor (IL-1R) respond to the cytokine and activate a transcription factor. Toll is a receptor involved in the establishment of dorsal-ventral polarity in *Drosophilla*. It is proposed that N and L6 probably activate plant defenses by a mechanism similar to those used by Toll and IL-1R.

Serine-Threonine Kinase

This domain may be present in the same *R* protein like Xa21 or it may exists independently. This is a central motif for kinase-mediated signal transduction in gene-for-gene interaction. This motif initiates signal transduction pathway by phophorylation/dephophorylation mechanism. The first *R* gene cloned (*Pto*) was a serine-threonine kinase. Its LRR counterpart (*Prf*) has also been identified later.

Functions of *R* Proteins

Despite significant insights into *R* gene structure, much remains to be elucidated about the molecular mechanisms by which *R* proteins recognize pathogens and transduce this information in the plant cell to initiate defense responses. For example, although *R* proteins were anticipated to play a direct role in recognizing pathogen-derived ligands, this has so far been reported only in case of Pto. Similarly, the cloning of a protein kinase *R* gene, *Pto*, and an LRR gene, *Prf*, from the same disease resistance pathway indicated that members of these two classes might represent different steps in a signalling pathway common to other *R* gene pathways. No clear evidence has emerged, however, to enable the placement of these two classes of proteins in a single pathway. Finally, the shared motifs among *R* proteins suggested that they might utilize common downstream signaling components to activate similar defense responses.

Recognition Specificity of *R* Proteins

A hallmark of *R* genes is their recognition-specificity for corresponding *avr* genes. Because of the extracellular life style of most plant pathogens, it was expected that *R* proteins would encode extracellular receptor-like proteins. This might be the case for some *R* proteins (*e.g.*, Xa21, Cf proteins); however, in the majority of cases *R* genes appear to encode intracellular proteins. The discovery of bacterial phytopathogenic type III secretion systems that would allow direct injection of pathogen proteins into the plant cell suggested that recognition might occur inside the plant cell. Transient expression of many Avr proteins inside plant cells has confirmed this speculation. Because LRRs are known to play a role in protein–protein interactions there has been much speculation that these regions are directly involved in recognition of Avr proteins. In support of this view, LRRs have been shown to have considerable variation among members within clustered gene families and there is now extensive genome sequence data to indicate that unequal crossing over, gene conversion, and divergent selection on genes within families has occurred to create new recognition specificities. Recently, the product of the rice blast *R* gene *Pita* has been shown to interact with the AVR-Pita protein in the yeast two-hybrid system.

Evidence for the direct interaction of an *R* and Avr protein has come from work with the *Pto–avrPto* system. Pto confers resistance to *Pseudomonas syringae* strains that express the AvrPto protein, and encodes a serine–threonine protein kinase that lacks any obvious receptor-like domain. Nevertheless, two laboratories have demonstrated that Pto interacts specifically with the AvrPto protein in the yeast two-hybrid system and that mutations, which disrupt this interaction, lead to loss of recognition in the plant cell. Pto recognition specificity for AvrPto resides in a threonine at position 204. Interestingly a threonine at this position is conserved in a large number of protein kinases and it has been proposed that phosphorylation of this residue might lead to a conformational change of the kinase which allows AvrPto binding. Overexpression of Pto activates a variety of defense responses targeted against diverse pathogens. A downstream role for LRR genes is also supported by the recent report that a mutation in the *RPS5* bacterial LRR *R* gene suppressed resistance to several avirulent *Pseudomonas* strains and to strains of the downy mildew pathogen.

Localization of *R* Proteins in the Plant Cell

If *R* proteins are directly involved in recognition of Avr proteins, then it is possible they are localized to the plasma membrane to intercept the incoming pathogen protein. Unfortunately, there is little information to date about the localization of most *R* proteins in the plant cell. In the best studied case, using an epitope-tagging strategy, the RPM1 protein (of the LZ–NBS–LRR class) was found to be enriched in the plasma membrane fraction and was removable by treatments that release peripherally-associated membrane proteins. This localization contrasts with that predicted by the analysis of RPM1 structural motifs. A plasma-membrane localization for the RPM1 protein supports a direct role in recognition for RPM1 although interaction of RPM1 with either of its corresponding Avr proteins, AvrB and AvrRpm1, has not been observed. The lack of a signal peptide or transmembrane domain in RPM1, similar to *R* proteins in several other classes, suggests that these proteins might be tethered to membranes *via* an interaction with other proteins. Interestingly, a two-hybrid screen has uncovered an RPM1-interacting protein that is a good candidate for fulfilling such a role.

Involvement of *R* Proteins in Signal Transduction

There have been significant developments in our understanding of signaling steps that lie downstream of *R* genes although much still remains to be learned. The similarity of *R* proteins to components of the *Drosophila* Toll and mammalian IL-1R (interleukin-1 receptor) pathways has been previously noted, and the complexity of these pathways is perhaps a harbinger of what is to be discovered downstream of *R* proteins. Despite the structural similarities between the TIR-like *R* proteins and Toll/IL-1R, and between Pto and the IL-1R-associated kinase (IRAK), it still remains to be determined if mechanistic similarities exist. Extensive searches of publicly available plant expressed sequence tag databases have not uncovered genes with obvious similarities to mammalian genes encoding MyD88, IL-1R accessory protein, TRAF6, NF-?B and other components downstream of the IL-1R receptor, and it seems likely that plants have evolved several unique proteins to fulfill signaling roles downstream of *R* genes. Nevertheless, as in mammalian systems, a role for reactive oxygen intermediates and nitric oxide in plant defense signaling has been established and there is now evidence in plants for other mammalian-like second messengers of nitric oxide signaling including cyclic GMP and cyclic ADP-ribose. In addition, in common with IL-1R signaling, mitogen-activated protein kinases (MAPKs) have been identified that are activated by *R* gene pathways.

Role of the NBS-LRR Motifs in Signaling

With the exception of the Xa21 protein, which contains a protein kinase domain, the primary amino acid sequence of other LRR–NBS *R* proteins does not suggest an obvious mechanism for the initiation of signal transduction. The most thought-provoking proposition to be put forth about the function of NBS-like *R* proteins is based on the noted similarity between them and the CED-4 and APAF-1 proteins, which activate proteases involved in mammalian apoptosis. Heterodimerization between CED-4 and Apaf-1 and their respective proteases, CED-3 and caspase-9, results from the interaction between homologous domains present in the amino terminal portions of the proteins. By analogy, it has been proposed that NBS–LRR proteins might rely on their TIR or LZ regions to form heterodimers with downstream proteins. In this model, the LRRs confer recognition specificity, whereas the NBS serves to activate downstream effectors.

Signaling via the Pto kinase

Pto and Xa21, by virtue of being protein kinases, offer the best opportunities to elucidate the role of *R* proteins in signal transduction *via* protein phosphorylation. Both Xa21 and Pto are active protein kinases and, at least for Pto, kinase activity appears to be required for its role in disease resistance. Pto interacts with several plant proteins in the yeast two-hybrid system and to date two of these, Pti1 (a serine–threonine kinase) and Pti4 (a DNA binding protein) have been found to be specific phosphorylation substrates for Pto. The Pti4/5/6 proteins resemble the ethylene response element binding proteins (EREBPs) originally isolated from tobacco. Like the EREBPs, Pti4/5/6 bind the GCC box *cis* element that is present in the promoter of many pathogenesis- related (PR) genes and other genes that are often expressed in plants upon exposure to ethylene or pathogens. Thus, the identification of the *Pti4/5/6* genes defines the first direct link between an *R* gene and expression of defense-related genes. Using purified proteins, Pti4 has been found to be phosphorylated *in vitro* by Pto on at least four sites. Experiments are under-way to examine the possible role of these phosphorylation events in nuclear localization, binding to the GCC box, or in transactivation of target gene expression. In addition, the *Pti4/5/6* genes are regulated differently in response to ethylene, jasmonic acid (JA), salicylic acid (SA), wounding, and pathogens. Thus, it seems possible that their proteins serve to integrate different signals that induce common target genes. Characterization of the *Pti1/4/5/6* genes suggests that multiple pathways diverge directly from the Pto kinase. Because *Pti1* overexpression accelerates the HR, one pathway might lead to the HR and perhaps to other defense responses. Additional, distinct pathways appear to be mediated by the interaction of Pto with the putative transcription factors Pti4/5/6. The *Pti1/4/5/6* genes were identified by a yeast two-hybrid screen and to date no mutations in these genes have been identified in tomato. Thus additional evidence for their role in disease resistance is being sought by transgenic analyses and biochemical approaches.

Role of MAPKs in *R* Gene Signaling

A key role for protein phosphorylation in *R* gene signaling is clearly implicated by the isolation of the Pto, Xa21, and Pti1 kinases. However, there are relatively few studies that have examined protein phosphorylation events further downstream of *R* genes. Two recent studies now provide compelling evidence for the activation of MAPKs by *R* gene pathways and, in addition, suggest that these kinases are convergence points for many different defense responses in plants. In tobacco, two previously identified MAPKs, a 48 kD salicylic acid induced protein kinase (SIPK) and a 44 kD wound-induced protein kinase (WIPK), were found to be activated during infection by tobacco mosaic virus. Activation of WIPK occurred only in tobacco plants containing the tobacco mosaic virus *R* gene *N* indicating that WIPK might play a role in gene-for-gene mediated defense against virus infection. Activation of WIPK required both serine–threonine and tyrosine phosphorylation and also an increase in its mRNA and protein levels. Although many defense responses mediated by the *N* gene are dependent on SA, the activation of WIPK was not affected in plants lacking this signaling molecule. Additional evidence that SIPK and WIPK lie downstream of *R* genes comes from studies with the *Cf-9 R* gene and its corresponding gene *Avr9*. Treatment of tobacco suspension cells expressing *Cf-9* with the Avr9 peptide resulted in rapid activation of both SIPK and WIPK. Similarly to the observations with the *N* gene system, the WIPK transcript was found to increase in an *Avr9*-dependent fashion in *Cf-9*-containing tobacco cells. Because WIPK and SIPK are known to be activated by many stimuli including non-specific elicitors, wounding, and SA, these reports indicate that some *R* gene pathways converge on components shared with general stress-responsive pathways. Despite the clear involvement of the *N* and *Cf-9* genes in activating MAPKs, it remains to be determined whether these kinases are directly involved in disease resistance and, if so, what role they play.

Genomics of Disease Resistance Genes

Genomic approaches are beginning to impact on disease resistance, particularly on resistance genes and the genes encoding downstream regulators regarding their structural organization and expression. An interesting observation on number and organization of disease resistance genes was made from genomic sequencing of *Arabidopsis* and rice. The complete *Arabidopsis* sequence permits a comprehensive analysis of the diversity of NB-LRR *R*-gene sequences in one plant. Annotation revealed ~150 sequences with homology to the NB-LRR class of *R* genes. *R* homologues are unevenly distributed between chromosomes, with 49 on chromosome 1, 2 on chromosome 2, 16 on chromosome 3, 28 on chromosome 4, and 55 on chromosome 5 of *Arabidopsis*. Despite the fact that many previously isolated *R* genes seem to reside in local multigene families, there are 46 singleton *Arabidopsis R* gene homologues, 25 doublets, 7 loci with three copies, and individual loci with four, five, seven, eight and nine NB-LRR-encoding genes. Thus, *Arabidopsis* has ~100 *R* loci distributed over all the chromosomes (Dangl and Jones, 2001; Khurana *et al*., 2001). Analysis of rice BAC end-sequences representing ~5 per cent of the rice genome reveals probable presence of 750-1500 NBS-encoding genes in rice (Meyer *et al*., 1999). All NBS-encoding genes in rice encode non-TIR type NBS, TIR-type genes have not been detected in genomic or expressed sequence tag (EST) sequences from any grass species (Meyer *et al*., 1999).

So far, only a few clusters of resistance genes have been sequenced. The complete sequencing of the *RPP5* cluster in *Arabidopsis*, the *Cf-4/9*, *Pto* cluster in tomato, and partial sequencing of the *Dm3* cluster in lettuce revealed highly duplicated regions containing little more than resistance-gene homologs. The *RPP5* cluster contains 8-10 homologs spread over ~90 Kb, interspersed with protein kinase pseudogenes and retrotransposones. The *Cf-4/9* cluster contains five resistance genes spread over 36 Kb and interspersed with fragments of lox genes, which may have played a role in the duplication of the region. Five *Pto* homologs are found to be spread over 60 Kb along with a single NBS-LRR gene, *Prf*, that is necessary for the function of at least two members of the *Pto* cluster. The *Dm3* region in lettuce is the largest resistance gene locus characterized at the molecular level so far; at least 24 resistance gene homologs are spread over at least 3.5 Mb. There was no evidence for functional genes in the *Dm3* region, other than homologs of *Dm3* and transposon related sequences (Khurana *et al*., 2001).

Genes that have altered expression in compatible and incompatible plant-pathogen interactions have been targeted for characterization by microarray analysis (Baldwin *et al*., 1999). These analyses will provide comprehensive data on expression profiles, both for genes already implicated in plant-pathogen interactions as well as for many genes that were not previously known to be involved in resistance or susceptibility. Moreover, the analysis of global expression profiles of a large number of plant genes in response to infection and treatment with signal molecules will provide a base to identify commonalties among defense pathways. Recently, changes in the expression patterns of 2,375 ESTs were examined simultaneously by cDNA microarray analysis in *Arabidopsis* after inoculation with an incompatible fungal pathogen *Alternaria brassicicola* or treatment with defense-related signaling molecules salicylic acid, methyl jasmonate or ethylene. A comparison of expression profiles revealed induction of *126* genes by multiple treatments that are likely to be regulated by the same or overlapping defense signaling pathways indicating the existence of a substantial network of regulatory interactions and coordination during plant defense among the different defense signaling pathways (Michelmore, 2000). However, the function of individual gene has to be confirmed on a gene-by-gene basis through high-throughput reverse genetics approaches, viral-induced gene-silencing or viral over-expression as well as the use of gene knockout libraries and promoter-trap strategies.

Knowledge on Plant-Pathogen Interactions Put to Work

The growing understanding of the diverse molecular mechanisms underlying the reactions of plants to pathogen infection holds promise for future progress in the development of strategies to engineer plants for enhanced resistance (Stuiver and Custers, 2001; Rommens and Kishore, 2000). Visible manifestation of the progress in this domain are cloning and characterization of many genes for disease resistance and their downstream effectors including SAR, and bacterial pathogenecity factors, elicitors, toxins etc. Viable strategies that are either employed to engineer disease resistance or envisaged to be potential against pathogen (especially for fungal and bacterial pathogen) attack are discussed here.

Plant genomics offers new tools so that novel disease resistance gene of broad-spectrum resistance can be looked for. Analysis of EST libraries demonstrated that plants such as *Arabidopsis* and soybean express hundreds of potential *R*-genes. Many such genes have already been mapped to genetically characterized resistance loci in a variety of plant systems, including *Arabidopsis*, potato, soybean, lettuce, maize, wheat etc. Obviously, the optimized discovery of *R*-genes in both domesticated and exotic germplasm will be of paramount importance in the future and these genes can be rapidly transferred into advanced commercial germplasm. One such gene is peeper *Bs2* conferring resistance against *Xanthomonas campestris*. The Avr protein recognized by *Bs2* is not only produced by pathovars of *X. campestris* that infects pepper but also by pathovars that infect hosts such as tomato, brassica and citrus. *Bs2* has recently been cloned and expression of this gene in tomato confers resistance against bacterial spot disease caused by *X. campestris* pv. *vesicatoria*.

Research has been initiated to broaden and optimize the activity of R-protein through genetic manipulation of *R*-gene. Wild-type tomato *Pto* gene confers resistance only to certain races of *Pseudomonas syringae* pv. *tomato* (Pst) that contain the Avr gene *AvrPto*. Replacement of the weak endogenous promoter of *Pto* with strong CaMV promoter resulted in not only a further increased resistance to Pst (*avrPto*) but also a partial control of unrelated pathogens such as *X. campestris* and *C. fulvum*. Studies on structure and function of R-protein demonstrated that LRR region is involved in *R*-gene specificity. Importantly, exchange of an LRR region resulted in one case in recognition of a different spectrum of pathogens than that of the originally used *R*-genes (Ellis *et al.*, 1999). Further studies on the various domains of *R*-genes may make it possible to optimize the efficacy and durability of *R*-genes.

Even more effective disease resistance genes may be isolated from resistant nonhost plant species. Because of the durability of nonhost resistance over time, it is commonly speculated that nonhost resistance could be exploited to improve disease resistance within host species. The nonhost resistance of tobacco against the potato pathogen *Phytophthora infestans* is correlated with the ability of tobacco to respond hypersensitivity to an elicitor of this pathogen. The product of a tobacco gene that shares similarities with *R* genes was reported to interact with a 10 kDa peptide elicitor produced by *P. infestans*, named INF1. Similarly, a second factor required for active nonhost responses is a 100 kDa parseley plasma membrane protein that binds to a cell wall glycoprotein elicitor of the soybean pathogen *P. sojae*. This parsley elicitor receptor is currently being isolated biochemically. The transfer of nonhost genes such as the tobacco and parsley elicitor receptors to susceptible hosts may have a tremendous impact on the ability to control *Phytophthora* species and other aggressive pathogens (Khurana *et al.*, 2001).

Knowledge on molecular mechanism of elicitor-receptor binding and consequent induction of hypersensitive reaction in host plants provides the researchers a new tool of elicitor-dependent artificial induction of broad-spectrum defense response. In one of the approaches, broad-spectrum disease resistance has been generated by the introduction of an elicitin gene coupled to a pathogen-inducible promoter. Transgenic tobacco plants were generated harbouring a fusion between the pathogen-inducible tobacco *hsr 203J* gene promoter and a *P. cryptogea* gene encoding the highly active elicitor cryptogein (Keller *et al.*, 1999). Under non-induced conditions the transgene was silent but upon infection by virulent pathogen, cryptogein production was stimulated which coincided with the fast induction of several defense genes at and around the infection sites. Induced elicitor production resulted in a localized necrosis that restricted further growth of the pathogen. The transgenic plants displayed enhanced resistance to fungal pathogens that were unrelated to *Phytophthora* species, such as *Thielaviopsis brasicola, Erysiphe cichoracearum* and *Botrytis cinerea* (Keller *et al.*, 1999).

Plant pathogens often produce non-specific toxins that are involved in the development of disease symptoms. Engineering plants for resistance to toxins produced by pathogens represents an elegant way to prevent the damage caused by infection. A good source for resistance genes is the toxin-producing organisms themselves, as they have to avoid the toxic effects on their own metabolism. Tabtoxin produced by *P. syringae* pv. *tabaci* during infection provokes clorosis by the inhibition of glutamine synthetase, resulting in the accumulation of toxic levels of ammonia. In *P. syringae* pv. *tabaci*, the gene *ttr* (tabtoxin resistance) encodes an acetylating enzyme that inactivates the tabtoxin. The cloned gene has been expressed in tobacco and the resulting transgenic plants developed complete resistance to the bacteria (Anzai *et al.*, 1989). Phaseolotoxin, produced by *P. syringae* pv. *phaseolicola*, is responsible for chlorosis on plants through the irreversible inhibition of the chloroplast enzyme ornithine carbamoyltransferase (Octase) which results in a block of arginine biosynthesis. Resistance of the bacterium to its own toxin is in part attributed to the presence of an Octase isozyme, encoded by the gene *argK*, that is insensitive to the toxin. Transformation of tobacco with *argK*, engineered for targeting to the chloroplast, rendered the plants insensitive to chlorosis induced by purified toxin. Transgenic plants showed an HR-like reaction, but no disease symptoms, after infection with *P. syringae* pv. *phaseolicola* (de la Fuente-Martinez *et al.*, 1992). Transgenic sugarcane was developed against leaf scald disease following this approach. The systemic, xylem invading pathogen, *Xanthomonas albilineans*, causes leaf scald disease of sugarcane. It produces a family of low molecular weight toxins (albicidins) that selectively block prokaryote DNA replication and cause the characteristic chlorotic symptoms by blocking chloroplast development. Transgenic sugarcane expressing an albicidin detoxifying gene (*alb*) (from *Pantoea dispersa*, bacterium that provides biocontrol against leaf scald disease) under the control of maize ubiquitin promoter developed resistance to leaf scald disease (Zhang *et al.*, 1999).

The genes involved in SAR response have also been targeted as potential candidate for manipulation towards resistance management. One such key SAR signaling gene is *Myb1*, which is induced by tobacco mosaic virus in resistant tobacco plants. *Myb1*, encoding a transcription factor that binds to a promoter element of the pathogenesis-related gene *PR1a*, has evaluated for its utility to enhance disease resistance levels. Modification of *Myb1* expression levels in transgenic tobacco plants was shown to increase resistance against both TMV and *Rhizoctonia solani* pathogen (Yang and Klessig, 1996). A gene, *Pad4*, encoding a protein with homology to lipases, has been cloned from *Arabidopsis* and shown to enhance signals that activate plant defenses. Inactivation of *pad4* leads to extreme susceptibility against a wide variety of pathogens, including *Erysiphe orontii, Peronospora parasitica* and *P. syringae* (Reuber *et al.*, 1998). Thus overexpression of *Pad4* in transgenic plants may enhance disease resistance signaling and may prove exceptionally interesting for the development of broad-spectrum disease resistance. Genetic analysis of SAR in *Arabidopsis* led to the cloning of the *NIM1/NPR1* (noninducible immunity/nonexpressor of PR) gene, mutations in which SAR induction is abolished. The NIM1/NPR1 protein shows similarity to NF-KB and I-KB factors controlling numerous cellular responses in mammalian systems, consistent with a key regulatory role. Understanding the function of *NIM1/NPR1* and its homologues will be fruitful in elucidating the mechanism of SAR establishment in *Arabidopsis* and in crop plants.

Progressive harnessing these genes for engineering disease resistant crops may depend upon the ability to separate lesion and disease resistance phenotypes and work with some *Arabidopsis lsd* mutants suggests that this may be possible. Similarly, other pathways leading to expression of defense genes (thionin etc.) other than PR genes are also be the focus of present research. For example, infection of *Arabidopsis* with necrotrophs such as *Alternaria brassicola* leads to induction of thionin and defensin-like genes such as *PDF1.2*, but does not result in PR-1 induction. Overexpression of the thionin gene in *Arabidopsis* leads to partial resistance against *Fusarium oxysporium*, indicating that non-SAR pathways may be useful for disease resistance engineering (Khurana *et al.*, 2001).

Conclusions

The isolation and preliminary characterization of *R* genes has now been achieved. Questions still remain now are (i) site of localization of *R* gene product in plant cell, (ii) whether the *R* gene products truly serve as the elicitor-binding receptors, (iii) how *R* gene products communicate with downstream components of signal transduction cascade etc. In future there are very exciting possibilities for structure-function research into the mechanisms of pathogen detection and signal initiation, ranging from sequence comparison, domain swapping, and site-specific mutagenesis to determination of crystal structures. Isolation of *R* genes and their corresponding avr gene in pathogens has fulfilled the structure-function requirement for the operation of gene-for-gene interaction. However, it has now opened up a new chapter on cellular and molecular mechanism of signal transduction in disease resistance. Commonalties among defense pathways identified by functional genomics help manipulation of resistant trait based on genes or regulators conferring broad-spectrum resistance. With the success of the first wave of transgenics for disease resistance, development of second generation will be more precise in regulation of transgene expression in endowing durability and to eliminate adverse affect on plant growth and yield (Khurana, 1998, 2002).

REFERENCES

Ainsworth, G.C. (1981). *Introduction to the History of Plant Pathology*. Cambridge University Press, Cambridge, 315 pp.

Albersheim, P. and Anderson-Prouty, A.J. (1975). Carbohydrates, proteins, cell-surfaces and the biochemistry of pathogenesis. *Annu. Rev. Plant Physiol.*, **26**: 31-52.

Anzai, H., Yonetyama, K. and Yamaguchi, I. (1989). Transgenic tobacco resistant to bacterial disease by the detoxification of a pathogenic toxin. *Mol. Gen. Genet.*, **219**: 492-494.

Bent, A.F. (1996). Plant Disease Resistance genes: function meets structure. *Plant Cell*, **8**: 1757-1771.

Biffen, R.H. (1905). Mendel's law of inheritance and wheat breeding. *J. Agril. Sci.*, **1**: 4-48.

Black, W., Mastenbroek C., Mills, W.R. and Petersen, L.C. (1953). A proposal for an international nomenclature of races of *Phytophthora infestans* and genes controlling immunity in *Solanum demissum* derivatives. *Euphytica*, **2**: 173-178.

Crute, I.R. and Pink, D.A.C. (1996). Genetics and utilization of pathogen resistance in plants. *Plant Cell*, **8**: 1747-1755.

Dangl, J.L. and Jones, J.D.G. (2001). Plant pathogens and integrated defense responses to infection. *Nature*, **411**: 826-833.

de la Fuente-Martinez, J.M., Mosqueda-Cano, G., Alvarez-Morales, A. and Herrera-Estrella L. (1992). Expression of a bacterial phaseolotoxin-resistant ornithyl transcarbamylase in transgenic tobacco confers resistance to *Pseudomonas syringae* pv. *phaseolicola*. *Biotechnology*, **10**: 905-909.

Ellingboe, A.H. (1982). Genetical aspects of active defense. In *"Active Defense in Plants, Proc. NATO Conf.*, Cape Sounion, Greece, Wood, R.K.S. (Ed.), pp. 179-192, New York, Plenum.

Ellis, J.G., Lawrence, G.J., Luck, J.E. and Dodds, P.N. (1999). Identification of regions in alleles of the flax rust resistance gene *L* that determine differences in gene-for-gene specificity. *Plant Cell*, **11**: 495-506.

Farrer, W. (1898). The making and improvement of wheats for Australian conditions. *Agric. Gaz, NSW*, **9**: 131-168.

Flor, H.H. (1942). Inheritance of pathogenicity in *Melampsora lini*. *Phytopathology*, **32**: 653-669.

Gabriel, D.W., Loschke, D.C. and Rolfe, B.G. (1988). Gene-for-gene recognition: the ion channel defense model. In *"Molecular Genetics of Plant–microbe Interaction"*, Palacios, R. and Verma, D.P.S. (Eds.), pp. 3-14, APS Press, St. Paul, MN, USA.

Hammond-Kosack, K.E. and Jones, J.D.J. (1996). Resistance gene-dependent plant defense responses. *Plant Cell*, **8**: 1773-1791.

Johal, G.S. and Briggs, S.P. (1992). Reductase activity encoded by the HM1 disease resistance gene in maize. *Science*, **258**: 985-987.

Johnson, T. and Newton, M. (1940). Mendelian inheritance of certain pathogenic characters of *Puccinia graminis tritici*. *Can. J. Res.*, **C18**: 599-611.

Keller, H., Pamboukdjian, N., Ponchet, M., Poupet, A., Delon, R., Verrier, J-L., Roby, D. and Ricci, P. (1999). Pathogen-induced elicitin production in transgenic tobacco generates a hypersensitive response and nonspecific disease resistance. *Plant Cell*, **11**: 223-35.

Khurana, S.M. Paul, Pattanayak, D., Chakrabarti, S.K. and Naik, P.S. (2001). Molecular approaches in plant disease management. *Annu. Rev. Plant Pathology*, **1**: 83-121.

Khurana, S.M.Paul. (1998). *Pathological Problems of Economical Crop Plants and Their Management*. Jodhpur, India: Scientific Publishers. 656 pp.

Khurana, S.M.Paul. (2002). Application of molecular biotechnology for crop protection. In: *Hi-Tech Horticulture*, pp. 267-282, (eds., Chadha, K.L., Choudhary, M.L., and Prasad, K.V.), Horticultural Society of India, New Delhi.

Martin, G.B. (1999). Functional analysis of plant disease resistance genes and their downstream effectors. *Curr. Opin. Plant Biol.*, **2**: 273–279.

Martin, G.B., Brommonschenkel, S.H., Chunwongse, J., Fray, A., Ganal, M.W., Spivey, R., Wu, T., Earle, E.D. and Tanksley, S.D. (1993). Map-based cloning of a protein kinase gene conferring disease resistance in tomato. *Science*, **262**: 1432-1436.

McNew, G.L. (1967). Landmarks during a century of progress in use of chemicals to control plant diseases. In *"Plant Pathology: Problems and Progress 1908-1958"*, Holton, C.S., Fischer, G.W., Fulton, R.W., Hart, H. and McCallan, S.E.A., (Eds.), pp. 42-54, Central Book Depot, Allahabad.

Meyer, B.C., Dickerman, A.W., Michelmore, R.W., Sivaramakrishnan, S., Sobral, B.W. and Yong, N.D. (1999). Plant disease resistance genes encode members of an ancient and diverse protein family within the nucleotide-binding superfamily. *Plant J.*, **20**: 317-332.

Michelmore, R. (2000). Genomic approaches to plant disease resistance. *Curr. Opin. Plant Biol.*, **3**: 125-31.

Person, C., Samborski, D.J. and Rohringer, R. (1962). The gene-for-gene concept. *Nature*, **194**: 561-562.

Reuber, T.L., Plotnikova, J.M., Dewdney, J., Rogers, E.E., Wood, W. and Ausubel, F.M. (1998). Correlation of defense gene induction defects with powdery mildew susceptibility in *Arabidopsis* enhanced disease susceptibility mutants. *Plant J.*, **16**: 473-85.

Rommens, C.M. and Kishore, G.M. (2000). Exploiting the full potential of disease-resistance genes for agricultural use. *Curr. Opin. Biotechnol.*, **11**: 120-125.

Schenk, P.M., Kazan, K., Wilson, I., Anderson, J.P., Richmond, T., Somerville, S.C. and Manners, J.M. (2000). Coordinated plant defense responses in *Arabidopsis* revealed by microarray analysis. *Proc. Natl. Acad. Sci. USA*, **97**: 11655-11660.

Stakman, E.C. and Piemeisel, F.J. (1917). Biologic forms of *Puccinia graminis* on cereals and grasses. *J. Agril. Res.*, **10**: 429-495.

Stuiver, M.H. and Custers, J.H.H.V. (2001). Engineering disease resistance in plants. *Nature*, **411**: 865-868.

Vanderplank, J.E. (1984). *Disease Resistance in Plants*, Second Edition, Academic Press, Inc., New York, pp. 194.

Yang, Y. and Klessig, D.F. (1996). Isolation and characterization of a tobacco mosaic virus-inducible *myb* oncogene homolog from tobacco. *Proc. Natl. Acad. Sci. USA*, **93**: 14972-14977.

Zhang, L., Ku, J. and Birch, R.G. (1999). Engineered detoxification confers resistance against a pathogenic bacterium. *Nat. Biotechnol.*, **17**: 1021-1024.

10

Entomogenous Fungi and Their Further Prospects as Mycoinsecticides

M.S. PATIL

Professor & Ex-Head,
Department of Botany, Shivaji University, Kolhapur – 416 008 (M.S.), India

Entomogenous fungi or 'fungi entomogeni' are mostly specialized parasites of insects, causing epizootics, a few are weak parasites, and species of *Septobasidium* are symbiotic with scale insects in nature mostly on terrestrial habitats. Many are also known in aquatic habitats. Insect population numerically is very large with a great diversity and representing 30 million species and their diseases are quite common, widespread and severe enough to eliminate almost whole insect population in a given habitat provided when humid environment, dense population, susceptibility and effective dissemination of a pathogen. Almost all groups (= classes) of fungi infect to insects except Myxomycetes and Basidiomycetes. Among these, the members of the whole class–Laboulbeniomycetes (1340 species of 128 genera) are the obligate parasites of true insects (Hexapoda) and also of Arachnida (mites) and Diplopoda (millipedes), second large group, the order Entomophthorales (256 spp. belonging to 10 genera, Balazy, 1993). More than a dozen of the genera from Pyrenomycetes, more than two dozen genera from Deuteromycotina. Entomogenous fungi are of two types: parasitic and opportunists (saprobic) parasitic forms are again of two types: Exoparasites and Endoparasites. Most entomogenous fungi are endoparasites but sporulating outside followed by death except *Cicadina* Peck on gregarious cicadas. Mode of infection by the propagule to the insects–eggs, larvae, pupae and adult is rather difficult to observe but commonly via germ tube–directly, through enzymatic actions or going in the body along with food during feeding on plants/substrate. De Geer (1782) was the first to describe a fungus attacked house fly (*Musca domestica* L.) and later on named by Ferd Cohn (1855) as *Empusa muscae* Cohn but corrected by Fresenius (1856) as *Entomophthora muscae* (Cohn) Fresenius. Now there are hundreds of species belonging to 35 genera excluding the class Laboulbeniomycetes. Plants are vital to life and provides ca. 80 per cent food. Plants are damaged enormously in field and forest by three major groups: Pathogens (13.6 per cent), pest (11.5 per cent) and weed (9.5 per cent) and also in storage affecting quantitative and qualitative losses (Mukhopadhyaya, 1992) in order to protect the cultivated crops is of prime important or to keep the crops healthy/disease free, pest free by many different agronomical and agricultural practices. Among the methods to protect the crop plants and also to control diseases, pests and weeds, pesticides are very commonly used. It is known that nearly 35000 pesticides products are used throughout the world involving 1500 basic chemical ingredients (Gangawane, 1999). In India, production of pesticides during the years 1990-95 was more than one lakh tonnes and ca. 800 pesticidal compounds have been registered under Pesticide Act, 1968 (Veerbhadra Rao and Sharma, 1991). Excess use of pesticides have created a lot of problems of environmental damage/imbalance, residual effects, development of resistance in pests and pathogens and thus, many pesticides are out of use. The major pests so far include 10,000 species of insects and diseases caused by 8000 species of fungi (Deshpande, 1999). Control of insect pest, particularly by their natural enemies comprising–parasites, predators and pathogens (viruses, MLOs, Bacteria and fungi) in agroecosystem is a continuous process. Among the first to examine the possibilities of using pathogenic fungi for insect control was Metchnikoff (1879) who isolated, from diseased wheat cockchafers (*Anisoplia austriaca* Habst.), the green muscardine: *Metarrhizium anisopliae* (Metch.) Sorokin, which he reared on sterilized beer mash. However, Rorer (1918), who worked out the practical aspect and *M. anisopliae* was used in sugarcane field to control pest of sugarcane: Frog hopper–*Tomaspis varia* in W. Indies and Trinidad. In India, Kamat *et al.* (1952) worked on same pathogen for to control Pyrilla of sugarcane in Maharashtra as an agent of biological control.

Now a days, many classes of organisms having pesticidal properties have been commercially produced on large scale under different trade names *e.g.* Elcar (HaNPV) released in 1976 by EPA (Carter, 1976), Sporeine (BT) in 1950, Biofungicide MTR-35 etc. of three classes: (i) *Bioinsecticides*: Bacteria, Fungi, Viruses, Protozoa, Nematodes; (ii) *Biofungicides*: Bacteria and fungi; (iii) *Mycoweedicides*: Fungi to control insects, fungi and weeds.

Among the most promising taxa of fungi belonged to Hyphomycetes and Zygomycetes *viz.*, *Beauveria*, *Entomophthora*, *Metarrhizium*, *Nomuraea*, *Spicaria*, *Neozygites*, *Verticillium* species. To culture the entomogenous fungi is a most difficult task. The species of these genera show narrow to wide host range; *Beauveria* and *Metarrhizium* species are the most promising and produced on a large scale and marketed by variety of trade names (Biofungicide MTR-35, Boverin, Biotrol–FBB, Beauvericin, Bassianin, Cytochalasin C and D, Destruxins B etc.) due to their primarily soil-borne nature, wide host range, toxin producing ability, easily culturing, heavily sporulating ability etc. It is said that *M. anisopliae* infects, naturally to more than 200 species of insects (Roberts, 1973). More than 200 different pests of plants have been successfully controlled (Bhagat, 1991). The ability of entomogenous fungi has been greatly appreciated by their production of CDES (Cuticle dissolving enzymes) on large scale by submerged fermentation using locust cuticle as a sole source of carbon. Extracellular proteolytic enzymes–esterase, endopeptidase, aminopeptidase and carboxypeptidase are produced within 24 hrs. by *M. anisoliae* and *B. bassiana*. These two taxa exhibit immense diversity, a key factor multi-million dollar business. There are many limitations which will be overcome by knowing overall biology of a fungus and insect host. It is indeed highly essential to know regional insects as pests, their pathogens and damage to crop plants and use of local pathogens is highly advantageous than a exotic/foreign which is generally found unsuitable, non-adaptive to local conditions. Thus, study of entomogenous fungi is a most fascinating topic for research and their further use as biological control.

A Brief History and Milestone Discoveries/Events in Microbial Pest Control

Specialized animal parasites of great infective potential with simultaneous distinctive selectivity specific to host, play an important role to reduce population density at field level *i.e.* mostly invertebrate–pest.

1. The earliest known report to use large yellow ant with long legs (*Oecophylla smaragdina*) by Chinese citrus-growers into orange field (grove) to protect wiry-worm fruits and in Yemen to protect date palm (*Areca dachylifera* L.) trees in 16th and 18th century.
2. The father of Charles Darwin, Erasmus Darwin was first to point out the possibility of biological control in 1800.
3. Metchnikoff (1878) was the first to observe bacterial diseases in insects caused by *Bacillus salutaris* to beetle larvae of *Anisopliae austriaca* and Husz (1928) who practically applied *Bacillus thuringiensis* as an agent of biological control isolated from *Ephestia kuehniella* Zeller, however, it was isolated earlier by Barlier in 1911 and tested on alfalfa caterpillar–*Colias philodice-eurytheme* Biosduval by Steinhaus (1951), who really recognized the detrimental effects of pesticides on environment and human health.

The fungus which infects silkworm larvae and quite destructive in silk-worm-rearing industry all over the world and it was first noted in Germany, a mascardine disease in 1821, which it was identified in 1835 by Bassi Balsamo as *Beauveria bassiana* Balsamo, a white mascardine pathogen, and M. anisopliae (Metchn.) sorokin, a green-mascardine pathogen to silk worm larvae. There are many more fungal pathogens to silkworm and produced variety of forms of diseases, *e.g.* species of *Aspergillus, Nornoraea, Spicaria, Isaria, Paecilomyces* species. Plants, insects and fungi are the most dominant organisms in the earth surface of plants serve and primary host for the parasitism of fungi and insects; insects are also second largest group of the organisms next to plants *i.e.* 30 million species (Hackett, 1989; Clark, 1982) and also served as the host to fungi.

Kirby and Spence (1826) who established the pathological relationship of the insect and pathogen to produce the insect diseases. This parasitism differs in each species of a group of major entomogenous fungi as well as its host range to be worked out by using biotest methods. Large number of strains or pathotypes exist which appear morphologically similar but differed in the host selectivity, pathogenesis, physiology, and biochemical properties.

The most remarkable efforts of Rover (1918), Berlier (1911) and Husz (1928) who practically used *Metarrhizium anisopliae* and *Bacillus thuringiensis* to control the pests *viz.* frog hopper of sugarcane *Tomaspis varia* and alfalfa caterpillar: *Colias philodice-eurytheme* respectively and a new field of biological control of pests, what Steinhaus (1951) anticipated that a bacterium which forms endospores would be a better an agent of biological control than non-sporing and claimed the faster action comparing with a virus. And a first commercial product 'sporein' was produced in France in 1938 (Luthy *et al.*) and its use to treat flour prevent the infestation from Mediterranean flour moth–*E. kuehniella* Zeller (Jacob, 1950).

Kamat *et al.* (1952) have also tried and worked out the feasibility of *M. anisopliae* against Pyrilla of sugarcane of Deccan in Maharashtra; unfortunately, the work was discontinued. Vassiliev (1914) and

Barger and Swain (1918) observed in the field 60-70 per cent mortality of the larvae *Anisoplia austriaca* by green muscardine pathogen (i.e. *M. anisopliae*) and 80 per cent of walnut aphid–*Chromophis juglandicola* Kott. in California (America). Today's serious problem of control of mosquitoes, vectors of many human diseases and also carrier is a challenge but many fungi are known as pathogens mostly to the larvae *viz. Coeomomyces* species, *Lagenidium giganteum* + *Culicinomyces clavisporac* and *Fusarium* species. Umphlett and Medelin (1968) have studied this problem much in details and Iyengar (1962), who studied a promising species–*C. indicus*, which infects to *Anopheles gambiae* Ayengar and established the following parameters for success infection in aquatic habitat.

1. Free evaporating body of water
2. PH–8.0
3. Water temperatures: 28–32°C in day time and should be less below 25°C and above 34°C.

Even though, it is a promising system, but a complicated situation and multiple requirements, it becomes very difficult to use the species of *Coelomomyces* (a wall-less endoparasite of mosquito larvae–Coelomomycetaceae–Blastocladiales–Chytridiomycetes) as an agent of biological contrast in practice. So many attempts other than fungus as an agent of biological control have been tried *e.g. Reesimermis nielgeni* Tsai and Grundmann, a parasitic nematode of mosquitoes *i.e.* mermithid nematode parasites of mosquitoes are world wide in distribution and more than 20 species have been reported from at least 63 host species, however, all are obligate parasites and thus, difficult to culture but their potential has been greatly appreciated.

One host mosquito larva: 10 infective stage (*R. nielseni*) after a week of exposure nematode parasitized the larva and move to shallow tray to collecting container–to sand-water bath with excess water in which after excess water has been drained out, the mermithids–moulted–mated and oviposited during 1-4 months and then eggs were collected from sand. The cost of production of pre-parasitic juveniles is 7-10 cent/million. In rice field of California (USA) 1150 nema/M^2 area resulted 80-85 per cent parasitisation of *Anopheles freebornii* Aitken.

This encouraging data attracted the attention of United States Environment Protection Agency (USEPA) to work further on many other species of mosquitoes. The exact role of high degree of parasitism, suppression of mosquito population, not exactly known and thus require further research. This case is mosquito (larvae) verses fungus–*Coelomomyces/Lagenium/Fusarium* and provides strategy and feasibility at field lever–a system is workable or non-workable or financially feasible or not, and to find alternatives because as Poinar (1975) recognized some 3142 insect–nemode relationships in nature and more than 450 viruses have been described from 500 orthropod species (Ignoffo, 1974). Narsimhan (1970) and Narsimhan and Thirumalachar (1977) gave an excellent account on biological control using bacteria and fungi against insect pests in India. As well as many excellent research books and articles have been published time to time on this matter *i.e.* microbial–control of pests, a most promising and urgently needed problem to replace insecticides not to develop problems of environment disturbance. Abalis (1981) studied the biochemical and pharmacological problems of the insecticidal chemicals *viz.* cyclodipeptides destruxins and bassianotides produced by *Metarrhizium anisopliae* and *Beauveria bassiana*; production of conidia, a source of inoculum commercially on large scale by Daaust *et al.* (1983) and Quintela (1994).

Sometimes more than mechanism involves in the nature of infection or parasitism *e.g. Trichoderma harzianum* antagonizes other fungi by necrotrophic parasitism while others (mycoparasitic) exhibit biotrophism. The same species *T. harzianum*, commonly a mycoparasitic species can also parasitize the elm bark beetle–*Scolytus*, a dual pathogenicity towards insect and fungi is quite novel and also shared by *Verticillium lecanii*, a pathogen of coffee green bug and also on other fungi, rusts, mushrooms etc.

Such strains within a species as above, can be separated on the basis of morphological and biochemical characteristics. This dual pathogenicity has been attributed to the hydrolytic enzymes which facilitate the entry of a pathogen.

Exoskeleton of the adult insect or cuticle of the larvae is a first line of defence and protective in function and complex chemically. It is made of wax, lipids, proteins and chitin, in which major component is chitin, 25-50 per cent on dry weight basis. A fungal cell wall components are two: skeletal and matrix. Chitin and R-glucans are the major while S-glucans and manoproteins are the components of matrix.

So, insects and fungi share a key structural polymer, chitin, a -1, 4-linked N-acetylglucosamine polymer (Shaikh and Deshpande, 1993).

The protein degrading enzymes (CDEs) mainly include proteases, lipases and chitinases and enzymes such as chitinases, proteases and glucanases are the main components of mycolytic enzyme (ME) complex (Deshpande, 1999; Vyas *et al.*, 1989 and Mendonsa *et al.*, 1996).

The composition of CDEs and ME is not drastically different, and chitinases contribute significantly in the process of degradation of protective covers. This leads to the selection of such agents which are the producers of CDEs and MEs *i.e.* chitinases (Lorito *et al.*, 1993).

Large-Scale Production

Large-scale production to use as biocontrol agents aims at the maximum possible number of propagules (infective units) infecting to their hosts insects. And it is expected that fungal propagules as a bio-insecticides must be alive (viable) at the time of their use. The expected minimum self life is of 18 months at 20°C. The biomass produced needs storing ideally without loosing its viability.

For this, both liquid and solid fermentation systems can be used commonly to produce desirable quantity of biomass. Liquid fermentation is found to be the most suitable process. Of course both processes having certain advantages and disadvantages. Propagules/spores or conidia produced by solid state fermentation found to be advantageous

A. Mass Multiplication and Formation of Entomopathogenic Fungi Against Insects

Propagules

The fungal kingdom is a diverse assemblage of organisms with a great variety of structural types as a result of their reproduction or multiplication. These are the units of infection produced during their life cycle either on natural substrate and habitats and on artificial culture media: Hyphal bodies (*Entomophthora* species), conidia (almost all Hyphomycetes: *Beauveria, Metarrhizium, Isaria, Spicaria, Norumaea, Verticillium, Paecilomyces, Trichoderma* spp. etc.), zoospores (aquatics: *Coelomomyces, Lagenidium* spp.), oospores/zygospores (*Entomophthora virulenta, Pyth. Oligrandum*), blastospores (*Neozygites* spp.), capilliconidia (*N. parvispora* and many species of Entomophthoraceae), Chlamydospores (*Trichoderma* and *Gloeocladium* spp.), yeast like cells, mycelial and blastospores (in dimorphic forms: many entomopathogenic fungi, *Paecilomyces fumosoroseus*).

Best Growth Media

Variety of types of media have been successfully used to produce large scale inoculum of a biological agent (fungus) on commercial bases-either solid, liquid or diphasic (both solid and liquid) media. Commonly, for the *Metarrhizium anisopliae*, *Beauveria bassiana*, *Aschersonia aleyrodes* a mixture of organic bases and porous materials in a proportion 2-3: 7-8 at 25°C for a period of 4 weeks with 30 per cent water retention capacity. The organic materials of low costs, readily available, may be given as a priority as: husk, rice-bran, rape seed meal, fish meal, feather meal, crab shell, charcoal, zeolite etc. Second type of raw materials as a major part of the culture media seeds whole or crushed or molasses with bran/husk–moistened.

Common Entomogenous Fungal Groups and Genus

Class-Chytridiomycetes (200 sporic fungi)

1. *Coelomyces* (07 spp.) pathogenic to mosquito larvae and larvicidal.
2. *Myiophagus* spp.

B. Class-Oomyces (Zoosporic Fungi)

3. *Lagenidium* spp. (*L. giganteum*) Mosquito larvae: *Culex tarsalis* (U.S.A.)
4. *L. callinectas* in crab eggs.
5. *Culicinomyces clavisporus*–on mosquitoes.

C. III. Class: Zygomycetes

Order: Entomophthorales: Entomophthoraceae

6. *Erynopsis* (3 spp.): Coleoptera: Cantharidae; Diptora: *Nemalocera* (Chironomidae).
7. *Massospora* (13 spp.): Gregarious Cicadas (12 Genera); a specialized obligatory, biotrophic pathogen of adults, sporulating in living labdomen–deterioration.
8. *Neozygotyes* (11 spp.): Aphids, Thrips and scale insects.
9. *Entomophthora* (13 spp.): on insects (terrestrial and aquatics).
10. *Entomophaga* (14 spp.): Insects and harvest spiders.
11. *Batkoa* (5 spp.): Nematoceros flies; hoppers, crane flies, Diptera: Tipulidae; Heteroptera: Pyrrhocoridae.
12. *Strongwellsea* (2 spp.): Fannia: Muscidae & Anthomyideae (Diptera).
13. *Zoophthora* (79 species): Insects.
14. Form genus: *Tarichium* (38 spp.) on insects and mites.

IV. Class: Triochomycetes (ca. 30 genera):

Obligate-symbionts of Arthropods: fresh water and terrestrial–Diptera larvae, may fly nymphs, beetles, cryustaceas: crabs, isopods, amphipods and millipods.

V. Class: Laboulbeniomycetes (128 G and 1340 spp.)

Studies of the Laboulbeniales are difficult for a mycologist who is usually a student of Botany due to lack of knowledge of Entomology–Arthropods, which is indispensable (Tavers, 1985 and Majewski, 1994).

VI. Ascomycotina

1. *Ascosphaera*
2. *Calonectria*
3. *Cordyceps* (ca. 100 spp.)
4. *Hypocrella*
5. *Myrangium*
6. *Nectria*
7. *Ophiocordyceps*
8. *Podonectria*
9. *Sphaerostilbe*
10. *Torrubiella*
11. *Ceratocystis* spp.–bark beetles (Ambrossia fungi).

VII. Hymenomycetes: Phragmobasidiomycetidae: Septobasidiales: Septobasidiaceae:

1. *Septobasidium* (165 spp.) on scale insects of the following genera: *Aspidiotus*, *Chionaspis*, *Chrysamphalus*, *Lepidosaphes*, *Mytilapsis*, *Perlatoria*, *Schizostachyum*, *Diaspis* and *Cercococcus* spp.
2. *Uredinìella* (one sp.)

VIII. Plectomycetes

1. *Aspergillus* (5 spp.): Insects, beetles, spiders, silk worms, etc. = *Sterigmatomyces* spp.

Plant Pests

Many insects, mites and nematodes feed on plants *i.e.* phytophagus, thus causing injuries, resembling those caused by plant pathogens–viruses, MLOs, RLOs, bacteria and fungi. Major groups of insect pests are as follows:

I.
A. Collembolla: Springtails–wingless (aptera) insects causing small holes in leaves.
B. Orthroptera: Grasshoppers, locusts, crickets, cockroaches with biting mouth parts.
C. Isoptera: Termites
D. Thysanoptera: Thrips (as pests and vectors)
E. Hemiptera: Plant bugs: Capsid bugs; Aphids; white flies, scale insects, mealy bugs, leaf hoppers (Jassids).
F. Lepidoptera: Moths and butterflies.
G. Coleoptera: Beetles.
H. Hymenoptera: Saw flies, wasps, bees, ants etc.
I. Diptera: Two winged flies: Gall widge, crane flies, fruit flies, carrot flies, wheat bulb fly, cabbage root fly etc.

II. Nematodes

Phytophagous (feeding on plants).

III. Common Pests

Vegetables, fruits, ornamentals, forest plants, plantation crops, storage pests, orchards etc.

1. *Plesiocoris rugicollis* Fall–Capsid bug–pest of willow and apple (Fryer, J.C.F., 1929).
2. *Mayetiola destructor* Say–Hessian Fly–of wheat
3. *Cephus cinchus* Nort.–wheat stem sawfly–of wheat.
4. *Aphis rumicis* L.–Common bean aphid on *Rumex*
5. *Brevicoryne brassicae* L.–Common bean aphid on *Brassica*.
6. *Acyrthosiphum pisum* Harris–Pea aphids–
7. *Myzus persicae* Sulzer–Peach aphid–
8. *Euproctis chrysorrhoea* L.–brown tail moth–of chickweed.
9. *Pieris brassicae* L.–Cabbage butterflies
10. *P. rapae* L.–Cabbage butterflies
11. *Plutella maculipennis* Curt.–Diamond back moth–on cruciferous plants.
12. *Leptinotarsa decemlineata* Say–Colorado potato beetle.
13. *Ceratitis capitata* Wild.–Mediterranean fruitfly–*Citrus*.
14. *Pyrausta rubilalis* Hbn.–European corn borer–Maize.
15. *Cecidophyopsis ribis* Nal.–'Big Bud' of black current mite–*Ribes* spp.
16. *Tomaspis saccharina* Dist.–Frog Happer of sugarcane–causes blight of sugarcane.
17. *Kokothrips robustus*Uzel.–Pea thrips–Peas
18. *Thrips tabaci* Lind.–Tobacco thrip.–Tobacco
19. *Phylloxera vitifoliae* Fitch.–Vine pest–Grapes.
20. *Pemphigus betae* Doane–Sugar beet Root Aphid
21. *Lycophotia margaritosa* Haw.–Cutworm of soil.
22. *Euxoa auxiliaries* Grote–Army cut worm–in soil.
23. *Eutettix tenallus* Baker–Beet leaf hopper.
24. *Epilachna varivastis* Muls.–Mexican bean beetle.
25. *Pseudococcus citri* Risso–Coffee Mealy bug.
26. *Chlorops pumilionis*–Gout fly of barley
27. *Oscinella frit* L.–Frit fly of oat
28. *Lymantria dispar* L.–gypsy moth (Boisgirand, 1840)
29. *Icerya purchasi* Maskell–Cottony cushion or fluted scale
30. *Promecotheca reichei* Baly–Fiji Hispid beetle of coconut palm.
31. *Saissetia oleae* Bernard–The black seale–on Olea
32. *Pseudococcus* spp.–Ciotrus mealy bugs.
33. *Trialeuroides vaporariorum* West–green house white fly.

34. *Encarsia partenopea* Masi–a chaleid.
35. *Tetranychus urticca* Koch–glass house red spider mite in cucumber houses.
36. *Pseudococcus lilacinus* Ckll.–mealy bug of coffee–Kenya
37. *Pseudococcus njalensis* Laing–vector to transmit swollen short disease of cacoa–
38. *Panonychus ulmi* Koch–Fruit tree red spider mite.
39. *Lepidosaphes ulmi* L.–Oyster shell scale or orchards.
40. *Spilonota ocellana* D & S.–eye spotted bud moth on orchards
41. *Laspeyresia pomonella* L.–Codling moth on orchards.
42. *Anisoplia austriaca* Hobst.–Wheat cockchafers
43. *Psylla mali* Schmidb.–Apple sucker in Europe
44. *Dialaurodes* spp.–white fly on fruit trees.
45. *Lepidosaphes beckii* Newm.–Purple scale on citrus.
46. *Melolontha melolontha* L.–Cockchafer grub.
47. *Bemisia tabaci*–white fly on Tobacco and also vector for legume viruses.
48. *Schistocerca pallens* Thumb.–Locust
49. *Colia philodice eurythemae* Biosduval–alfalfa caterpiller.
50. *Ephestia kuchniella* Zaller–Mediterranean flour moth.
51. *Junonia coenia* Hubner–buckeye caterpillar.
52. *Pristiphora erichsonii* Htg.–Larch saw fly.
53. *Tineola bisselliella* Hum.–Clothes moth
54. *Limantria monacha*-nunn moth
55. *L. dispar* L.–gypsy moth.
56. *Neodiprion sertifer* Geoffr.–European pine saw fly.
57. *Diprion hercyniae* Htg.–European spruce saw fly.
58. *Heliothis zea*-cotton ball worm
59. *H. virescens* F.–tobacco bud worm.
60. *Drosophila melanogaster* L.–fruitfly.
61. *Tribolium confusum* Duv.–flour beetle.
62. *Anthosomus grandis* Boh–Boll weevil.
63. *Pyrausta nubilalis*–European corn borer.
64. *Oncopeltus fasciatus* Dallas–milkweed bugs.
65. *Clysia ambiguella* Hl.–Vine moth
66. *Polychrosis botrana* Schiff.–Vine moth.
67. *Hoplocampa testudinea* Klug.–apple sawfly.
68. *Aspidiotus perniciosus* Const.–San Jose scale.
69. *Lygus pabulinus* L.–Capsid bug
70. *Archips argyrospila* Wlk.–Fruit tree leaf roller

71. *Calandra granaria*-Grain weevil
72. *Operophthera brumata* L.–Winter moth.
73. *Pieris brassicae*-Large cabbage butterfly.
74. *Helikothrips haemorrhoidalis* Bouche–Thrips
75. *Kaliosyphinga ulmi* Sund.–European elm sawfly leaf minor.
76. *Aphis fabae* Scop.–a polyphagus aphid/plant lice
77. *Lacanobia oleracea* L.–Tomato moth.
78. *Synanthelon pictipes*–lesser peach tree borer
79. *Synenthedon tenuis* Butler–Clearing moth of Japanese persimmon orchards.
80. *S. sequoia–Sequoia* pith moth
81. *Aegeria tibialis*–Cotton wood crown borer.
82. *Lobesia botrana*–grapevine moth.
83. *Pectinophora gossypiella* Saunders–Pink ballworm moth.
84. *Dendoctonusd brevicomis*–Western pine bottle.
85. *Aphis gossypii* Glov.–cotton aphid.
86. *Operophtera brumata*-winter moth
87. *Erannis defolaria* L.–Mottle winter moth.
88. *Alsophila aescularia* Schiff.–March moth.
89. *Erioischia brassicae*–Cabbage root fly.
90. *Melanoplus differentialis* Thos.–Grasshopper.
91. *Pieris brassicae*–Large white cabbage butterfly.
92. *P. rapae*-Small white cabbage butterfly.
93. *Contarinia pyrivoora* Riley–Pear midge
94. *Dacus oleae Gmel.*–Olive fly.
95. *Euteffix tenellus*-Leaf hopper (Jassid) (curly tip of beet virus vector).
96. *Cicadula sexnotata* Fall.–Leaf hopper (Jassid) (curly tip of beet virus vector) for AYD.

Promising species of fungi (for pest control):

1. *Aschersonia aleyroides* Webber (Coel.)
2. *Beauveria bassiana* (Bals.) Vuill (Hyph.)
3. *B. bassiana* (Bals.) Vuill. Strain Gk 2016.
4. *B. bronguiartii* (Sacc.) Petch.
5. *B. densa* (Link) Picard.
6. *B. tenella* (Delacroix) Siemaszko.
7. *B. sulfurescens* (non-pathogen to insects, but produced entomotoxic glycoprotein).
8. *Batko apiculata* (Thaxter) Humber (Entom.)
9. *C. lecanii* Zimmermann
10. *Coelomomyces africanus* (blastocladiales–Coelomomycetaceae, mosquito larvicidal).

11. *C. anophelesciae*
12. *C. indicus* Iyengar (= *C. Indiana* Iyengar, nom. nud.).
13. *C. dodgei*
14. *C. psorophorae*
15. *C. punctatus* Couch
16. *C. stegonyiae*
17. *Conidiobolus obscurus* (Hall & Dunn) Remaud. & Keller
18. *Entomophaga pyriformis* (Thoizon) Balazy (Entom.)
19. *Hirsutella lecaniicola* (Jaap) Petch (Hyph.)
20. *H. thompsonii*
21. *Lagenidium giganteum* (Oomycetes)
22. *M. album*
23. *Metarrhizium anisopliae* (Metch.) Sorokin (Hyph.)
24. *M. flavoviridae*
25. *Neozygites floridana* (Wieser & Muma) Remaud. & Keller (Ernto.)
26. *N. microlophi* Keller
27. *N. fresenii* (Nowakowski) Remaud. & Keller.
28. *N. lecanii* (A. zimmermann) Balazy = *Enpusa lecanii* A. Zimm. (basionum) = *Derexiomyces lecanii* Narasimhan and Thirumalachar
29. *N. parvispora* (Mac Leod, Tyrrell & Carl) Remaud. & Keller. = *Ento parvispora* Mac Leod *et al.*
30. *Paecilomyces fumosoroseus* (Hyph.)
31. *Spiccaria rileyi* (Farlow) Maublac). = Nomuraea rileyi (Farlow) Samson
32. *Spicaria* spp.
33. *Zoophthora aquatica* (Anderson and Ringo ex Anders. and Anagno Stakis) Balazy
    = *Ent. aquatica* Anders and Ringo.
34. *Z. aleurodis* Balazy and Manole.
35. *Z. anfuiensis* (Li) Humber.
36. *Z. aphidis* (Hoffmann) Remaudier and Hennebert
    = *Ent. aphidis* Hoffmann (basionym).
37. *Z. neoaphidis* (Remaud and Hennebert) Balazy
    = *Erynia neoaphidis* Remaud and Hennebert (basionym).
38. *Z. phytonomi (Arthur) Batko*
    = *Ent. phytonomi* Arthur
39. *Z. radicans* (Brefold) Batko
    = *Emp. radicans* Brefeld, 186, 1870; (basionym)
    = *Ent. radicans* Brefeld, 345, 1877.
40. *Verticillium lecanii*

Out of these above taxa only a few species/strains have been exploited commercially to produce myco-insecticides especially species of *Metarrhizium*, *Beauoeria*, *Verticillium*, *Spicaria/Nomurae*, *Paecilomyces* and *Cephalosporium* and sold in market with different trade names. However, which strain is used is a trade secrete.

However, today's market of biopesticides represents less than 1.0 per cent of a total worldwide pesticide market approaching 30 billion dollar annual sales. This 1 per cent biocide corresponds to 160-165 million dollar. More than 90 per cent of this represented by a product derived from single insecticidal bacterium–*Bacillus thuringiensis* Barliner (Bt.) with 3000 strains in last 2-3 years in America. This is mainly due to:

1. Weak penetration into market;
2. Performance factor *i.e.* slow action and less persistentness;
3. Production difficulties on large scale;
4. Lack of understanding of mode of action.

REFERENCES

Agarwal, G.P. (1990). Entomogenous Fungi in India and Management of insect pests, *Indian Phytopath.*, **43(2)**: 131-142.

Aregger, E. (1992). Conidial production of the fungus *B. brongniartiion* bouley and quality evaluation during storage at 2°C. *J. Invertbr. Pathol.*, **41**: 143-150.

Balazy, S. (1993). *Flora of Poland: Fungi (Mycota)*. Vol. 24: Entomophthorales, Polish Acad Sci., Inst. of Botany, Krakow, 356 pp.

Bell, J.V. (1974). *Insect Diseases* (ed.) Cantwell, G.E.), Marcell Dekker, N.Y., 185 pp.

Bhagat, R.C. (1991). Fungi as microbial pesticides. *Everyman Sci.* (February-March), p. 21.

Charnley, A.K. (1989). In: *Biotechnology of Fungi for Improving Plant Growth* (Whipps, J.A. *et al.*, ed.), Cambridge Univ. Press, U.K., 85 p.

Copell, S.C., *et al.* (1977). *Biological Insect Pest Suppression*. Springer, Verlag, 153 pp.

Couch, J.N. (1938). The genus *Septobasidium*, Univ. N. Carolina Press, Chapel Hill, Carolina, 430 pp.

Deshpande, M.V. (1998). In: *Biotechnology–A Challenge for 2005*; Proceedings of the Symposium, All India Biotech. Asso., New Delhi.

Deshpande, M.V. (1998). For Health, Wealth and Sustainable Environment. In "*Microbes*" (ed. Verma, A), Malhotra Publ. House, New Delhi, 281 pp.

Deshpande, M.V. (1998). In: Novel Biopesticides, CSIR Golden Jubilee Ser., CSIR, New Delhi.

Daoust, R.A. *et al.* (1983). Studies on prolonged storage of M. anisopliae conuidia: Effect of Temp. and R.H. on conidial viability and virulence against mosquitoes. *J. Invertibrate Pathol.*, **41**: 143-150.

Evan, S.C. (1988). Co-evolution of entomogenous fungi and their insect hosts. In "Co-evolution of fungi with plants and anbimals" (eds. Pirozynski, K.A. *et al.*).

Frigo, S.M. *et al.* (1986). Culture media and conidiation in M. anisopliae (Metsch.) Sorokin. *Anais-da-Escola Universidale de Sao-Paulo*, **43**: 285-293.

Guintela, E.D. (1994). Production of *M. anisopliae* (Metsch.) Sorokin on coarse grain rice. *Anais–da–sociedade–Entomologica-d-Brasil*, **23**: 557-560.

Jenkins, N.E. and C. Prior (1993). *Mycol. Res.*, **97**: 1489.

Letge, J.P. *et al.* (1977). Media suitable for industrial production of *Ent. virulenta* zygospores. *Biotechn. Bioengineer*, **19**: 1269-1284.

Letge, J.P. *et al.* (1988). Aphid pathogens. In "Aphids–their biology, natural enemies and control" (Helle, W., ed.). *World Crop Pests*, **28**: 323-336.

Madelin, M.F. (1966a). Fungal Parasites of insects. *Ann. Rev. Entomol.*, **11**: 423-448.

Moorhouse E.R., *et al.* (1993). Selection of virulent and persistent M. anisopliae isolates to control black vine weevil (*O. sulcatus*) larvae on glasshouse *Begonia*. *J. Invertibr. Pathol.*, **62**: 47-52.

Muzumdar, D. *et al.* (1995). Mass production of *B. bassiana* and its potentiality on rice hispa. *Indian Phytopath.*, **48(3)**: 275-278.

Narsimhan, M.J. (1970). Entomogenous fungi and possibility of their use for biological control. *Indian Phytopath.*, 23: 16-26.

Narsimhan, M.J. and M.J. Thirumalachar (1977). Entomogenous fungi attacking scale insects on coffee leaves in India. Abstr. II. *Int. Mycol. Cong. Tampa, Florida*, **2**: 446.

Petch, T. (1948). A revised list of British Entomogenous fungi. *Trans. Brit. Mycol. Soc.*, **81**: 286-304.

Roberts, D.W. *et al.* (1976). The use of fungi for microbial control of insects. In: *Microbiological Control of Insects and Mites* (Burges *et al.*, ed.), A.P. N.Y., 125-149 pp.

Rombach, M.C. *et al.* (1988). Production of *B. bassiana* (Deuteromycotina: Hyphomycetes) in different liquid media and subsequent conidiation on dry mycelium. *Entomophaga*, **33**: 315-324.

Sharma Shashi *et al.* (1988). Mass multiplication and formulation of entomopathogenic fungi and their efficacy against white-grubs. *J. Mycol. & Pl. Pathol.*, **29(3)**: 299-305.

St. Leger, R.I. *et al.* (1979). *Trends Biotechnol.*, **15**: 83.

Steinhaw, E.A. (ed.) (1963). *Insect Pathology*, Vol. II. A.P. N.Y. 689 pp.

Umphlett, C.J. (1968). Ecology of *Coelomomyces* infection of mosquito larvae. *Mycol. Stud.* pp. 108-114.

Vimla Devi, P.S. (1994). Conidia production of the entomopathogenic fungis: *Nomuraea rileyi* and its evaluation for control of *Spodoptera litura* (Fab.) on *Riccinus communis*. *J. Invertibr. Pathol.*, **63**: 145-150.

Williamson, M. (1991). Biocontrol Risk. *Nature*, **353**: 394.

Zimmerman, G. (1986). Insect pathogenic fungi as pest control agents. In: *Biological Plant and Health Protection*, (ed. J.M. Franz., Gustav Fischer Verlag, Stuttgart, pp. 217-231.

11

Integrated Disease Management in Rice

C.S. REDDY

Principal Scientist, Directorate of Rice Research, Rajendranagar, Hyderabad – 500 030, India

"At best, the world food outlook for the decades ahead is grave; at worst, it is frightening." Thus stated Forrest F. Hill, Member of the Ford Foundation, before the Trustees of the Foundation in 1959, which was the world food situation at that time. Precisely twenty years later, while expressing concern on the situation of rice crop, Robert F. Chandler Jr., the First Director General of International Rice Research Institute, Philippines, mentioned, "So dependent upon rice are the Asian countries that throughout history a failure of that crop has caused widespread famine and death". Since then, there has been a revival of interest among the researchers in finding the reasons for this failure and solutions to save the crop on which many of the rice based countries depend.

Rice (*Oryza sativa* L.), the principle staple food crop in India, holds the key to our country's ability to produce enough food for our people. It is grown in 44.62 million hectares with annual production of 93.08 million tons (http://agricoop.nic.in/statistics 2003). To sustain and exist as a nation we must produce 100 million tons by 2010 and 140 million tons by 2025 (Mishra, 2002). Therefore, the major concern in coming years is to increase the productivity from the present level of 2.066 t/ha to more than 3 t/ha. To achieve this goal, the losses due to abiotic and biotic stresses have to be tackled.

Among the biotic stresses, it is often mentioned that diseases caused by plethora of microorganisms, take a heavy toll of the crop in the humid tropical rice growing environment. The scars left behind, by the extent of human suffering and famine caused by the Bengal famine during pre independent India, are still vivid in the minds of the people. When administration and civil supplies failed, the severe crop loss caused to rice by *Helminthosporium* triggered the famine in eastern parts of

India. While an average yield losses span from 5 to 15 per cent over large areas, total crop failure due to pests and disease epidemic is regularly encountered in some or the other pockets of the country.

In modern India, disease spectrum and intensity are changing continuously because of dynamic nature of rice culture. Before the advent of high yielding varieties, the major disease problems were blast and brown spot. After the introduction of high yielding nitrogen responsive varieties such as TN 1, Padma, IR 8 and Jaya, bacterial blight and rice tungro virus occurred in epidemic proportions in the Indo-Gangetic plains and coastal Indian states. In mid 1980s, sheath blight became economically important in isolated regions of Andhra Pradesh, Assam, Kerala, Punjab, Tamil Nadu and West Bengal. Later, consequent to the introduction of highly susceptible varieties such as IR 50, Swarna and BPT 5204, blast disease re-emerged as a major production constraint, and altered cropping intensity led to the development of tungro epidemics in peninsular India.

As it stands today, the diseases like blast, sheath blight, bacterial blight and tungro are the limiting factors in the successful cultivation. Now, these diseases are continue to be the major problem in many rice growing areas of the country, though at several stages, false smut, grain discoloration and sheath rot are of local importance and location specific in nature.

Disease management is one of the major functional components in rice production system in India. It plays a vital role in checking the diseases from flaring up to epidemic proportions or at least in slowing down the epidemics. Although management of the diseases can be achieved either by the use of resistant varieties or chemical control or manipulation of cultural practices, one method alone may not avoid danger of sustaining heavy losses. Therefore attempts were made to integrate several types of control measures to reduce yield losses.

A logical disease management program has to be normally based on the knowledge of the current disease profile in a given ecosystem, and management strategies need to consider the impact of disease severity on crop yield. The crop under field conditions often infected with more than one disease. In this context, Integrated Disease Management being a well knot unit which utilizes the cultural, biological, chemical means of management assumes greater significance.

The details of the relative importance of rice diseases, their distribution pattern in the country, primary source of inoculum, secondary spread, mode and form of survival of pathogen, favourable factors for disease epidemics, and available technology as on today for the management of rice diseases are discussed below and presented in Tables 11.1 to 11.4.

Blast

Blast disease aptly named to the damage it causes, is caused by *Magnaporthe grisea* (anamorph *Pyricularia grisea*). It can occur at all stages of growth causing heavy and at times a total loss in yield. Farmers in Andhra Pradesh, Himachal Pradesh, Tamil Nadu and in many parts of India, had to abandon popular susceptible cultivars because of the loss incurred from blast attack.

On leaves, symptoms develop from a small, pin head, water-soaked grayish dot that turns within 2 to 3 days into a spindle-shaped spot of ashy appearance and of 1 cm long and 0.5 cm wide. Occurrence of numerous such spots results in drying and death of the leaf. When the node is affected, it turns black and breaks at the nodal junction. The most disastrous stage of the disease is the panicle blast. Its attack results in improper grain filling, poor milling recovery and also chaffy earheads similar to the damage of white ears by stem borer. Hence, careful surveillance will help in arresting the disease in its initial stages. If not, the disease moves faster from field to field by producing myriad number of spores that are disseminated by wind in all directions. These spores upon falling on rice plant, initiate the disease. The repeated cycles of spore production and infection continues throughout the crop growth and under favourable conditions, the green lush crop growth is turned to burnt up appearance, and all this can happen within a fortnight leaving nothing to harvest.

The factors that influence blast epidemics are:

1. Susceptible variety
2. Availability of inoculum to initiate the disease
3. Excessive application of nitrogen fertilizer

Table 11.1: Relative Importance of Rice Diseases and Distribution Pattern in India

Disease	*Pathogen*	*Plant Parts Affected*	*Distribution (States)*
Major Diseases			
Blast	*Pyricularia grisea*	Leaves, nodes, panicle base and seed	Andaman Islands, A.P., Assam, Bihar, Gujarat, Haryana, H.P., J&K, Karnataka, Kerala, M.P., Maharashtra, Orissa, T.N., U.P., W. Bengal, and all North Eastern states.
Sheath blight	*Rhizoctonia solani*	Sheath, stem, leaves and panicles	Andaman Islands, A.P., Assam, J&K, Kerala, Orissa, Punjab, T.N., U.P., W. Bengal.
Bacterial blight	*Xanthomonas campestris* pv. *oryzae*	Leaf, sheath and panicle	Andaman Islands, A.P., Assam, Bihar, Gujarat, Haryana, Kerala, M.P., Maharashtra, Orissa, Punjab, T.N., U.P., W. Bengal
Tungro	Virus transmitted by *Nephotettix* spp.	Leaf, stem and sheath	A.P., Assam, Bihar, Karnataka, Manipur, Orissa, Punjab, T.N., U.P., W. Bengal.
Emerging Diseases			
Sheath rot	*Sarocladium oryzae*	Sheath (especially flag leaf sheath), panicle, seed	Andaman Islands, A.P., Assam, Bihar, Kerala, Karnataka, Orissa, T.N., U.P., W. Bengal.
False smut	*Ustilaginoidea virens*	Panicles	Andaman Islands, A.P., Assam, Bihar, Gujarat, Haryana, Himachal Pradesh, Karnataka, Maharashtra, Orissa, Punjab, Uttaranchal, U.P. W. Bengal, N.E. States.
Minor Diseases			
Brown spot	*Helminthosporium oryzae*	Leaves, sheath and panicle	A.P., Assam, M.P., Orissa, Punjab, T.N., U.P. W. Bengal.
Stem rot	*Sclerotium oryzae*	Stem	Haryana, Punjab, Tamil Nadu.
Glume discoloration	*Sarocladium oryzae, Helminthosporium oryzae,* and saprophytic fungi	Seed	Andaman Islands, A.P., Assam, Bihar, Gujarat, Haryana. H.P., J&K, Karnataka, Kerala, M.P., Maharashtra, Orissa, T.N., U.P., W. Bengal, and all North Eastern states.
Bakanae	*Gibberella fujikuroi*	Seed, stem	A.P., Haryana, Punjab.
Bunt	*Tilletia barclayana*	Seed	Haryana, Punjab.
Bacterial leaf streak	*Xanthomonas campestris* pv. *oryzicola*	Leaves	A.P., Assam, Gujarat, Haryana, Karnataka, M.P., Maharashtra, Orissa, T.N., U.P., and W. Bengal.

Table 11.2: Primary Source of Inoculum, Secondary Spread, and Mode and Form of Survival of Pathogen

Disease	*Primary Source of Inoculum/ Survival Through*	*Secondary Spread Through*	*Over Wintering/Survival of Pathogen/Mode of Survival*
Blast	Seed, infected leaves, stubbles, straw	Air	Mycelium, conidia, straw, stubbles, seed and grass hosts
Sheath blight	Soil, stubbles, weeds, soil	Contact with infected leaves, weeds	Sclerotia, mycelia, straw, stubbles, weeds and soil
Bacterial blight	Seed, rhizosphere of weeds, infected leaves, stubbles	Irrigation water, rain splashes	Cells, straw, stubbles and grass hosts
Tungro	Infected stubbles	Green leaf hoppers	Stubbles and grass hosts
Sheath rot	Seed	Air, insects	Mycelium, conidia, straw, stubbles, flag leaf, sheaths and seed
False smut	Soil, seed	Air	Chlamydospores
Brown spot	Seed, leaves, straw	Air	Mycelium, conidia, straw, stubbles, and seed
Stem rot	Soil, stubbles	Irrigation water	Mycelium, conidia and infected plants
Glume discoloration	Seed, soil, stubbles	Air	Conidia and infected panicles
Bakanae	Seed, soil, stubbles	Air	Conidia and infected material
Bunt	Soil, seed	Air	Teleutospores
Bacterial leaf streak	Seed	Irrigation water, rain splashes	Cells and infected plant material

4. Cloudy and drizzle weather or dew resulting in continuous leaf wetness for more than 10 hours
5. Night temperature between 19 to 21°C
6. Relative humidity above 92 per cent.

If these conditions prevail, vigilant scouting has to be done for the appearance of initial symptoms and assessment of disease prevalence. In nursery and at active tillering stages, lower leaves have to be examined for the symptoms of the disease and at these crop stages, usually, the disease starts from lower to upper leaves.

When initial blast spots are seen as small roundish to slightly elongated, necrotic gray spots about 1 to 2 mm in diameter, immediate spraying with effective blasticides like ediphenphos 50 EC @ $500g^{ha-1}$ or tricyclazole 75 WP @ $250g^{ha-1}$ or broad-spectrum fungicides like carbendazim 50 WP @ $500g^{ha-1}$, has to be resorted to intercept further development of the disease. Nitrogen split dose can succeed the spray and it is advisable not to top dress with nitrogen fertilizer until the spray is given. Hence, nitrogen management with need-based application of fungicide can control the diseases on a susceptible variety. The commercial varieties resistant to blast disease are Sasyasree, Rasi, IR 36, IR 64, Vikas, Tulasi, Aditya, Srinivas and Swarnadhan.

Table 11.3: Favourable Factors for Epidemics

Outbreak Conditions	*Perpetuation*	*Cultural Practices*
Blast		
Low night temperature (15 to 24ºC), High RH (90 per cent and above), Prolonged dew deposition and cloudy weather.	Conidia and mycelia in straw. Continuous cropping, Weeds. Perfect stage.	Varietal susceptibility. High nitrogen use. Dense population. Delayed planting. Aged seedlings. Wet leaves. Non-removal of stubbles and weeds.
Sheath Blight		
Temperature range (23 to 26°C). High humidity (90 per cent and above) and cloudy weather. Heavy winds.	Sclerotia/mycelia in soil. Left over infected material and straw. Weeds.	Varietal susceptibility. High nitrogen use. Dense population. Insect injury and wounding. Non-removal of stubbles and weeds.
Bacterial Leaf Blight		
Temperature (25 to 32°C). Low sunshine hours. Cyclones/typhoons/ strong winds/intermittent rains. Humidity (90 per cent and more for 12-14 h/day).	Dormant at low temperature and humidity in left over material. Stubbles. Weeds. Continuous cropping.	Varietal susceptibility. Insect injury and wounding. Excess nitrogen especially as late top dressing, phosphate and potassium deficiency. Non-removal of stubbles and weeds. Pruning of leaves. Planting under shade.
Tungro		
Green leafhopper population. Cool and moist climate for vector population build up.	Infected stubbles. Link host (*Echinocloa colonium*) weeds. Migration of vectors.	Varietal susceptibility. Non-removal of stubbles and weeds.
Sheath Rot		
Rain/drizzle/prolonged dew period at flowering. High RH (90 per cent) and moderate temperature.	Infected stubbles.	Varietal susceptibility. Poor panicle exertion. High nitrogen usage. Presence of insects and mites in the field.
False Smut		
Temperature (25ºC to 30ºC). High RH (95 per cent). Moderate rainfall/high RH and dew drops particularly during flowering. Cloudy period and spell during anthesis.	True sclerotia and ascospores from high altitude area (*Clavlceps oryzae-sativae*) and Soil borne (?)	Varietal susceptibility. Excess nitrogen application, especially as, late top dressing.
Brown Spot		
Temperature (25 to 32ºC) RH over 90 per cent, heavy monsoon and cloudy days	Infected seed and plant material.	Nutrient deficiency, drought and poor soil conditions. Poor water management.
Stem Rot		
High humidity (90 per cent and above) and high temperature (30°C to 32°C) Aged plants.	Sclerotia/mycelia in soil. Left over infected material and straw. Weeds.	High nitrogen and phosphorus level. Low potassium, Flood/poor drainage, Close planting.

Table 11.4: Available Technology for the Management of Important Rice Diseases

Resistant Varieties	*Cultural Practices*	*Chemical Control Practices*
Blast		
Rasi, IR 36, IR 64, Sasyasree, Srinivas, Tikkana, Simhapuri, Parijatha, Salivahana, Gauthami.	1. Select healthy seed. 2. Apply moderate nitrogen levels (80–100 kg/ha) in 3 to 4 splits. 3. Avoid excess nitrogen; skip final nitrogen in blast infected fields. 4. Destroy stubbles/weeds, etc.	*Seed Treatment (/Kg seed)*: Tricyclazole 75 WP @ 2 g Carbendazim 50 WP @ 2 g *Spray (/L water)*: Tricyclazole 75 WP @ 0.6 g Ediphenphos 50 EC @ 1 ml Iprobenphos 48 EC @ 2 g Carbendazim 50 WP @ 1 g Thiophanate-methyl 75 WP @ 1 g Propiconazole 25 EC @ 1 ml
Sheath Blight		
Swarnadhan, Vikramarya, Radha, Pankaj, Manasarovar.	1. Avoid using infested seed. Select healthy seed. 2. Apply moderate nitrogen levels (80–100 kg/ha) in 3 to 4 splits. 3. Avoid excess nitrogen; skip final nitrogen in sheath blight infected fields. 4. Destroy stubbles/weeds, etc. 5. Check brown plant hopper, if any, which predisposes the crop to sheath blight infection	*Spray (/L water)*: Validamycin 3 L @ 2.5 ml Hexaconazole 5 EC @ 2 ml Propiconazole 25 EC @ 1 ml Iprobenphos 48 EC @ 2 g Carbendazim 50 WP @ 1 g Thiophanate-methyl 75 WP @ 1 g Bitertanol 25 WP @ 2 g Iprodione 50 WP @ 2 g Chlorathalonil 75 WP @ 3 g
Sheath Rot		
Bala, Cauvery, Kakatiya, Janaki, Tella Hamsa, Sabarmati.	1. Select healthy seed. 2. Apply moderate nitrogen levels (80–100 kg/ha) in 3 to 4 splits. 3. Avoid excess nitrogen, skip final nitrogen in sheath rot infected fields. 4. Avoid insect/mechanical injury to the crop and also check other diseases, which actually predispose the crop to sheath rot.	*Seed Treatment (/Kg seed)*: Mancozeb 75 WP @ 2.5 g Captan 50 WP @ 2 g *Spray (/L water)*: Carbendazim 50 WP @ 1 g Ediphenphos 50 EC @ Thiophanate-methyl 75 WP @ 1 g Propiconazole 25 EC @ 1 ml Hexaconazole 5 EC @ 2 ml
Brown Spot		
Rasi, Jagannath, IR 36.	1. Select healthy seed. 2. Apply total amount of nitrogen in split doses. 3. Give adequate and balanced nutrition to the crop. Correct soil deficiency of K_2O, Ca, Zn, Mn or low pH, by suitable amendments.	*Seed Treatment (/Kg seed)*: Mancozeb 75 WP @ 2.5 g Iprodione 50 WP @ 2 g *Spray (/L water)*: Mancozeb 75 WP @ 2.5 g Iprodione 50 WP @ 2 g Carbendazim 50 WP @ 1 g Chlorathalonil 75 WP @ 2 g

Contd...

Table 11.4–Contd...

Resistant Varieties	*Cultural Practices*	*Chemical Control Practices*
Stem Rot		
Jagannath, Sabarmati, Pankaj.	1. Apply total amount of nitrogen in split doses. 2. Avoid excess nitrogen application. 3. Skip final nitrogen application in stem rot infected fields. 4. Check brown plant hopper damage to the crop, which in turn reduces the stem rot incidence. 5. Destroy infected stubbles, infected plants, weeds, etc.	*Spray* (*/L water*): Iprobenphos 48 EC @ 2 ml. Carbendazim 50 WP @ 1 g Thiophanate-Methyl 70 WP @ 1 g
Bacterial Blight		
Ajaya, IR 64, Sona Mahsuri, Swarna, Radha, Janaki, Sujata, Asha, Udaya, PR 4141, Prasad, Gobind, Co 43, Biraj, Suraj.	1. Avoid using infected seed. Select healthy seed. 2. Apply moderate nitrogen levels (80–100 kg/ha) in 3 to 4 splits, in bacterial blight prone areas. 3. Avoid excess nitrogen; skip final nitrogen application in bacterial blight infected fields. 4. Avoid field to field irrigation. 5. Avoid shade in the field. 6. Avoid insect damage to the crop. 7. Destroy stubbles/weeds, etc.	No chemical control methods are recommended.
Tungro		
Vikramarya, Nidhi, Radha, Annapurna, Triveni.	1. Destroy tungro infected plants, stubbles, weeds, etc. 2. Avoid ratooning. 3. Introduce non-rice crop such as pulse/oilseed crops in tungro endemic areas. 4. Avoid electrical lighting near paddy nurseries.	*Nursery* (*Granular application*): Carburfuran 3 g @ 25 Kg/ha Phorate 10g @ 12.5 Kg/ha Fipronil 0.4 g @ 25 Kg/ha *Main Field* (*Spray*): Monocrotophos 36 EC (1 L/ha) Carbaryl 50 WP (1.5 Kg/ha) Phosphomidon 85 WSC (600 ml/ha) Fipronil 5 EC (1 L/ha)

Sheath Blight

Sheath blight disease caused by the fungus *Thanetophorus cucumeris* (*Rhizoctonia solani*) is widely prevalent but seldom assumes epidemic proportions. The disease affects all plant parts above water-line *viz.*, sheaths, inter-node, upper leaves and panicles. On sheaths, the spots are greyish-grey with ellipsoid or ovoid shape. The leaf of the affected sheath dries up and under humid conditions, white threads of fungal body can be seen over the surface of leaf sheaths. The infected leaves and internodes turn grey to straw colour with lateral brown bands resembling snake-skin.

On the infected portions, initially white pin-head sized bodies appear which later turn to brown and dark brown, with flattened base called sclerotia, visible to the naked eye. Many such sclerotia aggregate and appear in a large compact mass of dark brown colour. They adhere to plants, but are easily detached and are carried by water stream, and upon contacting plants, initiate the disease at the point of attachment.

Under favourable conditions of high humidity, intermittent rainfall and sunshine, the disease spreads to top portions of the plant and chokes the emergence of the panicle and if panicle has emerged, it invades spike-lets causing improper grain filling or sterility. In our country especially in southern part, this pathogen does not produce any propagative air borne spores, hence its spread is slow either through sclerotial bodies or by physical contact between healthy and affected plants.

The pathogen has a wide host range and occurs on all grasses and broad leaved weeds grown on rice bunds causing similar symptoms and producing sclerotial bodies. These sclerotial bodies fall in paddy water and initiate infections on rice crop. Even if leaves of rice plants come in contact with infected weeds on bunds, they pick-up infection and spread the disease. Hence, keeping bunds clean of weeds will help in checking the disease spread from primary sources.

The disease spread is very slow and a change in environment from humid to dry weather will alleviate the crop from disease. Otherwise, if the humid weather is likely to persist for prolonged time and disease is noticed, a spray of either validamycin 3 L or hexaconazole 5 EC @ 1000ml^{ha-1} or thiophanate-methyl 70 WP or carbendazim 50 WP @ 500g^{ha-1} will prevent the spread of the disease.

Sheath Rot

Sheath rot incited by *Sarocladium oryzae,* has become widespread in Andhra Pradesh, Kerala, Orissa, Tamil Nadu, Bihar, West Bengal and North Eastern States, causing significant losses in yield. The disease appears during heading to maturity stages and the pathogen generally attacks the upper most leaf sheath enclosing the young panicles: Oblong or irregular spots with chocklate brown coloration may develop on the boot leaf sheath. If the pathogen attacks before the panicle emergence, the young panicles may not emerge and rot completely or emerge only partially. Grains inside the choked panicles and on the partially emerged panicles may be chaffy, light to dark-brown and covered by a white to light-pink mat of mycelium and spore mass. If the pathogen attacks after the panicle emergence, the grains may be partially or completely filled and the glume discoloration is caused. Generally, it was observed that susceptibility of a variety is associated with poor panicle emergence.

Application of carbendazim, twice at fortnightly intervals commencing from boot leaf stage would be effective. Other products found useful in reducing the sheath rot infection are thiophanate-methyl, ediphenphos, hexaconazole, propiconazole and mancozeb.

False Smut

False smut is more prevalent in seasons favourable for good growth and high yields, and the farmers consider its incidence as an indication of good harvest. However, severe outbreaks have been reported from Andhra Pradesh, Assam, Bihar, Gujarat, Eastern Uttar Pradesh, Haryana and Punjab. A heavy incidence of disease was observed in 1990 in Punjab and damaged the crop up to 70 per cent. The yield loss was maximum in PR 106 and its breeding lines, followed by Jaya and PP 103. In the years of severe incidence, losses equaled bacterial blight and sheath rot.

The symptoms of false smut are visible only after flowering. The pathogen, *Ustilaginoidea virens,* grows in the ovary and transforms it into large, yellowish and velvety green balls, which become greatly enlarged at later stage. The spore balls are covered by a membrane in the early stages, which bursts with further growth and the loose velvety pseudomorphs become visible.

Generally, the disease is common and serious in temperate and cooler mountainous areas than in hot tropical regions. It occurs more frequently in upland rice areas than in lowland areas and when a spell of wet weather synchronizes with the heading time. High moisture favours the development of the disease. The number of rainy days during flowering period influences the disease incidence more than the amount of rainfall.

Though little is known about the varietal resistance to false smut, varieties like Bala, Cauvery, Sabarmati, Prakash and Pankaj were found not infected when they were grown in between severely infected Mahsuri plots. Among the fungicides, copper oxychloride, mancozeb and chlorothalonil effectively reduced the incidence of the disease.

Brown Spot

In general, brown spot does not cause much loss in yield. But occasionally, when epidemics occur, 25-53 per cent of the produce might be lost. Although this disease is much more widely distributed than blast, except during epiphytotic years, it is unlikely to cause appreciable loss in yield. Serious damage is generally associated with other primary problems such as nutrition and physiological disorders.

The disease occurs more or less every year in mild to severe form in many upland and rainfed low land rice growing areas. It is widely distributed thorough out the country, especially along the eastern parts of Uttar Pradesh, Gangetic Plains of north Bihar, Chattisgarh plains of Madhya Pradesh, Tarai region of West Bengal, eastern parts of Orissa, Assam, Andhra Pradesh and Tamil Nadu.

The fungus, *Helminthosporium oryzae,* causal agent of brown spot of rice, may also cause glume discolouration, reducing the market value. On the leaves, first the spots appear as small brown margins and grayish centre. These spots frequently run together so that the entire leaf surface withers and dries up. The spots may also be distinct, isolated and scattered through out the leaf surface. On the glumes, the disease appears as black or brown spots, and in severe cases the greater portion of the glume may be covered by these spots.

Since the disease is seed-borne, seeds should be obtained from disease free areas. If affected seeds are to be sown, fungicidal seed treatment will be useful in reducing the disease on seedlings. If the disease is observed in the later growth stages of the crop, iprodione (2 g/L) or mancozeb (2.5 g/L) can be sprayed for controlling the disease. Among released varieties, Rasi, IR 36 and Jagannath are resistant to the disease.

Other methods suggested and tried for the control of this disease include, field sanitation, crop rotation, proper fertilization, good water management and use of soil amendments.

Bacterial Leaf Blight

Bacterial blight caused by *Xanthomonas campestris* pv. *oryzae,* shot into prominence after the introduction of nitrogen responsive, high yielding variety T(N)1, in mid-sixties. It also resulted in the withdrawal of this variety from cultivation due to its vulnerability under high nitrogen management.

Bacterial blight is a monsoon disease and its incidence and severity is very much influenced by rainfall, number of rainy days, susceptibility of the cultivar and nitrogen fertilizer application. Severe epidemics recorded in 1979 and 1980 in north western India, reducing the grain yields drastically.

The disease occurs in two phases–"kresek" or wilt phase which usually occurs during active tillering stage, results in the death of tillers and the hill resembling the damage of 'dead hearts' caused by stem borer. This phase is more common and often occurs in epidemic form in Punjab, Haryana and western Uttar Pradesh, and occasionally in Bihar. The other one is leaf blight phase which is common throughout the rice growing areas of the country appears from maximum tillering to heading stages. Sometimes the disease is mistaken by farmers as only drying due to unfavourable conditions and expects the crop to recover. But this is not to be, as under favourable conditions the disease spreads quickly causing drying of leaves, thus weakening the plant, resulting in poor panicle emergence with half filled discoloured spike lets.

The disease initially starts from either one or both sides of leaf margin, mostly at upper half of the leaf. The margins will become water-soaked and turns yellow to straw color in a wavy pattern. As the disease progresses, the drying spreads downwards and inwards of leaf blade causing drying and death of the leaf.

Rain splashes coupled with wind aid in dissemination of the bacterium from field to field. Injuries to rice leaves caused by rain storms, strong winds and intercultural operations form entry points for the pathogen to enter besides other natural pores present in rice leaves. Once inside the leaf, the bacteria multiply faster producing millions of bacterial cells, besides causing blighting of the leaf. On the surface of affected leaves, milky dew drops may be observed in the early morning in humid and warm conditions. These milky dew drops which ooze out of the affected areas disseminate by rain splashed wind, repeating the cycle a number of times up to flowering stage causing heavy losses in yield. Field to field irrigation from infected to non-infected areas also aids in the spread of the disease.

Since, there are no effective chemicals to protect the crop; the only recourse is development of resistant varieties and nitrogen management. If the favorable weather persists and disease is at a flare-up stage, it is advisable to withdraw application of nitrogen fertilizer at that stage of crop growth to minimize the effect of disease on yield.

In the absence of chemical control, actual breeding for disease resistance at many rice research stations has been carried out fruitfully resulting in resistant commercial cultivars. Some of the commercially cultivated resistant varieties are Ajaya, PR 114, Swarna, MTU 4870, HKR 120, IR 36, IR 64 and Saket 4.

Rice Tungro Disease

Among the virus diseases that affect rice, tungro disease is widely prevalent, particularly, in the north eastern and eastern coastal region of peninsular India. The worn 'tungro' in *tagallo,* a language in the Philippines, means degenerated growth, causing leaf discoloration which varies from light green to orange yellow leaves, reduction in tiller number and stunted plant growth. The panicle emergence is delayed resulting in small and sterile panicles. In some cultivars, the panicles do not emerge at all.

The virus spreads by green leafhoppers (GLH), *Nephotettix virescence* and *N. nigropictus,* for which rice is the preferred host. Viruliferous green leafhoppers introduce the virus into rice leaves when they probe to suck nutrients. Thus, a tungro infected plant suffers from damage caused by both virus and the insect. The disease can occur at any stage from nursery onwards. The disease spread depends upon the availability of virus infected plants; the insect vector population and the susceptibility of the variety.

The importance of weed hosts and self-sown rice plants cannot be overlooked. Weeds also harbor the virus and the vectors also feed on weeds. With the growth of the rice crop, which is a preferred host for GLH, these migrate to rice crop and those that fed on virus-infected weeds/self sown rice, initiate the disease. The initial disease on rice crop is seen along the weedy border of rice fields and spreads into the, main field thereon. The self-sown or highly susceptible varieties could prove disastrous has been proved by the outbreak of tungro in IR 64, in 1990 in coastal Andhra Pradesh. Even moderately susceptible varieties succumbed to disease due to large areas of infected crop of a susceptible variety and abundant vector population.

To control tungro virus, the ideal method is to grow resistant varieties, especially in the endemic areas to cut down the primary infection and primary infected areas, so that moderately resistant varieties can be harvested with minimum damage and little loss in yield. At present, only two varieties *viz.*, Vikramarya and Nidhi were found resistant.

The other strategy is to monitor the insect populations and indexing for the rice tungro disease, and keeping the bunds and vicinity free of weeds. If insect population is found on increase (two insects/hill), it should be controlled by the application of insecticides such as carbofuran 3g or phorate 10g @ 0.75 or 1 kg a.i.$^{ha-1}$ or spraying monocrotophos 36 EC or phosalone 35 EC @ 0.5 kg a.i.$^{ha-1}$.

Care should be taken that the nursery is free from tungro virus. In case the infected plants are observed, insecticide application should be resorted immediately to eliminate the vectors. Otherwise, the infected plants form an excellent source in the main field for the spread of the disease by vectors.

General Control Measures

For general precautionary measures to prevent or minimize diseases in the field, the common management practices to be adopted are:

1. Select the seed from healthy fields.
2. Raise nursery in elevated areas.
3. Adopt judicious nitrogen management when susceptible variety is planted.
4. Monitor the crop and weather conditions.
5. Keep the bunds and vicinity areas weed free.
6. Adopt need based application of fungicides by spraying at initial stages of disease with recommended dosages.

REFERENCES

Chandler, R.F., Jr. (1979). *Rice in the tropics: A guide to the development of national programs.* Boulder, Colo., Westview Press, p. 1.

Hill, F.F. (1959). In: *An Adventure in Applied Science* (1982), Robert F. Chandler, Jr. pp. 1, Chap. 1.

http://agricoop.nic.in/statistics2003/chap4a.htm#chap46a, accessed on 1st July, 2003.

Mishra, B. (2002). Rice in India. *Agrolook,* **3**: 10-15.

12

Management of Charcoal Rot of Soybean

SUDHA MALL

Institute of Environment Management and Plant Sciences, Vikram University, Ujjain – 456 010, India

ABSTRACT

Charcoal rot of soybean caused by *Rhizoctonia bataticola* poses annual threat year after year in dry areas of tropical countries. It is mainly a host-stress associated disease because plants growing in stress conditions are most susceptible to this pathogen. The various practices suggested to manage this disease are use of resistant varieties, crop rotation, inter-sequential cropping and seed treatment with fungicides but none of them were found to be very effective. Most important practice is to reduce host-stress and this is best done by heavy irrigation to maintain high soil moisture or to flood soil for 3–4 weeks before sowing. In drier areas water scarcity does not allow this practice, in such cases biological control with the help of local biocontrol systems like soil antagonists and hyperparasites are the most promising because this method of management is environmental friendly as well as economically viable.

Key Words: *Rhizoctonia bataticola, Soybean, Charcoal rot, Management.*

Soybean has established itself as one of the most important crop for oil, deoiled cake and low price source of high quality protein. Soybean has replaced sorghum, millet and other crops giving comparatively lower economic return. Government agencies have encouraged soybean cultivation as kharif crop in India (Shroff *et al.*, 1992). In Madhya Pradesh specially in western part, the area under soybean cultivation has increased significantly and in 1995 the state had 77 per cent of the total soybean crop area in India and appropriately it is named as 'Soybean State'.

Charcoal rot caused by *Rhizoctonia bataticola* Taub. Butler (*Macrophomina phaseoli*–pycnidial stage) is one of the most important disease causing heavy losses year after year. Charcoal rot of soybean is favoured at 30°C specially in dry soils. In epidemiology of disease host stress is more significant than pathogen inoculum in soil hence it is a stress disease. *Rhizoctonia bataticola* has multipotential capacity to infect soybean plant at all stages of growth resulting in cumulative loss which may be more than 30 per cent of the total crop yield and when calculated in terms of money it may be up to several crores on all India basis. But mostly infection takes place in seedling stage and continues throughout the plant life and infects all parts of the plant. The fungus causes general root rot infecting root and lower portion of stem and the infected tissue appears as dipped in charcoal dust, a diagnostic symptom of the disease.

Rhizoctonia infection reduces plant growth, height, volume and root weight by more than 50 per cent. In above ground symptoms the disease shows low plant vigour and premature yellowing and drop of top leaves which may be due to vascular plugging. Flat pods without seeds has frequently been reported in severely infected fields. Damage to root system is most evident during pod formation and filling stages when demand for water and nutrients are highest. Diseased plant matures several weeks earlier which further contributes yield loss.

Light and electron microscopic studies have revealed that penetration of root surface occurs most frequently between epidermal cells, aided both by chemicals and mechanical pressure. After penetration hyphae ramifies in cortical regions. The disease is most severe if infected seeds are sown when seedlings show 70–80 per cent mortality. Pathogen survives in soil year after year as sclerotia in soil or in infected crop residues because of this *R. bataticola* can not be eliminated although it is a poor saprophyte and weak pathogen. Though the disease has been reported from India in 1936 (Likhite, 1936) but detailed account appeared only after intensification of soybean cultivation when disease acquired significance due to its wide spread occurrence (Khare *et al.*, 1970, Choudhary and Karnakar, 1978) in Madhya Pradesh and other states. Ten of the top soybean producing countries are Argentina, Bolivia, Brazil, Canada, China, India, Indonesia, Italy, Paraguay and United States, all these together produce 97.3 per cent of world soybean, they also have disease problem and the estimated loss in 1998 was $ 6 billion. Charcoal rot was one of the most damaging disease in all these countries and intensive researches are going on to minimize losses due to diseases (Wrathen, 1999).

Another important aspect of modern agriculture is the use of chemical pesticides. These pesticides besides affecting target component also have non-target effect on host, soil-borne pathogens and soil microflora (Altman and Campbell, 1977, Rodriguez *et al.*, 1980) and thus influencing the incidence and development of diseases. In case of soybean also herbicide like basalin (common name–Fluchloralin) has been found to enhance charcoal rot disease by predisposing host for *Rhizoctonia bataticola* infection (Nagori and Mall, 1992). While analyzing the effect of basalin on interaction between *R. bataticola* and rhizosphere microflora, Nagori (1992) demonstrated that basalin inhibits microbes in rhizosphere of soybean while supports competitive colonization of *R. bataticola* and thus enhancing charcoal rot infection and development. The effect was also observed in improvement of pectolytic enzyme production by the pathogen. Effect of herbicide and other pesticides on disease incidence has also been reported in other studies (Bowen *et al.*, 1986, Sinclair, 1989). Further in recent years high concentration of pesticide, residues have been reported in food grains, fruits, vegetables, milk and milk products etc. These and various other researches on pesticide residual toxicity has been successful in creating awareness among pesticide users regarding hazards associated with indiscriminate and excessive use of pesticides.

Management

It is evident from above introductory account of charcoal rot of soybean that the disease is most severe in plants under stress. In such cases most important management programme to minimize yield loss is by introducing agricultural practices those reduce host stress and maintain host vigour.

Charcoal rot management practices suggested by Bhatnagar and Tiwari (1995) are:

1. Growing resistant varieties
2. Follow up rotation
3. Seed treatment with captan and thiram.
4. Mixed and intercropping.
5. Irrigation of fields to maintain high soil moisture or flooding the field 3–4 weeks before sowing.

As regards resistant varieties there is no variety that can serve the purpose as the varieties recommended for Madhya Pradesh are susceptible to the disease under elevated temperature and dry weather conditions in crop season. Although soybean crop loss can be reduced by rotating with small grains as wheat and barley for 1–2 years. But it is still difficult to eliminate *R. bataticola* from soil because it survives for years in crop refuse and can infest other hosts as well. While seed treatment can reduce seed and seedling rot but it is less effective in charcoal rot. In mixed and companion cropping for successful production crop combination must be selected carefully so that maximum land and sunrise can be utilized. Companion crop for soybean JS 71-05 for Malwa region is pigeonpea where paired rows of soybean and one or paired rows of pigeon pea are sown. However this practice has been of little benefit because of long survival and resistant nature of sclerotia. Heavy irrigation and flooding of the field for 3–4 weeks before planting though reduces stress on soybean plants and also cause loss of sclerotia but this practice has severe limitations in dry areas. Even early withdrawal of monsoon also causes enhancement of disease in infected plants. In Brazil a comparison between conventional and no-till cropping system showed that the latter was better alternative to suppress charcoal rot because it preserves soil moisture and reduce disease progress. Under such conditions biological control of the disease should be explored by making soil inhospitable to pathogen with help of soil microbes. In this endeavor interactions already occurring in soil are picked up and exploited *in situ* as per need (Mall, 1999). In addition, biological control is also one of the most important part of integrated disease management where biological agents are used to reduce pesticide spray. In this effort species of microorganisms of local strains are found to be very successful. Growth of *R. bataticola* and also the disease incidence could not be suppressed by providing inoculum of *Streptomyces alleus* and *S. griseus* in case of cotton. But growth of *R. bataticola* can be checked in soil column when it comes in contact with *Trichoderma lignorum* (Norton, 1954) but Dhingra and Khare (1973) could not found any antagonistic property of *T. lignorum* against *R. bataticola* which probably could be due to alkaline nature of soil. In perti dish hyphae of *Trichoderma* spp. coil around the hyphae of *R. bataticola* resulting in coagulation of protoplasmic contents of cell followed by dissolution of cell wall. *Trichoderma* species which were found to be very effective hyperparasite may also give promising result in this case. *Aspergillus niger* produce some antibiotic but it is not mycoparasite and is not able to suppress *R. bataticola* in soil (Dhingra and Khare, 1973). Further soil infested with *Bacillus cerreus* inhibited the pathogen both in amended and non-amended soil (Norton, 1954). In on line literature on soybean

Sinclair has studied antagonistic and synergistic effect between various fungi and bacteria associated with soybean seeds for biological control but none was found to be effective in suppressing charcoal rot in field experiments to economically beneficial extent.

It can be concluded from above account that management to control charcoal rot of soybean is still best done by removing host stress in soil by improving soil moisture but it is possible only in areas where sufficient water is available. Biotechnological approach using biocontrol agents can be the only environmentally safe way to minimize the disease specially in drier areas.

References

Altman, J. and Campbel, C.L. (1977). Effect of herbicides on plant diseases. *Ann. Rev. Phytopath,* **15**: 361-385.

Bhatnagar, P.S. and Tiwari, S.P. (1955). *Technology for increasing soybean production in India*. National Research Centre for Soybean (ICAR) Khandwa Road; Indore.

Bowen, J.E., Yorinori, J.J. and Sinclair, J.B. (1986). Effect of herbicide on soybean disease development and seed qualities in state of Parona, *Fitopathologia Brasileie.* **11**: 205-216.

Choudhary, A.K. and Karnakar, S.K. (1978). The influence of plant age on susceptibility of soybean. (*Glycine max* L Meml) to *Macrophomina phaseolina* (Tass) Goid. *Indian Agriculturist.,* **22**: 181-183.

Dhingra, O.D. and Khare, M.N. (1973). Biological control of *Rhizoctonia bataticola* on Urdbean. *Phytopath. Z.,* **76**: 23-29.

Khare, M.N., Kulkarni, S.N., Sharma, H.C. and Kumar, S.M. (1970). Diseases of soybean in Madhya Pradesh. 2nd *Workshop Conference of Soybean,* Jabalpur.

Likhite, V.N. (1936). Host range of Gujarat cotton root rot. *Proc. Assoc. Econ. Biol.* Coimbatore, **3**: 18-20.

Mall, S. (1999). Managing soil for suppression of root diseases: A renewed approach. In: *Modern Approaches and Innovations in Soil Management,* Ed. D.J. Bhagyaraj, Ajit Verma, K.K. Khanna, H.K. Kheri, Rastogi Publ. Meerut.

Nagori, M.S and Mall, S. (1992). Effect of basalin on interaction between charcoal rot disease of soybean and soil microflora. *AMSE Proc.,* **32**: 15.

Norton, D.C. (1953). Linear growth of *Sclerotium bataticola* through soil. *Phytopathology,* **43**: 633-636.

Norton, D.C. (1954). Antagonism in soil between *Macrophomina phaseolina* and selected soil inhabiting organisms. *Phytopathology,* **44**: 522-524.

Shroff, V.N., Saksena, K.K. and Tiwari, S.S.L. (1992). Soybean ki unnat kheti (Hindi) *Soybean Processors Assoc. of India,* 72 pp.

Sinclair, J.H. and Hackman, P.A. (1989). *Compendium of soybean diseases.* A.P.S. St. Paul N.M. 106 pp.

Sinclair, J.B. (1999) *Soybean research documents Online.* Google. Com.

13

Application of Genetic Engineering for Disease Management in Vegetable Crops

T.S. THIND[1], J.K. ARORA[1], H.J.S. DHALIWAL[1], PREM RAJ[1], C. MOHAN[1] & M.I.S. GILL[2]

[1] *Department of Plant Pathology,* [2] *Department of Horticulture, Punjab Agricultural University, Ludhiana – 141 004, India*

The green revolution is one of the biggest success stories of India cited globally, which enabled the country to convert its nightmarish "begging bowl status" to self sufficiency. With the introduction of hybrid seeds and other advanced farming techniques in staple foods, production in India has jumped from 50 million tonnes a year in 1950 to about 200 million tonnes a year for the last several years. But with the burgeoning populations the challenge is to increase the yield by developing high yielding varieties and protect the crop from biotic (diseases and pests) and abiotic stresses. Crop losses caused by pests and diseases are estimated to be about 30 per cent on an average. The great Irish famine resulting in successive failures of potato crop in Ireland due to *Phytophthora infestans* in 1840's had turned the attention of the world to the importance of plant diseases. In India, the Bengal famine due to damage of rice crop caused by *Helminthosporium oryzae* left devasting effects. Now globally, certain plant diseases have become a havoc in several countries and their attack is being managed by several methods inducing agronomic practices, resistant varieties and use of agrochemicals. But indiscriminate use of toxicants poses many harmful effects on ecosystems and appears to be not a viable and long term strategy apart from increasing the input cost. In this context biotechnology appears to much needed strength and stability for economical and sustainable management of plant diseases.

Global Status of Genetically Modified Plants

Biotechnology is being commercialized very fast. In 1996, 1.7 million hectares were planted with transgenic crops. This increased to 11 million hectares in 1997 and about 28 million hectares in 1998 (Table 13.1). In the first three years of commercialization (between 1996 and 1998), eight countries excluding China-five industrial and three developing countries have contributed to more than a fifteen fold increase in the area covered by transgenic crops globally. The number of countries growing commercialized transgenic crops increased from five in 1997 (USA, Argentina, Canada, Australia and Mexico) to eight in 1998 when South Africa, Spain, and France grew transgenic crops for the first time.

Table 13.1: Global Area (million hectares) of Transgenic Crops in 1996, 1997 and 1998

Year	** Total Area*	*Area in Industrial Countries*	*Area in Developing Countries*
1996	1.7	1.7	0.0
1997	11.0	9.5	1.5
1998	27.8	23.4	4.4

* Excluding China; James (1998)

The main transgenic crops which are being commercialized include soybean (52 per cent), corn (30 per cent), cotton (9 per cent), canola (9 per cent) and potato (~1 per cent). Transgenics have been produced for herbicide tolerance (71 per cent), insect resistance (28 per cent) and quality traits (1 per cent). However, research is being done to genetically modify most plants with a high economic value such as cereals, fruits, vegetables and floricultural and horticultural species.

Status of Plant Biotechnology in India

In India research on transgenic crops has been going on steadily by the concrete efforts of DBT, ICAR and private entrepreneurs but none of the transgenic plants is commercialized on vast scale. The Department of Biotechnology has set up several committees such as ISBC (International Bio-safety Committee), RCGM (Review Committee on Genetic Manipulation) and GEAC (Genetic Engineering Approval Committee) which comprise experts from various fields. These committees help DBT in preparation of regulation guidelines and ensure that all the prescribed bio-safety measures are followed in laboratory experiments before granting permission for field trials and commercial cultivation (Manjunath, 2000).

India has lagged much behind in commercializing the genetically modified (GM) crops. In May, 2002 three cotton hybrids having Bt gene for control of American boll worm have been cleared by Genetic Engineering Approval Committee (GEAC). The increase in yield is to the extent of 10-15 per cent with IPM and there is a considerable saving in pesticides. India has identified 12 major field crops *viz.* rice, wheat, maize, sorghum, cotton, pigeon pea, chick pea, mung and urd bean, potato, mustard, soybean and pearl millet which need transformation for specific diseases like bacterial blight, rust, wilt, downy mildew and viruses and pests like shoot fly, pod borer and American boll worm (Hunsigi, 2004).

Biotechnology helps to modify the genetic material of plants to enhance their tolerance to pests and diseases that, in turn, increases the yield and improve its quality and nutritional value. Different methods used in biotechnology for plant disease management include tissue culture techniques, somaclonal variations and genetic transformation.

This article aims to discuss use of genetic engineering techniques for the management of disease problems in some vegetable crops.

Genetic Transformation

Genetic engineering is the most potent biotechnological approach which deals with transfer of special gene assemblies through various transformation techniques. Transgenic plants carry an additional stably integrated foreign gene. The cloned gene having desirable traits is introduced without co-transfer of undesirable genes from the donor. The transgenic plants are normal plants in which the genes carrying the required traits are introduced by genetic engineering. This transformation method provides access to greater gene pool as the gene may be coupled from bacteria, fungi, animal, human beings, unrelated plants and even from chemical inducers in the laboratory (Chahal and Gosal, 2002).

Methods of Gene Transfer

On the basis of mode of transfer, transformation of genes is accomplished by two methods:

1. Vector mediated
2. Direct transfer

Vector Mediated

In this method the transgene is combined with a vector like *Agrobacterium tumefaciens* and *Agrobacterium rhizogene*.

Direct Transfer

The transgenes are physically delivered to the target tissues by different techniques *viz.* microinjection, electroporation of protoplast, liposome encapsidation, DNA injection into intact plants, incubation of seeds with DNA, particle bombardment, silicon carbide fibre method (Klein *et al.*, 1987; Grimsely *et al.*, 1987; Kloti *et al.*, 1993; Christou, 1994 and Sanford, 1988).

Development of Transgenic Plants Resistant to Fungal Pathogens

Fungal pathogens are principle causes of destructive diseases of many economically important crops and can inflict serious economic losses varying from 15 to 30 per cent (Grover and Gowthaman, 2003). Majority of transgenics produced against plant pathogens include single gene manipulation as well as combined effect of two genes. In all cases, genes regulating pathogen related proteins, antimicrobial proteins, antimicrobial compounds, enzymes controlling production of active oxygen species and phytoalexins have been exploited to produce fungal resistant transgenic plants. The strategy for the production of fungal resistant transgenics can be classified into two categories *viz.*

1. Antifungal protein mediated resistance
2. Antifungal compound mediated resistance

Antifungal Protein Mediated Resistance

Pathogenecity Related Proteins

These proteins are induced during hypersensitive response or related situation, by wounding, application of elicitor or a chemical and have to play active role in natural defense mechanism. On the basis of serological properties PR proteins have been grouped into five categories (PR1 to PR5) and act as direct antimicrobial proteins or enzymes and some may act indirectly like oligosaccharides and peroxidases which catalyse the cross linking of macromolecules in the cell wall (Dixon *et al.*, 1995). Chitinases and glucanases (PR2 and PR3 proteins) are fungal cell wall hydrolysing enzymes. Chitinases were isolated and characterized from several source like bacteria, fungi and insects and shown to inhibit the fungal growth by degrading newly synthesized chitin at the hyphal tip and can be grouped into four classes–basic polypeptides, acidic polypeptides, bifuctional chitinases and acidic chitinases. The first fungal resistant transgenic was developed in 1991 by Broglie *et al*. constituting expressed bean chitinase gene in tobacco and *Brassica napus* and the plants showed enhanced resistance to *Rhizoctonia solani*. Four synthetic peptides Pep 6, 7, 11 and 20 inhibit the growth of *Phytophthora infestans* and *Alternaria solani* in potato and one synthetic hexapeptide inhibits the growth of *Pencillium italicum* and *Penicillum digitatum* in citrus during decays in post harvest infection (Lopez *et al.*, 2000). Since then, there have been a number of reports on transgenics developed by constituting expressing PR-Protein gene (Table 13.2).

Table 13.2: PR-Protein Genes used in making Fungus-Resistant Transgenic Plants

Plant Species	*PR*	*Donor*	*Fungus Tested*	*Reference*
Canola	PR 3	Bean	*Rhizoctonia solani* *Pythium aphanidermatum*	Broglie *et al.* (1991)
Carrot	PR 5	Tobacco	*Erysiphe heraclei*	Stuvier *et al.* (1996)
Potato	PR 5	Potato	*Phytophthora infestans*	Zhu *et al.* (1996)
Rapeseed	PR 3	Tobacco, tomato	*Phoma lingam* *Scleotinia sclerotiorum*	Grioson *et al.* (1996)
Tomato	PR 2	Tobacco	*Fusarium oxysporum* f.sp. *lycopersici*	Jongedijk *et al.* (1995)
	PR 3	Tobacco	*Fusarium oxysporum* f.sp. *lycopersici*	Jongedijk et al.(1995)

Dasgupta *et al.* (2003)

Osmotin

Overexpressing osmotin exhibits antifungal activities against different pathogens in transgenic potato and tobacco plants. The transgenic potato crop with overexpressing osmotin showed resistance against *Phytophthora infestans* (Liu *et al.*, 1994).

Cystein Rich Protein

A separate group of antifungal polypeptides comprises chitin binding proteins, plant defensins, thionins and heveins all these have antifungal activities. Transgenic tomato plants expressing heve in gene showed less symptoms as compared to control plants when infected with *Trichderma hamatum* (Lee and Raikhel, 1995). Over expression of endogenous thionins enhanced resistance in Arabidopsis against *Fusarium oxysporum* f.sp. *mathiolac* and the transgenic plants exhibit less infection, less chlorophyll loss and disrupted hyphal growth (Epple *et al.*, 1997). Among defensins RS-AFP2 (*Raphanus sativus* antifungal protein-2) is the well studied plant defensins (Broekaert *et al.*, 1995). Cystein rich antimicrobial protein (Ace-Amp-1) from onion over expressed in geranium lead to increased resistance to *Botrytis cinerea* (Bi *et al.*, 1999). A gene for cystein rich defensins from *Alpha alpha* seed alfAFP (*Alpha*

alpha antifungal peptides) when expressed under the control of 35S promoter in transgenic potato governs the resistance to *Verticillum dahlia, Alternaria solani* and *Phytophthora infestans* (Gao, 2000).

Non-Plant Antifungal Proteins (Lysozymes)

Lysozymes are ubiquitous enzymes with specific hydrolytic activities and can be differentiated on the basis of the substrate specificity. The HEWL gene has been expressed in transgenic potato and tobacco plants that exhibit antimicrobial activity against several chitin containing fungi like *Botrytis cinerea, Verticillum albo-atrum* and *Rhizoctonia solani.*

Antifungal Compound Mediated Resistance

After the pathogen exposure or under stress conditions low molecular weight, anti microbial compounds such as phtyoalexins are produced which play an important role in plant resistance to fungal and bacterial pathogens. Expression of stillbene synthase gene from grapevine catalyse the production of stillbene type phytoalexins (resveratrol) and enhance the resistance to infection by *Botrytis cinerea*. Production of active oxygen species such as super oxide, anion radicals, hydroxial radicals and hydrogen peroxides (oxidative burst) play an important role in plant defense responses to pathogen infection. Transgenic potato plants expressing an H_2O_2 generating fungal genes for glucose oxidase were found to have cleavated levels of H_2O_2 that enhanced the protection to late blight of potato (Wu *et al.*, 1995). Some examples of transgenic plants having fungal resistance are mentioned in Table 13.3.

Table 13.3: Transgenic Plants Generated in Various Crops for Fungal Resistance

Plant	*Gene/Protein*	*Target Pathogen*	*Reference*
PR-Proteins			
Cucumber	Chitinase from rice	*Botrytis cinerea*	Tabei *et al.* (1998)
Brassica napus	Bean chitinase	*Rhizoctonia solani*	Broglie *et al.* (1991)
Potato	B-1, 3-endoglucanase from soybean	*Phytophthora infestans*	Borkowska *et al.* (1998)
Antifungal Proteins			
Potato	Osmotin from tobacco	*Phytophthora infestans*	Liu *et al.* (1994)
Genes/Protein that induced artificial cell death			
Potato	Barnase and barstar	*Phytophthora infestans*	Strittmatter *et al.* (1995)
Potato	Glucose oxidase from *Aspergillus niger*	*Phytophthora infestans*	Wu *et al.* (1995)

(Pattanayak and Kumar, 2000)

Development of Transgenic Plants Resistant to Viral Diseases

Viruses have become predominant causes of disease manifestation in many crop plants. These have been reported to cause huge losses by reducing yield and produce quality of field, vegetable, fruit and ornamental plants (Bent, 1996). In contrast to bacterial and fungal diseases, plants that are infected with viruses can not be cured because of parasitic interaction at genetic level and hence difficult to control through use of chemicals. With the advancement in biological sciences, virologist and molecular geneticist are able to construct virus gene based on nucleotide sequence derived from the virus itself or from virus associated nucleic acid. There are mainly two approaches for developing genetically engineered resistance depending on the source of gene used. The gene can be either from the pathogenic virus itself or from any other source. The former approach is called as pathogen derived resistance (PDR). In PDR a part or complete viral genome is introduced into the plant, which subsequently interferes with one or more essential steps in life cycle of the virus (Dasgupta *et al.*, 2003). Other approaches include expression of foreign gene (Non pathogen derived resistance) is based on the utilizing of the host resistance gene which prevents virus multiplication and spread by interacting with incoming viral genome or their expression product or by killing the host cell (Kundu and Mandal, 2001).

Engineering Strategies

Coat Protein Mediated Resistance (CPMR)

The phenomenon has been termed as genetically engineered cross protection. Introduction of viral coat protein gene into the plant makes the plant resistant to virus from which the gene for the CP was derived. The resistance manifested as delayed appearance of symptoms as well reduced titre of virus. The mechanism of CPMR may be protein mediated or RNA mediated. The protein mediated resistance operates in two stages, firstly in the initially injected cell where virion disassembly might be inhibited. Secondly, in experiment with Alfafa mosaic virus and Potato virus X, CPMR proposed that transgene CP might interfere with viral origin of assembly with viral nucleic acid necessary for replication. RNA mediated protection was based on the cellular RNA degradation mechanism that gets activated when higher level of a particular transcript accumulated within the cell leading to the degradation of specific transcript or co-suppression with viral genome (Bendahmane and Gronenborn, 1997). CPMR was first demonstrated for TMV in tobacco, subsequently virus resistant transgenics have been developed in tomato, melon, rice, papaya and sugarbeet. A transgenic variety of yellow squash called Freedom11 has been released in USA having resistance to viral infection (Shah *et al.*, 1995). CPMR is the extensively exploited mechanism for production of transgenics resistant to viral diseases. Some examples of coat protein mediated transgenics resistance to viruses in crops are given in Table 13.4.

Satellites

Satellites are small RNA molecules which show little, if any, sequences homologies with the parent virus to which they are associated. Tomato exhibited a non nerogenic satRNA sequence any developed faint symptoms of CMV ìnfection. This technique is used as biocontrol agent in field trials in several countries. (Tien *et al.*, 1991; McGarvey *et al.*, 1993). The satellite strategies are limited because many viruses have no satellite and also have some disadvantage. Some satellite virus combinations result into more severe symptoms, so sufficient caution will have to be exercised in adopting this technology.

Dysfunctional Movement Protein

This phenomenon was first to be used to engineer resistance against TMV in tobacco by producing modified movement proteins (MP) which are particularly active as transgenes. After entering into plant cell, a virus usually multiplies in cytoplasm or nucleus and then moves to the adjacent cells through plasmodesmata. The major viral protein involved in this process is the movement protein. A defective MP competes with its wild type during the infection process. Transgenic plants expressing a defective MP would block all the sites of the viral genomes thereby effectively minimizing binding of the functional MP. Since the viral genome is now coated with dysfunctional MP, intra and inter cellular movement of this virus is curtailed (Parthasarathy *et al.*, 2001). The resistance conferred by dysfunctional MP is also broad spectrum in nature. The transgenic potato plants expressing a movement protein of PLRV were found to restrict the movement of PVX, PVY and PLRV (Tacke *et al.*, 1996).

Table 13.4: Coat Protein Mediated Transgenic Response to Viral Diseases in Crops

CP Gene	*Virus Group*	*Challenge Virus*	*Transgenic Plant*	*Reference*
AIMV	Alfamo	AIMV	Tomato	Tumer *et al.* (1987)
PVS	Carla	PVS	Potato	Mackenzie *et al.* (1991)
SqMV	Como	SqMV	Squash	Pang *et al.* (2000)
CMV	Cucumo	CMV	Cucumber	Gonsalves *et al.* (1992)
			Squash	Tricolli *et al.* (1995)
			Tomato	Xue *et al.* (1994)
TYLCV	Gemini	TYCV	Tomato	Kunik *et al.* (1994)
PLRV	Luteo	PLRV	Potato	Kawchuk *et al.* (1990)
PVX	Potex	PVX	Potato	Hoiekama *et al.* (1989)
PVY	Poty	PVY	Potato	Malonoe *et al.* (1994)
PhMV	Tymo	PhMV	Tomato	Vidya *et al.* (2000)

Kundu and Mandal (2001)

Antisense RNA

Antisense gene is produced by inverting the orientation of coding region in relation to the promoter so that the natural sense strand of gene becomes antisense and is transcribed. The presence of antisense genes produced antisense and sense RNA which are complimentary to the sequences of each other and thus pair as double stranded RNA so that mRNA is not available for translation leading to silencing of normal genes or cause post-transcriptional down regulation of gene activity. The technology is useful for those viruses which are restricted to particular tissue such as phloem limited PLRV and the viruses which replicate in the nucleus, such as ssDNA containing gemini viruses. Bendahmane and Gronenborn (1997) produced transgenic expressed antisense RNA against tomato yellow leaf curl virus (TYLCV).

Transgenics with Non-Pathogen Derived Resistance

In non-pathogen derived resistance, the genes derived from either the host plant or any other nonpathogenic source are utilized. Strategies under this are as follows:

1. Oligoadenylate synthetase
2. Antiviral plant bodies
3. Natural resistance gene
4. Ribosomal inactivation protein (RIP)

Oligoadenylate Synthetase

Use of enzymes of mammalian interferon system is the latest in the development of virus resistant transgenic plants effective against a broad range of plant viruses. Interferons are proteins secreted by mammalian cells during viral infection. They induce the synthesis of other proteins which ultimately lead to the inhibition of viral multiplication. One of these is the RNA degradation pathway referred as

2′–5′ oligoadenylate synthetase (2–5 Aase) system. This pathway is activated by dsRNA, the usual replicating intermediate of RNA viruses (poorly activated by cellular dsRNA's). The synthesized 2′–5′ oligodenylates 2–5A in turn activate ribonuclease L (RNAseL) a single stranded RNAse, which degrade viral and cellular RNAs. Degradation of cellular RNAs results in cell death and thus it limits virus replication and spread (Beachy, 1996). Turve *et al.* (1993) first reported the constitutive expression of 2–5 Aase in transgenic potato plants that were protected against PVX infection under field conditions.

Anti-Viral Plant Bodies

These are specific anti-viral antibodies in plants, commonly known as plant bodies. Monoclonal antibodies against various gene products of TSWV have been introduced into tomato to generate continued resistance to both TSWV and root knot nematode (Wilson, 1993). There is possibility of interaction of the antibody with the Ca^{2+} binding on the coat protein that could disrupt the uncoating of the virus or the assembly of the viral progeny.

Natural Resistance Genes

Some plants show natural resistance to certain viruses. This resistance to viruses may be the result of inhibition of virus replication, virus movement or hypersensitive reaction (HR). In HR virus is restricted within the cells of primary infection, due to necrotic death of the cells around the primary infection site and resistance is encoded by the single dominant gene in most of the cases. Transgenic tomato plants expressing the N gene of *Nicotiana glutinosa* controls the HR in response to infection by tobacco mosaic virus (TMV). The plants that express N gene are resistant to all strains of TMV. Feller *et al.* (1998) obtained resistance in potato transgenic expressed as genome-linked proteinprotease (N1a) genes of three poty viruses.

Ribosomal Inactivating Proteins

Several plants have been found to contain antiviral proteins, commonly termed as ribosome-inactivating proteins (RIPs). RIPs inhibit the translocation step of translation by catalytically removing a specific adenine base from 28S ribosomal RNA. They are synthesized either as pre or pre-pro proteins and targeted to vacuoles. Because of their specific intracellular localization, RIPs do not affect the endogenous 28S RNA.

It is supposed that RIPs enter cells together with the viruses and exert the damage to the host ribosome or possibly viral RNA. Some RIPs not only inhibit local virus multiplication in RIP-treated leaves but also block viral multiplication systemically. Hence RIPs release a signal that induces systemic resistance to viruses (Dasgupta *et al.*, 2003).

The genes for RIPs have been isolated from a number of plant sources. The cDNAs for PAP (Pokeweed), MAP (*Mirabilis jalapa*) trichoxanthin (*Trichoxanthes kirilowi*) dianthin (*Dianthus caryophyllus*) momorcharin (*Momordica charantia*), CA-SRI (*Clerodendron aculeatum*), ricin (*Ricinus communis*), etc. have been isolated and characterized. These cDNAs for PIP (*Phytolasca insularis* protein) have also been used to transform plants and in many cases the transgenic plants have shown broad–range antiviral activities *e.g.* in potato.

Development of Transgenic Plants Resistant to Bacterial Diseases

Genetic engineering for bacterial resistance has met with relatively little success. Genes encoding antibacterial proteins that have been cloned and expressed in plants in an attempt to confer resistance to bacterial diseases are mentioned in Table 13.5.

Table 13.5: Examples of Successful Improvement of Resistance to Bacterial Diseases Through Genetic Engineering

Protein	*Origin*	*Transgenic Species*	*Resistance to*	*Level of Resistance*
Production of Antibacterial Proteins of Non-Plant Origin				
Shiva–1	Giant silk moth	Tomato	*Ralstonia solanacearum*	Partial
Lysozyme	T4 bacteriophase	Potato	*Erwinia carotovora*	Partial
Tachyplesin	Horseshoe crab	Potato	*Erwinia carotovora*	Partial
Inhibition of Bacterial Pathogenecity or Virulence Factors				
Phaseolotoxin-insensitive OC Tase[a]	*Pseudomonas syringae* pv. *phaseolicola*	Bean	*Pseudomonas syringae* pv. *phaseolicola*	Total
Enhancement of Natural Plant Defense				
Pectate lyase	*Erwinia carotovora*	Potato	*Erwinia carotovora*	Partial
Glucose	*Aspergillus niger*	Potato	*Erwinia carotovora*	Partial

(Mourgues *et al.*, 1998)

Production of Antibacterial Proteins of Non-Plant Origin (Lytic Peptides and Lysozymes)

Lytic peptides are small proteins with an amphipathetic, a-helical structure whose function is to form pores in bacterial membranes. Lysozymes are ubiquitous enzymes with specific hydrolytic activities directed against the bacterial cell-wall peptidoglycan. Three different lysozyme genes (egg-white, T_4 bacteriophage and human lysozyme) have been expressed in plants. Transgenic potato plants producing T_4 bacteriophage-lysozyme expressed partial resistance to *Erwinia carotovora* pv. *atroseptica in vitro* and green house experiments (Mourgues *et al.*, 1998). Similarly, the tachyplesin gene, which codes for a lytic peptide isolated from the horse-shoe crab, has also been expressed in transgenic potatoes resulting in a reduced amount of the tuber rot caused by *Erwinia carotovora*.

Inhibition of Bacterial Pathogenicity

Inhibition of bacterial toxins by cloning and expression of a gene whose product either inactivates or is insensitive to bacterial toxin. For example *arg*K gene in *P. syringae* pv. *phaseolicola* synthesizes the ornithine carbamoyl transferase that is resistant to its own phaseolotoxin and has been successfully employed to develop transgenic bean plants (Anzai *et al.*, 1989).

Enhancement of Natural Plant Defenses

Enhanced Production of Elicitors

Transgenic potato plants expressing a pectinolytic enzyme of *Erwinia carotovora* (known to release cell-wall sugar oligomers that act as elicitors) were significantly less susceptible to the bacteria than untransformed tubers (Wagner, 1987).

Reactive Oxygen Species (Oxidative Burst)

The accumulation of hydrogen peroxide and related reactive oxygen species like oxygen radicals immediately after pathogen recognition trigger the defense related responses including hypersensitive cell death. In potato, expression of glucose oxidase gene from *Aspergillus niger* resulted into increased level of resistance to *Erwinia carotovora* (Mourgues *et al.*, 1998).

Table 13.6: Strategies for Virus Resistance and Some Examples of Virus Tolerant Transgenic Plants

Transgene Product	*Stage of Infection Cycle*	*Mode of Action*	*Virus*	*Transgenic plant*	*Reference*
Pathogen Derived Resistance					
Coat protein	Uncoating	Competition for RNA	PEMV	Pea	Chowrira *et al.*, 1998
Movement protein	Transport	Interference with transport	PLRV, PVX & PVY	Potato	Tacke *et al.*, 1996
Antisense RNA	Translation	Blocks viral RNA and prevents translation	PLRV	Potato	Kawchuk *et al.*, 1991
Statellite RNA	Assembly	Competes for capsids	CMV	Tomato Pepper	Kim *et al.*, 1997
Plant Genotype					
RIP	Translation	Inhibits translocation	PVX	Potato	Wang *et al.*, 1998
2', 5' oligoadenylte	Replication	Degrade ds-RNA	PVX	Potato	Truve *et al.*, 1993
Antibody	Assembly	Competes for viral protein against which it is generated	TSWV	Tomato	Wilson, 1993

Pattanayak and Kumar (2000)

Conclusion

Biotechnology offers a very promising alternative to synthetic inputs for disease management and a good supplement to the conventional plant breeding methods for disease management. Biotechnology should be combined with other advanced agricultural technologies (crop diversity, sound crop rotation etc.) that offers exciting and environmentally responsive way to meet the consumers demands for sustainable agriculture. The benefits of transgenic crops should reach the small and marginal farmers so that the second green revolution becomes a reality.

REFERENCES

Anzai, H.K., Yonesty, K. and Yamaguch, I. (1989). Transgenic tobacco resistant to bacterial disease by the detoxification of a pathogenic toxin. *Mol. Gen. Genet.*, **219**: 492-494.

Beachy, R.N. (1996). The 2'-5' antiviral system in plants: A dose of mammalian medicine. *Nat. Biotechnol.*, **14**: 1538-1539.

Bendahmane, M. and Gronenborn, B. (1997). Engineering resistance against tomato yellow leaf curl virus (TYLCV) using antisense RNA L: *Plant Mol. Biol.*, **33**: 341-357.

Bent, A.F. (1996). *Plant Cell.*, **8**: 1757-1771

Bi, Y.M., Cammue, B.P.A., Goodwin, P.H., Krishna Raj, S. and Saxena, P.K. (1999). *Plant Cell Rep.*, **18**: 835-840.

Broekaert, W.F., Terres, F.R.G., Cammue, B.P.A. and Osborn, R.W. (1995). *Plant Physiol.*, **108**: 1353-1358.

Broglie, K., Chat, I., Hilliday, M., Cressman, R., Biddle, P., Knowlton, S., Mauvais, C.J. and Broglie, R. (1991). Transgenic plants with enhanced resistance to the fungal pathogen *Rhizoctonia solani*. *Science*, **254**: 1194-1197.

Chahal, G.S. and Gosal, S.S. (2002). Principles and Procedures of Plant Breeding: Biotechnological and Conventional Approaches. Narosa Publ. House, New Delhi.

Christou, P. (1994). Applications to Plants. Pp. 71-99. In: Particle bombardment technology for gene transfer, Eds. Yang, N.S. and Christou, Oxford Univ. Press.

Dasgupta, I., Malathi, V.G. and Mukherjee, S.K. (2003). Genetic engineering for virus resistance. *Current Science*, **84**: 341-354.

Dixon R.A., Paiva, N.L. and Bhattacharya, M.K. (1995). Engineering disease resistance in plants: an overview. In: *Molecular Methods in Plant Pathology*, (eds) R.P. Singh and U.S. Singh (U.S.A., CRC Lewis Publishers) pp. 249-270.

Epple, P., Aple, K. and Bohlmann, H. (1997). *Plant Cell*, **9**: 509-520.

Feller, J.P., Collins, G.B. and Hunt, A.S. (1998). The Nla-proteinase of different plant potyvirus provides specific resistance to viral infection. *Crop Science*, **38**: 1309-1319.

Gao, A.G. (2000). *Nature Biotechnology*, **18**: 1307-1310.

Grimsley, N., Hohn, T., Davlew, J.W. and Hohn, B. (1987). Agrobacterium mediated delivery of infectious maize streak virus into maize plants. *Nature*, **325**: 177-179.

Grover, Anita and Gowthaman, R. (2003). Strategies for development of fungus-resistance transgenic plants. *Current Science*, **84**: 330-340.

Hunsigi, G. (2004). Genetically modified crops. *Kisan World*, **31**: 35-36.

James, C. (1998). Global review of commercialized transgenic crops. Ithaca, New York: International Service for the Acquisition of Agri-biotech application, **43**

Klien, T.M., Wolf, E.D., Wu, R. and Sanford, J.C. (1987). High velocity microprojectiles for delivering nucleic acids into living cells. *Nature*, **327**: 70-73.

Kloti, A., Iglesias, V.A., Wunn, J., Burkdardt, P.K., Datta, S.K. and Potrykes, I. (1993). Gene transfer by electroporation into intact scutellum cells of wheat embryos. *Pl. Cell Rep.*

Kundu, P. and Mandal, R.K. (2001). Transgenic approaches for producing virus resistant plants. *Proc. Indian natn. Sci. Acad.*, **67**: 58-30.

Lee, H.I. and Raikhel, N.V. (1995). *Braz. J. Med. Biol. Res.*, **28**: 743-750.

Liu, D., Raghottama, K.G., Hasegawa, P.M. and Bresson, R.A. (1994). Osmotin overexpression in potato delays development of disease symptoms: *Proc. Natl. Acad. Sci.* USA, **91**: 1888-1892.

Lopez-Garcia, B., Gonzales-Candelas, L., Perez-Paya, E. and Marcos, J.F. (2000). *Mol. Plant-Microbe Interact.*, **13**: 837-846.

Manjunath, T.M. (2000). Insect resistant transgenic crops: Their importance in integrated pest management. *Agro*, 32-33.

McGarvey, P.B. and Kaper, J.M. (1993). *Transgenic Plants* (Eds. Kung, S.D. and Wu, R.), Academic Press, New York, pp. 227-296.

Mourgues, F., Brisset, M. and Chevreau, E. (1998). Strategies to improve plant resistance to bacterial diseases through genetic engineering. *TIBTECH*, **16**: 203-210.

Parthasarathy, V.A., Bose, T.K. and Deka, P.C. (2001). Potato. In: *Biotechnology of Horticultural Crops*, **2**: 1-42.

Pattanayak, D. and Kumar, P.A. (2000). Plant biotechnology: Current advances and future perspectives. *Proc. Indian Natn. Sci. Acad.*, **66**: 265-310.

Sanford, J.C. (1988). The biolistic process. *Trends in Biotechnology*, **6**: 299-302.

Shah, D.M., Rommens, C.M.T. and Beachy, N. (1995). Resistance to diseases and insects in transgenic plants: Progress and applications to agriculture. *Trends in Biotechnology*, **13**: 362-368.

Strittmatter, G., Egen, M., Runtz, I., Janssens, J., Opsomer, C. and Botterman, J. (1994). Engineering of programmed cell death as a defence mechanism of potato against late blight disease. *Journal of Cellular Biochemistry (Supplement)*, **18A**: 91.

Tacke, E., Solamini, F. and Rhode, W. (1996). Genetic engineering of potato for broad spectrum protection against virus infection. *Nat. Biotechnol.*, **14**: 1597-1601.

Tien, P. and Wu, G., (1991). *Adv. Virus Res.*, **39**: 321-339.

Truve, E., Aaspollu, A., Houkanen, J., Puska, R., Mehto, M., Hassl, A., Teeri, T.H., Kelve, M., Seppanen, P. and Saarma, M. (1993). Transgenic potato plants expressing mammalian 2′-5′ oilgoadenylate synthetase are protected from potato virus X infection under field conditions. *Biotechnology*, **11**: 1048-1052.

Wagner, R.R. (1987). The biology of the *Rhabdoviruses*. Plenum Press, New York.

Wilson, T.M.A. (1993). Strategies to protect crop plants against viruses, pathogen derived resistance blossoms. *Proc. Natl. Acad. Sci.* USA, **90**: 3134-3141.

Wu, G., Shortt, B., Lawerence, E.B., Fitszsimmons, K.C. and Shah, D.M. (1995). *Plant Cell*, **7**: 1357-1368.

14

Plant Growth Promoting Rhizobacteria to Augment Crop Production

K.V.B.R. TILAK & C. MANOHARACHARY

Department of Botany, Osmania University, Hyderabad – 500 007 (A.P.), India

Soil is the natural growth media for living plants and microorganisms. Soil-plant ecosystem depends on microbial activities which in turn contribute towards improving soil health, environmental quality and crop production. Bacteria are the most dominant group of microorganisms in the soil and probably equal to one half of the microbial biomass present there in. The most abundant bacterial genera present in the soil are *Pseudomonas, Arthrobacter, Clostridium, Bacillus, Micrococcus, Flavobacterium, Chromobacter, Sarcina, Enterobacter, Corynebacterium, Mycobacter* etc. (Mishra 1996; Rangaswami and Bagyaraj 1998; Subba Rao, 2000).

Some bacterial species living in the rhizosphere can affect plant growth in either a positive or in a negative way. Rhizosphere bateria that favourbly affect plant growth and yield of commercially important crops are denominated as Plant Growth Promoting Rihzobacteria (PGPR) (Kloepper and Schroth, 1978). Beneficial effect of rhizospheric bacteria have most often been based on increased plant growth, faster seed germination, better seedling emergences, enhanced nodulation and nitrogen fixation in leguminuous crops and suppression of diseases. As a consequence, PGPR have been further divided into subsets like emergence promoting rhizobacteria (EPR); nodulation promoting rhizobacteria (NPR) and disease suppressing rhizobacteria (DSR) (Kloepper *et al.*, 1989). However, there can be considerable overlap among these subjects. For instance, PGPR that act through biological control can also enhance germination or growth. In recent years, many reviews have appeared dealing with PGPR isolation, screening, ecology, physiology and their use as agrobiotechnological inoculants (Glick, 1995; Lazarovits and Nowak, 1997; Saxena *et al.*, 2000; Lata *et al.*, 2002, Glick, 2003). Investigations have brought to light instances where biological activities are markedly enhanced in two or three membered associations of organisms (Tilak *et al.*, 1982; Varma and Mathur, 1989; Negi *et al.*, 1990; El-Giami, 1992; Thiagarajan and Ahamad, 1993; Gupta *et al.*, 1998 a, b). Such syntrophic associations are of ecological importance with implied agricultural significance. In the present review, an attempt has been made to discuss the significance of plant growth promoting soil microorganisms in improvement of crop production.

Mechanisms of Growth Production

The mechanisms by which PGPR promote plant growth have been reviewed by several workers (Kloepper *et al.*, 1989; Glick 1995; Gupta *et al.*, 2000; Lata *et al.*, 2002; Glick, 2003). It has been suggested that production of plant hormones (Glick *et al.*, 1995; Glick, 2003); enhancement of plant growth nutrient uptake (Boddey and Dobereiner, 1995; Kennedy *et al.*, 1997; De Freitas *et al.*, 1997) and suppression of phytopathogenic microorganisms (Saxena *et al.*, 2000) via siderophores (Scher and Baker, 1982), -1, 3-glucanase (Fridlender *et al.*, 1993), chitinases (Renwick *et al.*, 1991), antibiotics (Shanahan *et al.*, 1992) and cyanide (Flaishman *et al.*, 1996) production may be involved.

Plant Growth Regulating Substances

Many rhizosphere bacteria produce IAA (Indole acetic acid) in culture media specially in the presence of tryptophan in the rhizosphere and rhizoplane of many plant species of economically important crop plants.

Tien *et al.* (1979) found production of plant growth promoting substances in liquid culture by the N -fixing *Azospirillum brasilense*. The bacteria produced small amounts of IAA, gibberellin and cytokinn-like substances.

Barea *et al.* (1976) examined culture supernatant of 50 phosphate–dissolving bacteria, isolated from rhizospheres of crop plants for IAA, gibberellins and cytokinins. Twenty bacteria were found to synthesize all three types of plant hormones; 43 produced IAA, 29 produced gibberellins and 45 cultures produced cytokinin-like substances.

Prikryl *et al.* (1995) reported production of IAA and some other auxins in liquid culture of *Pseudomonas cepacia* and *P. fluorescens* isolated from maize and bean rhizophere.

Brown and Burlingham (1968) detected the presence of three gibberellin-like substances in the culture of *Azotobacter chroococcum* strain A6. When an inoculum of *Azotobacter* was added to seeds or roots, the later development of tomato plants was altered, possibly because it was taken up by the seedlings at a critical stage of vegetative reproductive primordial differentiation.

Arshad and Frankenberger (1988) while studying the production of ethylene (C_2H_4) by the soil fungi *Acremonium falciforme* showed that microbially produced C_2H_4 can effect plant growth of etiolated pea seedlings. Etiolated pea seedlings presented a classical triple response, which includes reduction in elongation, swelling of the hypocotyls and a change in the direction of growth (horizontal), when *A. falciforme* was used as an inoculum. Further studies by Frankenberger and Arshad (1990) showed production of plant growth regulators (PGRs) by many soil microorganisms in the presence of suitable precursors. In laboratory studies, the microbial biosynthesis of auxins, cytokinins and ethylene in soil was monitored.

In liquid culture of *Azospirillum brasilense* Cd, the concentration of IAA increased rapidly with the beginning of stationary phase. This suggested that the high increase in IAA production in stationary phase is the expression of an overall change in cell metabolism when the carbon source is exhausted (Omay *et al.*, 1993).

Rhizobium leguminosarum strains shown to improve growth of canola and lettuce can serve as PGPR. Their growth promotive effects involve the production of IAA and cytokinin (Noel *et al.*, 1996).

Malik *et al.* (1997) observed IAA production in five strains of PGPR. These PGPR when inoculated in Kallar grass and rice resulted in increased acetylene reducing activity (ARA).

During the last few years, a new mechanism of plant growth promotion involving ethylene has been proposed. It was demonstrated that a small number of soil bacteria contains the enzyme ACC deaminase (Klee *et al.*, 1991). This enzyme hydrolyzes ACC, the immediate precursor of ethylene in plants, to yield ammonium and -ketobutyrate. Bacteria possessing ACC deaminase activity may therefore grow using ammonium as the sole nitrogen source. The beneficial rhizobacteria *Pseudomonas putida* strain GR 12-2 stimulates root elongation of different plants. Later, it was found that this bacteria contains ACC deaminase (Jacobson *et al.*, 1994). Mutants of this strain lacking ACC desminase activity were not able to promote root elongation of canola seedlings, implicating this enzyme in the mechanism of root growth promotion by this organism. A model of the possible role of ACC deaminase in root elongation is shown by Glick *et al.* (1995); following the binding of the bacteria to the seed coat.

A review on the formation of hydrocarbons by microorganisms indicated that while ethylene is a common product of fungi, the soft rot bacterium, *Pseudomonas solanacearum* has been shown to produce the gas in pure culture. More recently its formation by other bacteria has been demonstrated.

Experimental results on the effect of wild-type and mutant PGPR (*P. putida*) in the rooting of mungbean cuttings showed that treated cuttings had a significant higher number of roots compared with cuttings rooted in water. It was also clear that ethylene is involved in the initiation and elongation of adventitious roots in mungbean cuttings (Mayak *et al.*, 1999).

Enhancement of symbiotic nitrogen fixation and plant growth have been reported due to inoculation of Vitamin B 12 secreting *Pseudomonas* strain 267 in growth medium (Derylo and Skorupska, 1993).

Asymbiotic Nitrogen Fixation

Root-associated bacteria capable of fixing nitrogen occur regularly in diverse soils which vary widely in nitrogen content and the capability to metabolize N_2 provide a strong competitive advantage in soils of low organic nitrogen content.

Non-symbiotic nitrogen-fixing bacteria are important in the natural environment. Long-term nitrogen-balance studies have shown them to be capable of significant rates of nitrogen accumulation (Day *et al.*, 1975; Neyra and Dobereiner, 1977). Several studies indicate N_2 fixation rates of about 30-40 kg ha^{-1} $year^{-1}$ by associative bacteria.

Sprent (1979) presented an extensive list of species currently known to fix nitrogen. Common genera capable of fixing nitrogen include *Azospirillum, Azotobacter, Bacillus, Clostridium, Derxia and Klebsiella*. These are commonly designated as "free-living" bacteria, since they are able to exist in the soil and reduce nitrogen without entering into a symbiotic association with plants.

There have been many research efforts to demonstrate significant biological N_2 fixation by free living diazotrophs in the rhizosphere and on the surface of roots (Boddey and Doberiner, 1994; Okon and Labandera-Gonzalez, 1994; Saxena and Tilak, 1994).

Nitrogen fixation activities measured in agronomically important plants (wheat, maize, rice, sorghum) inoculated and colonized by *Azospirillum* and *Azotobacter* under greenhouse conditions or in field have generally been low (Okon and Labandera-Gonzalez, 1994).

It has been reported that in some cultivators of sugarcane of Brazil and of kallar grass in waterlogged salty soils in Pakistan, substantial quantities of nitrogen (about 50 kg N ha^{-1}) can be fixed by free-living diazotrophs associated with the plant *Acetobacter diazotrophicus* and *Herbaspirillum seropedicae* are N_2–fixing bacterial endophytes that might be responsible for the proposed high N_2–fixing in sugarcane (Boddey and Dobereiner, 1994). Species of the genus *Azoacus* are endophytes frequently isolated from roots of kallar grass (Reinhold-Hurek and Hurek, 1993).

Antagonism Against Phytopathogenic Fungi

During the last two decades, several examples of rhizobacteria capable of providing substantial disease control in the field have been reported (Saxena *et al.*, 2000). Many bacterial genera have shown their potential for biocontrol both under *in vitro* and *in vivo* conditions.

The usefulness of *Bacillus* as a source of antagonist for many plant pathogens (Campbell, 1989) is well known. Several potent strains from different species of *Bacillus* have been tested on a wide variety of plant species for their ability to control several diseases. *Bacillus* has ecological advantages because it produces endospores that are tolerant to extreme environmental conditions such as heat and desiccation. The greatest interest is *Bacillus subtilis* A 13, was isolated from lysed mycelium of *Sclerotium rolfsii* by Broadbent *et al.* (1971). It was found to be antagonistic to several pathogens and as a seed inoculant, increased the yield of carrots by 48 per cent, oats by 33 per cent (Merriman *et al.*, 1974) and peanuts by 37 per cent (Turner, 1987). Since, 1983 it has been sold as a bioinoculant under the trade name QUANTUM-4000 (Turner, 1987). *Bacillus* spp. and actinomycetes share several features that make them attractive for biological control agents including their abundance in soil and the production of various biologically potential metabolites active against a range of fungi (Silo-Suh *et al.*, 1994).

Fluorescent pseudomonads have revolutionized the field of biological control of soil-borne plant pathogenic fungi. During the last 25 years, they have emerged as the largest potentially most promising group of plant growth promoting rhizobacteria involved in the biocontrol of plant diseases (Barbosa *et al.*, 1995; Gomes *et al.*, 1996; Wei *et al.*, 1996). Fluorescent pseudomonads have received the most attention for several compelling reasons. First, fluorescent pseudomonads readily colonize roots in nature, where they are frequently the most common of microorganisms (Weller, 1988). The simple nutritional requirements and the ability to use many carbon sources that exude from roots and to compete with indigenous microflora, may explain their ability to colonize the rhizophere (Mazzola and Cook, 1991). Additionally, pseudomonads are amenable to genetic manipulation. These characteristics make them useful vehicle for the delivery of antimicrobial and insecticidal compounds and plant hormones to the rhizosphere. O'Sullivan and O'Gara (1992) reviewed the traits of fluorescent pseudomonads such as production of antibiotics, hydrogen cyanide, siderophore which are involved in suppression of plant root pathogens.

There are numerous example of biocontrol of several devastating fungal plant pathogens of important crops by fluorescent pseudomonads and has bee reviewed from time to time (O'Sullivan and O'Gara, 1992; Weller, 1998; Anith *et al.*, 1998; Pal *et al.*, 2000; Tilak and Saxena, 2001; Satyawani, 2001).

Solubilization of Mineral Phosphates

A majority of agricultural soils contain large reserves of phosphorus, of which a considerable part is accumulated as a consequence of regular application. The phenomenon of fixation and precipitation of P in soil is dependent on pH. In acidic soils, P is precipitated as Al and Fe phosphates, whereas in calcareous soils high concentration of Ca results in P precipitation. The soil is indeed a habitat for diverse group of organisms which employ variety of solubilization reactions to release soluble phosphorus from insoluble phosphates (Salih *et al.*, 1989; Illmer *et al.*, 1995; Singh and Kapoor, 1998). The potential of these phosphate solubilizing microorganisms (PSM) have been utilized as bioinoculants for crops grown in soils low in available P and amended with rock phosphate or tricalcium phosphate (Chabot *et al.*, 1996; de Freitas *et al.*, 1997; Gaind *et al.*, 2001). Kucey (1987) reported increased biomass and uptake of P, Cu, Fe and Zn in wheat inoculated with *Penicillum bilaji*. A direct correlation between population of P-solubilising *Penicilium bilaji* and wheat dry matter, seed and P yield was also reported by him.

Organic P is catalysed through hydrolysis of C-O-P ester bonds by phosphatase or phytase, which are very important in the nutrition of plants. In phytate acid soils which are rich in organic P, aspergilli were found to have the highest activity of the fungi tested (Tarafdar *et al.*, 1988). These phosphatase producing *Aspergilus* strains when inoculated in acid soils of Western Rajasthan (India), increased the P uptake and biomass yield of wheat and chickpea (Tarafdar and Rao, 1996).

PGPR on Plant Growth and Yield

There is an enormous volume of literature on the application of bacteria for improvement of plant performance (Whipps and McQuilken 1993; Okon and Labandera-Gonzalez, 1994; Saxena and Tilak, 1994; Bashan and Holguin, 1997; Okon *et al.*, 1998; Malik *et al.*, 1996, 1999; Saxena *et al.*, 2000), but few bacteria like *Azotobacter* and *Azospirillum* have been developed as commercial products. The organisms under most scrutiny for potential use in agriculture are bacteria belonging to the genera *Pseudomonas* and *Bacillus species* (Powell and Rhodes, 1994).

Conclusions

The PGPR's are known for their growth promoting ability by virtue of their ability to produce phytohormones, siderophores, HCN, chitinases, volatile compounds or antibiotics which help the host plant to enhance its productivity and reduce the infection through phytopathogenic microorganisms. Identifying such important traits will allow more efficient selection of new PGPR strains.

Efficient strains with specific traits to perform under adverse conditions like soil acidity, alkalinity, salinity, moisture stress and temperature (high or low) and for disease/pest resistance need to be selected or developed and tested for their performance.

Combined inoculation of PGPR strains with diazotrophs, phosphate solubilizers VAM fungi and nutrient mobilizers can be regarded as alternative means for more rational agricultural programme in economizing the use of chemical fertilizers. The interaction of PGPR with beneficial soil microorganisms could be harnessed in the preparation of inoculants in order to derive maximum benefits of inoculation which can be of commercial value of agricultural crops grown in tropical soils of India.

REFERENCES

Anith, K.N., Tilak, K.V.B.R. and Khanuja, S.P.S. (1998). Molecular basis of antifungal toxin production by fluorescent *Pseudomonas* sp. strain EM-85–A biological control agent, *Curr. Sci.*, **77(5)**: 671-677.

Arshad, M. and Frankenberger Jr., W.T. (1988). Plant growth regulating substances in the rhizosphere: microbial production and functions, *Adv. Agron.*, **62**: 45-151.

Barbosa, M.A.G., Michereff, S.J., Mariano, R.L.R. and Maranhao, E. (1995). Biocontrol of *Rhizoctonia solani* in cowpea by seed treatment with fluorescent *Pseudomonas* spp, Summa. *Phytopath.*, **21**: 151-157.

Barea, J.M., Navarro, E. and Montoya, E. (1976). Production of plant growth regulators by rhizosphere phosphate-solubilizing bacteria, *J. Appl. Bacteriol.*, **40**: 129-134.

Bashan, Y. and Holguin, G. (1997). *Azospirillum*–Plant relations: environmental and physiological advances (1990-1996). *Can. J. Microbiol.*, **43**: 103-121.

Boddey, R.M. and Dobereiner, J. (1995). Nitrogen fixation associated with grasses and cereals: Recent progress and perspectives for the future, *Fert. Res.*, **42**: 241-250.

Broadbent, P., Baker, K.F. and Waterworth, Y. (1971). Bacteria and actinomycetes antagonistic to fungal root pathogens in Australian soils, *Aust, J. Biol., Sci.*, **24**: 975.

Brown, M.E. (1972). Plant growth substances produced by microorganisms of soil and rhizosphere, *J. Appl. Bacteriol.*, 35: 443-451.

Brown, M.E. and Burlingham, S.K. (1968). Production of plant growth substances by *Azotobacter chroococcum*, *J. Gen Microbiol.*, **53**: 135-144.

Campbell, R. (1989). *Biological control of microbial plant pathogens*, Cambridge University Press, Cambridge, pp. 218.

Chabot, R., Autoun, H. and Cescas, M.P. (1996). Growth promotion of maize and lettuce by phosphate-solubilizing *Rhizobium leguminosarum* bv. *phaseoli*, *Plant Soil*, **184**: 311-321.

Cook, R.J. (1993). Making greater use of introduced microorganisms for biological control of plant pathogens, *Ann. Rev. Phytopath.*, **31**: 53-80.

Dashti, N., Zhang, F., Hynes, R.K. and Smith, D.L. (1998). Plant growth promoting rhizobacteria accelerate nodulation and increase nitrogen fixation activity by field grown soybean [*Glycine max* (L.) Merr.] under short season conditions, *Plant and Soil.*, **200**: 205-213.

de Freitas, J.R., Benerjee, M.R. and Germid, J.J. (1997). Phosphate-solubilizing rhizobacteria enhance the growth and yield but not phosphorus uptake of canola (*Brassica napus* L.). *Biol. Fertil. Soil.*, **24**: 358-364.

Derylo, M. and Skenipska, A. (1993). Enhancement of symbiotic nitrogen fixation by vitamin-securing fluorescent *Pseudomonas*, *Plant Soil*, **154**: 211-217.

El-Giami, M.S. (1992). Interactions between *Azotobacter* spp.- Rhizobium sesbani in the rhizosphere of *Sesbania sesban* (l.) Merrill plants and its efficacy on growth and symbiotic nitrogen fixation, *Zbl. Mikrobiol.*, **147**: 112-118.

Flaishman, M.A., Eyal, Z., Zilberstein, A., Voisard, C. and Haas, D. (1996). Suppression of *Septoria tritici* bloch and leaf rust of wheat by recombinant cyanide-producing strains of *Pseudomonas putida*, *Mol-Plant-microbe-Interact*, **9 (7)**: 642-645.

Frankenberger Jr., W.T. and Arshad, M. (1990). Microbial production of plant growth regulating substances in soil. *In*: *Proceedings of Second International Workshop on Plant Growth Promoting Rhizobacteria*, Interlaken, Switzerland, Oct. 14-19, 1990, pp.162-171.

Fravel, D.R. (1988). Role of antibiosis in the biocontrol of plant diseases, *Ann. Rev. Phytopathol.*, **26**: 75-91.

Friedlender, M., Inbar, J. and Chet, I. (1993). Biological control of soil-borne pathogens by a -1, 3 glucanase-producing *Pseudomonas cepacia*, *Soil Biol. Biochem.*, **25**: 1211-1221.

Gaind, S., Mathur, R.S. and Tilak, JK. V. B. R. (2001). Phosphate solubilizing microorganisms. *In: Recent Advances in Biofertilizer Technology* (eds.A. K. Yadav, M. R. Motsara and S.Ray Chaudhury). Soc.Promotion & Utilization Resources & Technology, New Delhi. India, pp. 190-206.

Glick, B.R. (1995). The enhancement of plant growth by free-living bacteria, *Can J. Microbiol.*, **41**: 109-117.

Glick, B. R. (2003). Plant growth promoting bacteria. *In*: *Molecular Biotechnology–Pinciples and Applications of recombinant DNA* (eds. B. R. Glick and J. J. Pasternak). ASM Press, Washington D.C., USA pp. 436-454.

Glick, B.R., Karaturovic, D.M. and Newell, P.C. (1995). A noval procedures for rapid isolation of plant growth promoting pseudomonads, *Can J. Microbiol.*, **41**: 533-536.

Gomes, A.M.A., Peiscoto, A.R., Mariano, R.L.R. and Michereff, S.J. (1996). Effect of bean seed treatment with fluorescent *Pesudomonas* spp. on *Rhizoctonia solani* control, *Arquivos-de Biologiae Technologia*, **39**: 537-545.

Gupta, A., Gopal, M. and Tilak, K.V.B.R. (2000). Mechanism of plant growth promotion by rhizobacteria, Review Article, *Indian J. Exptl., Biol.*, **387**: 856-862.

Gupta, A., Saxena, A.K., Gopal, M. and Tilak, K.V.B.R. (1998a). Bacterization of green gram with rhizosphere bacteria for enhanced plant growth, *J. Sci. Indust. Res.*, **57**: 726-736.

Gupta, A., Saxena, A.K., Gopal, M. and Tilak, K.V.B.R. (1998b). Enhanced nodulation of green gram by introduced *Bradyhizobium* when co-inoculated with plant growth promoting rhizobacteria, *J. Sci. Indust. Res.*, **57**: 720-725.

Illmer, P. and Schinner, F. (1995). Solubilization of inorganic calcium phosphates–solubilization mechanisms, *Soil Biol. Biochem.*, **27**: 257-263.

Jacobson, C.B., Pasternak, J.J. and Glick, B.R. (1994). Partial purification and characterization of 1-aminocyclopropane-1-carboxylate deaminase from the plant growth promoting rhizobacterium *Pseudomonas putida* GR12-2, *Can. J. Microbiol.*, **40**: 1019-1025.

Kennedy, I.R., Pereg-Grek, L.L., Wood, C., Deaker, R. Gilchrist, K., Katupitiya, S., Ladha, J.K. and Brujin, F.J. (1997). Biological nitrogen fixation in non-leguminous field crops: Facilitating the evolution of an effective association between *Azospirillum* and wheat, *Plant Soil*, **194**: 65-79.

Klee, H.J., Hyford, M.D., Kretzner, K.A. Barry, G.F. and Kishore, G.M. (1991). Control of ethylene synthesis by expression of a bacterial enzyme in transgenic tomato plants, *Plant Cell*, **3 (11)**: 1187-1193.

Kloepper, J.W., Hume, D.T., Scher, F.M. Singleton, C., Tipping, B., Laliberte, M., Franley, K., Kutchaw, T., Simonson, C., Lifshitz, R., Zaleska, I. and Lee, L. (1989). Plant growth promoting rhizobacteria on canola (rapeseed). *Plant Dis*, **72**: 42-46.

Kloeppeer, J. W., Leong, J., Teintze, M. and Schroth, M. N. (1980). Enhanced plant growth promoting rhizobacteria. *Nature* (London). **286**: 885-886.

Kloepper, J.W. and Schroth, M.N. (1978). Plant growth promoting rhizobacteria on radish. *In: Station de pathologie Vegetable et phytobacteria logie, INRA (Anyers, ed.).* Proc. 4th Int. Conf. Plant Pathogenic Bacteria, **2**: 879-882.

Kozyrovska, N., Koutunovych, G., Giomodova, E., Kuharchuk, P. and Kordyum, V. (1997). Novel inoculants for an environmentally-friendly crop production. *In: Global environmental biotechnology.* Proceedings of the 3rd Int. Symp. of Int. Soc. for Biotechnology. Boston, USA (ed. D.L. Wise). Resources, Conservation and Recycling **18**: 79-85.

Kucey, R.M.N. (1987). Alteration of size of wheat root systems and nitrogen fixation by associative nitrogen-fixing bacteria measured under field conditions, *Can. J. Microbiol.*, **34**: 735-739.

Lata, Saxena, A.K. and Tilak, K.V.B.R. (2002). Biofertilisers to augment soil fertility and crop production, *In: Soil Fertility and Crop Production* (ed. K.R. Krishna). Science Publishers Inc., Enfield (NH). USA, pp. 279-312.

Lazarovits, G. and Nowak, J. (1997). Rhizobacteria for improvement of plant growth and establishment, *Hort, Science*, **32(2)**: 188-192.

Loper, J.E. and Schroth, M.N. (1986). Influence of bacterial sources of indole-3-acetic acid on root elongation of sugarbeet, *Plant Pathology*, **76**: 386-389.

Malik, K.A., Bilala, R., Mehnaz, S., Rasul, G. Mirza, M.S. and Ali, S. (1997). Association of nitrogen-fixing, plant-growth promoting rhizobacteria (PGPR) with Kallar grass and rice, *Plant and Soil*, **194**: 37-49.

Malik, R.D.S. Kavimandam, S.K. and Tilak, K.V.B.R. (1996). Soil microorganism increases agricultural production, *Indian Farming*, **45**: 15-16.

Malik, R.D.S. Kavimandan, S.K. and Tilak, K.V.B.R (1999). *Kurthia* sp.–A noval ecofriendly inoculant for rapeseed (*Brassica campestris* var. *toria*), *Ind. J. Exp. Biol.*, **37**: 92-93.

Mayo, K., Davis, R.E. and Motta, J. (1986). Stimulation of germination of spores of *Glomus versiforme* by spore associated bacteria, *Mycologia*, **78**: 426-431.

Mazzola, M. and Cook, R.J. (1991). Effects of fungal root pathogens on the population dynamics of biocontrol strains of fluorescent pseudomonads in the wheat rhizosphere, *Appl. Environ. Microbiol.*, **57**: 2171-2178.

Merriman, P.R., Price, R.D., Kollmorgen, J.F., Piggott, T. and Ridge, E.H. (1974). Effect of seed inoculation with *Bacillus subtills* and *Streptomyces griseus* on the growth of cereals and carrots, *Aust. J. Agric. Res.* **25**: 219-226.

Mishra, R.R. (1996). *Soil Microbiology*, CBS Publishers and Distributors, New Delhi, pp.156.

Negi, M., Sachdev, M.S. and Tilak, K.V.B.R. (1990). Influence of soluble phosphorus fertilizer on the interaction between the vesicular-arbuscular mycorrhizal fungus and *Azospirillum brasilense*, *Biol. Fertil. Soils.*, **10**: 57-60.

Neyra, C.A. and Dobereiner, J. (1977). Nitrogen fixation in grasses, *Adv. Agron.*, **29**: 1-38.

Noel, T.C. Sheng, C., Yost, C.K. Pharis, R.P. and Hynes, M.F. (1996). *Rhizobium leguminosarum* as a plant growth promoting rhizobacterium: direct growth promotion of canola and lettuce, *Can. J. Microbiol.*, **42**: 279-283.

O'Sullivan, D.J. and O'Gara, F. (1992). Traits of fluorescent *Pseudomonas* spp. involved in suppression of plant root pathogens, *Microbiol. Rev.*, **56**: 662-676.

Okon, Y. and Labandera-Gonzalez, C.A. (1994). Agronomic applications of *Azospirillum*: an evaluation of 20 years worldwide field inoculation, *Soil Biol., Biochem.*, **26**: 1591-1601.

Okon, Y. Bloemberg, G.V. and Lugtenberg, B.J.J. (1998). Biotechnology of biofertilization and phytostimulation, *In: Agricultural Biotechnology* (Ed. A Altman), Marcel Deker, Inc., New York, pp. 327-349.

Omay, S.H., Schmidt, W.A. Martin, P. and Bangerth, F. (1993). Indole acetic acid production by the rhizosphere bacterium *Azospirillum brasilense* Col. Under *in vitro* conditions, *Can. J. Microbiol.*, **39**: 187-192.

Pal, K.K. Tilak, KV.B.R., Saxena, A.K. Dey, R. and Singh, C.S. (2000). Antifungal characteristics of a fluorescent *Pseudomonas* strain involved in the biological control *of Rhizoctonia solani*, *Microbiol. Rev.*, **155**: 233-242.

Parmar, N. and Dadarwal, K.R. (1999). Stimulation of nitrogen fixation and incubation of flavonoid-like compounds by rhizobacterium., *J. Appl. Microbiol.*, **86**: 36-44.

Powell, K.A. and Rhodes, D.J. (1994). Strategies for the progression of biological fungicides into field evaluation, *BCP Monograph*, **59**: 307-315.

Prikryl, Z., Vancura, V. and Wurst, M. (1985). Auxin formation by rhizosphere bacteria as a factor of root growth, *Biologia Plantarum* (*Praha*). **27**: 159-163.

Rangaswami, G. and Bagyaraj, D.J. (1998). *Agricultural Microbiology*, 2nd ed. Prentice Hall of India Pvt. Ltd., New Delhi, p. 422.

Reinhold-Hurek, B., Hurek T., Gillis, M., Hoste, B. Vancannet, M., Kersters, K. and Deley, J. (1993). *Azoarcus* gen. Nov., a nitrogen fixing proteobacteria associated with the rots of Kallar grass (*Leptochloa fusca* (L.) Kunth.) and description of two species *Azoarcus indigens* sp. Nov. and *Azoarcus communis* sp. nov, *Int. J. Syst. Bacteriol.*, **43**: 574-588.

Renwick, A., Campbell, R. and Coe, S. (1991). Assessment in *vivo* screening systems for potential biocontrol agents of *Gaeumanomyces graminis*, *Plant Pathol.*, **40**: 524-532.

Salih, H.M. Yahya, A.I., Abdul, R.A. and Munam, B.H. (1989). Availability of phosphorous in a calcareous soil treated with rock phosphate or superphosphate as affected by phosphate dissolving fungi, *Plant and Soil*, **120**: 181-185.

Sarig, S., Kapulnik, Y. and Okon, Y. (1986). Effect of *Azospirillum* inoculation on nitrogen fixation and growth of several winter legumes, *Plant Soil*, **90**: 335-342.

Satyavani, S. (2001). Biocontrol potential to *Trichoderma* spp. and *Pseudomonas fluorescens* in the management of pigeonpea (*Cajaus cajan* (L) Millsp) wilt, Ph.D. Thesis submitted to Deptt. of Botany, Osmania University, Hyderabad, A.P., India.

Saxena, A.K. and Tilak, K.V.B.R. (1994). Interactions among beneficial soil microorganisms, *Ind. J. Microbiol.*, **34(2)**: 91-106.

Saxena, A.K. and Tilak, K.V.B.R. (1997). Interaction of soil microorganisms with vasicular-arbuscular mycorrhiza, In: *New Approaches in Microbial Ecology* (eds. J.P. Tewari, G. Saxena, I. Tewari, N. Mittal and B.P. Chamola). Aditya Books Pvt. Ltd., New Delhi, India.

Saxena, A.K. Pal, K.K. and Tilak, K.V.B.R. (2000). Bacterial biocontrol agents and their role in plant disease management, In: *Biocontrol Potential and its Exploitation in Sustainable Agriculture*, (Eds. R.K. Upadhyay, K.G. Mukerji and B.P. Chamola, Kluwer Academic/Plenum Publisher, New York, pp.25-37.

Scher, F.M. and Baker, R. (1982). Effect of *Pseudomonas putida* and a synthetic iron chelator on induction of soil suppressiveness to *Fusarium* wilt phathogens, *Phytopathology*, **72**: 1567-1573.

Shanahan, P., O'Sullivan, D.J., Simpson, P., Glennon, J. and O'Gara, F. (1992). Isolation of 2,4-diacetyl, phloroglucinol from a fluorescent pseudomonad and investigation of physiological parameters influencing its production, *Appl. Environ. Microbiol.*, **58**: 353-358.

Singh, S. and Kapoor, K.K. (1998). Effect of inoculation of phosphate solubilizing microorganisms and an arbuscular mycorrhizal fungus and mungbean grown under natural soil conditions, *Mycorrhiza*, 7: 249-253.

Silo-Suh, I.A., Lethbridge, B.J. Roffel, S.J., Clardy, H., He, J. and Handelsman, J. (1994). Biological control activities of two fungistatic antibiotics produced by *Bacillus cereus* strain UW 85, *Appl. Environ. Microbiol.*, 160: 2023-2030.

Sprent, J.I. (1979). *The Biology of Nitrogen-Fixing Organisms*, McGraw-Hill, London, p.196.

Subba Rao, N.S. (2000). *Soil Microbiology*, 4th ed. Oxford & IBH Publishing Co. Pvt. Ltd., New Delhi, pp. 407.

Subba Rao, N.S., Tilak, K.V.B.R. and Singh, C.S. (1985 b). Synergistic effect of vesicular-arbuscular mycorrhizas and *Azospirillum brasilense* on the growth of barely in pots, *Soil Biol. Biochem.*, **17**: 119-121.

Subba Rao, N.S., Tilak, K.V.B.R. and Singh, C.S. (1985b). Effect of combined inoculation of VAM and *Azospirillum brasilense* on pearl millet (*Pennisetum americanum* L). *Plant and Soil*, **84**: 283-286.

Subba Rao, N.S., Tilak, K.V.B.R. and Singh, C.S. (1986). Dual inoculation with *Rhizobium* sp. and *Glomus fasciculatum* enhances nodulation, yield and nitrogen fixation in chickpea (*Cicer arietinum* L.). *Plant and Soil*, **95**: 219-226.

Suslow, T.V. and Schroth, M.N. (1982). Role of deleterious rhizobacteria as minor pathogens in reducing crop growth, *Phytopathology*, **72**: 111-115.

Tarafdar, J.C. and Cleassen, N. (1988). Organic phosphorus compounds as a phosphorus source for higher plants through the activity of phosphatases produced by plant roots and microorganisms, *Biol. Fertil. Soil*, **5(4)**: 308-312.

Tarafdar, J.C. and Rao, A.V. (1996). Contribution of *Aspergillus* strains to acquisition of phosphorus by wheat (*Triticum aestivum* L.) and chickpea (*Cicer arietinum* Linn.) grown in a loamy sand soil, *Agric-ecosyst-environ. Appl. Soil. Ecol.*, **3(2)**: 109-114.

Thiagarajan, T.R. and Ahmad, M.H. (1993). Influence of a VAM on the competitive ability of *Bradyrhizobium* spp. for nodulation of cowpea (*Vigna unguiculata* L. Walp) in non-sterilise soil, *Biol. Fertil. Soils*, **12**: 112-116.

Tien, T.M. Gaskins, M.H. and Hubbel, D.H. (1979). Plant growth substances produced by *Azospirillum brasilense* and their effect on the growth of pearl millet (*Pennisetum americanum* L.) *Appl. Environ. Microbiol.*, **37(5)**: 1016-1024.

Tilak, K.V.B.R. and Subba Rao, N.S. (1987). Association of *Azospirillum brasilense* with peal millet (*Pennisetum americanum* (L.) Leeke). *Biol. Fertil. Soil.*, **4**: 97-102.

Tilak, K.V.B.R., Singh, C.S. and Subba Rao, N.S. (1982). *Azospirillum* and *Aztobacter* inoculation. Effect on yield of maize and sorghum, *Soil. Biol. Biochem.*, **14**: 417-418.

Turner, J.T. and Backman, P.A. (1991). Factors relating to peanut yield increases after seed treatment with *Bacillus subtilis*, *Plant Dis.*, **75(4)**: 347-354.

Verma, S. and Mathur, R.S. (1989). Biocoenotic association between nitrogen fixing and phosphate solubilising microorganisms, *Curr. Sci.*, **58**: 1099-1100.

Wei, g., Kloepper, J.W. and Tuzun, S. (1996). Induced systemic resistance to cucumber diseases and increased plant growth by plant growth promoting rhizobacteria under filed conditions, *Phytopath*, **86**: 221-224.

Weller, M. (1988). Biological control of soil-borne plant pathogens in the rhizosphere with bacteria, *Ann. Rev. Phytopathol.*, **26**: 379-407.

Whipps, J.M., McQuilken, M.P. and Budge, S.P. (1993). Use of fungal antagonists for biocontrol of damping-off and *Sclerotinia* diseases, *Pestic. Sci.*, **37(4)**: 309-313.

15

Role of 2,4-diacetylphloroglucinol (DAPG) for Plant Disease Control: Its Importance to Rice Bacterial Blight Suppression in India

P. VELUSAMY[1], G. DEFAGO[2], L.S. THOMASHOW[3] & S.S. GNANAMANICKAM*

[1] *Center for Advanced Studies in Botany, University of Madras Guindy Campus, Chennai – 600 025, India*

[2] *Institute of Plant Sciences/Phytomedicine, Swiss Federal Institute of Technology, ETH-Zentrum, Zurich, Switzerland*

[3] *USDA-ARS, Department of Plant Pathology, Washington State University, Pullman, WA, U.S.A.*

ABSTRACT

We have used established methods for assembling and characterizing plant-associated bacteria from parts of Andhra Pradesh, Karnataka, Kerala, Maharashtra and Tamil Nadu states of India. This present investigation is on *Pseudomonas* strains as biocontrol agents of bacterial

* *Corresponding author,* E-mail: gmanick@vsnl.com

blight of rice but the emphasis is really on those fluorescent *pseudomonads* that produce a valuable antibiotic called, 2,4-diacetylphloroglucinol (DAPG). There has been no previous report on the production of DAPG by plant-associated bacteria from India and therefore, this study assumes special importance as a major study on DAPG-production by Indian strains of *Pseudomonas* which also act as biocontrol agents for rice bacterial blight. In the present study DAPG production has been implicated as an antibacterial compound involved in the suppression of one of the most important and devastating bacterial crop diseases, the bacterial blight of rice. To provide positive confirmation to the role of DAPG as the mechanism that mediates the biological suppression of rice bacterial blight, a genetic analysis of *P. fluorescens* strain PTB 9 was carried out. Transposon (*Tn5*) insertion mutants of this strain which were altered for either total or partial loss of its *in vitro* antibiosis towards *X. oryzae* pv. *oryzae*, were generated and were evaluated in a greenhouse experiment to compare their biocontrol efficacy with that of the wild type strain. In the greenhouse experiment, the wild type strain afforded 59.5 per cent bacterial blight suppression while the mutants afforded only less than 50 per cent (range: 17.1 to 23.8 per cent) bacterial blight suppression. Therefore, this observation with Phl$^-$ mutants on bacterial blight suppression strongly supports a positive role for DAPG production by *P. fluorescens* strain PTB 9.

***Key Words**: Bacterial blight, Biological control, 2,4-diacetylphloroglucinol (DAPG), Rice.*

Some plant-associated strains of fluorescent *Pseudomonas* spp. produce several metabolites, such as phenazine-1-carboxylicacid (PCA), 2,4-diacetylphloroglucinol (DAPG), pyoluteorin (Ptn), pyrrolnitrin (Pln), and oomycin A. Among these, 2,4-diacetylphloroglucinol is an antibiotic produced by fluorescent *Pseudomonas* spp. of diverse geographic origin (Dowling and O'Gara, 1994; Keel *et al.*, 1996; Thomashow and Weller, 1995; Raaijmakers *et al.*, 1999). Production of DAPG by fluorescent strains of *Pseudomonas* has been known since 1970 (Reddy and Borovkov, 1970) along with its broad toxic activity against bacteria, fungi and plants (Reddy *et al.*, 1969; Broadbent *et al.*, 1976). It has been implicated as the mechanism involved in the biological control of some of the most important crop diseases such as the root-rot of wheat caused by *Fusarium oxysporum* f. sp. *graminis* (Garagulya *et al.*, 1974), black root rot of tobacco caused by *Thielaviopsis basicola* (Defago *et al.*, 1990; Keel, *et al.*, 1992), damping-off of sugarbeet caused by *Pythium ultimum* and *Rhizoctonia solani* (Nowak-Thompson *et al.*, 1994) and the "take-all" of wheat caused by *Gaeumannomyces graminis tritici* (Defago *et al.*, 1990; Keel *et al.*, 1992; Gardener and Weller, 2001). Strains of *Pseudomonas fluorescens* that produce DAPG also have had a key role in the natural biological control of "take-all" known as "take-all decline" (Raaijmakers *et al.*, 1999; Raaijmakers *et al.*, 1997). Evidence for an important role of DAPG in plant protection comes from studies on DAPG-negative mutants of *Pseudomonas fluorescens* strain into which DAPG biosynthetic plasmids have been transferred. By this approach, it has been demonstrated that DAPG contributes to the control of black root rot of tobacco (Keel *et al.*, 1992), take-all of wheat (Keel *et al.*, 1992), *Pythium* damping-off of sugarbeet (Fenton *et al.*, 1992).

The antibiotic DAPG is produced by many strains of fluorescent *Pseudomonas* spp. with biocontrol activity against many soilborne bacterial and fungal plant pathogens. Genes required for DAPG synthesis by *P. fluorescens* Q2-87 are encoded by a 6.5-kb fragment of genomic DNA that can transfer production of DAPG to nonproducing recipient *Pseudomonas* strains. The nucleotide sequence was determined for the 6.5-kb fragment and its flanking region of genomic DNA in strain Q2-87 (Bangera and Thomashow, 1999). Six open reading frames were identified, four of which *(phl ABCD)* comprised an operon that includes a set of three genes *(phl ACB)* conserved between eubacteria and archaebacteria and a gene *(phlD)* encoding a polyketide synthase with homology to chalcone and stilbene synthases from plants. The biosynthetic operon is flanked on either side by *phl E* and *phl F*, which code respectively for putative efflux and regulatory proteins. Expression in *Escherichia coli* of *phl A, phl C, phl B,* and *phl D,* individually or in combination, identified a novel polyketide biosynthetic pathway in which *phlD* is responsible for the production of monoacetylphloroglucinol (MAPG). *PhlA, PhlC,* and *PhlB* are necessary to convert MAPG to DAPG, and they also may function in the synthesis of MAPG (Bangera and Thomashow, 1999; Bangera and Thomashow, 1996). Cloning of a 6.5-kb genomic DNA fragment from *Pseudomonas fluorescens* Q2-87 conferred production of DAPG upon DAPG non-producing recipient biocontrol strains and increased the biocontrol efficiency (Bangera and Thomashow, 1999).

DAPG is a bacterial and plant metabolite (Bangera and Thomashow, 1996; Bangera and Thomashow, 1999; Keel *et al.*, 1992), phenolic in nature, probably of polyketide origin with a broad-spectrum of antifungal, antibacterial, antiviral and antihelminthic properties (Garagulya *et al.*, 1974; Keel *et al.*, 1992; Nowak-Thompson *et al.*, 1994; Levy *et al.*, 1992; Reddy and Borovkov, 1970; Tada *et al.*, 1990). The production of DAPG by fluorescent *Pseudomonas* spp. is stimulated by glucose in many strains (Duffy and Defago, 1999). In addition, zinc sulfate and ammonium molybdate have been reported to favor DAPG production in some strains (Duffy and Defago, 1999). Recently, Duffy and Defago (1997) have identified fusaric acid, which is produced by the phytopathogen *F. oxysporum*, as an effector repressing DAPG production in a biocontrol strain of *P. fluorescens*. This fungal toxin can lead to repression of bacterial DAPG production in tomato roots, thereby neutralizing biocontrol of *Fusarium* crown and root rot.

In this chapter, we describe the identification of strains of DAPG-producing fluorescent pseudomonads in India, their antagonism to the rice bacterial blight pathogen, and their efficacy in the biological suppression of bacterial blight. This is the first time that 2,4-diacetylphloroglucinol production is reported from India and is implicated in the suppression of a major tropical pathogen, *Xanthomonas oryzae* pv. *oryzae*.

Materials and Methods

Rice Seed, Pathogens and Bacteria

Rice seeds of indica rice variety IR24 (obtained from S. McCouch, Cornell University, Ithaca, NY) were used. Bacterial blight pathogen (*Xanthomonas oryzae* pv. *oryzae*) (*Xoo*) was isolated from naturally infected rice leaf tissues. The leaf bits were then ground in a sterile mortar and pestle with double sterile water. A loopful of this suspension was streaked onto peptone sucrose agar (PSA) plate and incubated at 28°C for 2 to 4 days. The plates were observed for the appearance of characteristic colonies of *Xanthomonas oryzae* pv. *oryzae* (*Xoo*) and the identity of the pathogen was confirmed by disease development upon inoculation onto rice plants.

More than 637 plant-associated bacteria were obtained from the crop rhizosphere of various part of India. The rhizosphere samples were labeled and stored at 4°C till isolations were made from them. Bacteria were isolated from the soil suspensions of these rhizosphere samples after serially diluting them and plating onto King's B agar medium. The plates were incubated at 28°C for 2 days. Single colonies exhibiting bluish to green fluorescence were picked under UV light (= 365 nm) and further purified on the same medium. Such purified bacterial strains were stored in 50 per cent glycerol at 4°C for further use.

Dual Plate Assays

Dual plate assays were performed for the identification of bacterial antagonists of rice BB pathogen *Xoo* on peptone sucrose agar (PSA) medium (Fig. 15.1). These assays led to the identification of 278 bacterial antagonists out of 637 screened. The antagonist strains showed zones of inhibition whose diameter ranged from 0.5 to 4.7 cm. (Table 15.1).

Table 15.1: List of Plant-Associated *Pseudomonas* Fluorescent Strains for Biological Control of Rice Bacterial Blight, *Xanthomonas oryzae* pv. *oryzae* (Xoo) in cv. IR24

Place of Collection	*Code Number of Strain in Dual Plate*	*Inhibition Xoo Assay*[a]	*Biological Suppression of BB disease (%)*		*Production of DAPG*[c]
			Nethouse[b]	*Field*[b]	
Karnataka	KAD7	1.8	56.87	53.93	+
Karnataka	IMV14	2.8	58.68	56.74	+
Karnataka	IMV2	2.8	36.78	33.18	+
Karnataka	BGR19	0.5	27.48	42.71	+
Kerala	PTB9	2.1	58.78	64.46	+
Kerala	MON1	3.3	19.70	37.13	+
Tamil Nadu	TVM8	1.2	16.32	23.51	+
Tamil Nadu	VEL17	2.3	23.33	54.08	+
Tamil Nadu	VEL10	0.7	17.03	17.52	+
Tamil Nadu	GDY4	2.1	30.73	41.44	+
Tamil Nadu	GDY7	2.3	51.57	45.30	+
Tamil Nadu	TRP5	1.7	33.68	34.54	+
Tamil Nadu	TRP18	2.3	53.15	48.34	+
Tamil Nadu	MDR9	2.1	30.68	23.29	+
Tamil Nadu	MDR7	1.3	57.11	54.43	+
Tamil Nadu	STR7	1.8	19.51	36.04	+
Tamil Nadu	VGP13	2.5	55.39	51.62	+
Tamil Nadu	MDR16	1.4	37.60	46.12	+
Tamil Nadu	PDY5	1.8	58.83	56.88	+
Tamil Nadu	VLB7	1.7	33.64	9.08	+
Tamil Nadu	KVR5	2.5	55.92	49.75	+
Tamil Nadu	TNI13	2.2	26.05	28.19	+
Tamil Nadu	KOV8	3.0	38.69	9.71	+
Tamil Nadu	RJP31	2.4	10.31	0.50	+
Tamil Nadu	KOV3	1.8	21.18	49.21	+
Tamil Nadu	PDU1	2.9	8.16	11.80	+
Tamil Nadu	PDU9	2.7	16.94	2.04	+
	Control	–	–	–	–

[a] 278 rhizosphere bacterial strains (out of 637 strains tested) inhibited the growth of *Xanthomonas oryzae* pv. *oryzae*.

[b] Results of a replicated field experiment conducted at the Regional Agricultural Research Station, Pattambi, Kerala. Each figure is a mean of 20 replications.

[c] Production of DAPG was identified through a PCR-based screening procedure which amplified a 745 bp DNA fragment in 27 out of 278 tested strains.

PCR-Based Screening for 2,4-diacetylphloroglucinol Production by Plant-Associated Bacteria

To confirm the production of DAPG, intervening sequence specific primers *Phl*2a 20-mer (5′-GAGGACGTCGAAGACCACCA-3′) and *Phl*2b 20-mer (5′-ACCGCAGCATCGTGTATGAG-3′) which were developed originally from the *PhlD* sequence of *P. fluorescens* Q2-87 by Raaijmakers *et al.* (1997) were used.

By this method, a total of 27 strains out of 278 antagonistic strains that inhibited the growth of *Xoo* showed a characteristic 745-bp DNA fragment amplification. This was confirmed from the PCR products of DAPG-producing strain (reference strain) whereas no PCR products were amplified from DNA of the non-producing strains. The predicted 745-bp fragment of the same size was amplified from DNA of 27 DAPG producing Indian strains (Fig. 15.2).

Evaluation of DAPG Producers for BB Suppression-*In vivo*

In a nethouse and a field experiment, DAPG producer strains of *P. fluorescens* (listed in Table 15.1) were applied as a seed coat to rice seeds of bacterial blight-susceptible cv. IR24. The seed coat contained 10^8 cfu/ml in 0.1 per cent carboxymethycellulose (cmc) which served as an adhesive when the bacterial treatments were given overnight (12-14 h). Subsequently, the seeds were planted in seed beds and the rice seedlings raised from bacterial treatments and a nontreated control were transplanted when they were 21 days old in 1 m x 1 m sized field plots. At the same time plants were also transferred into big pots in the nethouse for the purpose of assessing disease suppression. Both these experiments were carried out at the Regional Agricultural Research Station, Pattambi, Kerala. At the time of transplanting, the seedlings were given a root-dip in the bacterial cell suspension [(containing 10^8 cfu/ml) (or in sterile distilled water in the case of untreated control)]. When the seedlings were 35 and 45 d old, they received two additional foliar spray applications of the respective bacteria at 10^8 cfu/ml of spray fluid. A day after the second foliar spray application or when seedlings were 46 d old, they were inoculated with a cell suspension (10^6 cfu/ml) of the rice bacterial blight pathogen, *X. oryzae* pv. *oryzae* by the clipping method. The inoculated plants (both bacterial treatments and control) were evaluated for the development of bacterial blight lesions by measuring the lesion length (in cm) 15 days after the inoculation. Reductions in bacterial blight lesion length, compared with lesions of those in untreated control plants are furnished in Table 15.1.

Analysis of Biocontrol Mechanism(s) Involved in Inhibition of X. *oryzae* pv. *oryzae* by Antagonistic Bacteria

Further confirmation for DAPG production by this sub-set of 27 strains was afforded by ethyl acetate extracts of their culture fluids with an authentic sample of DAPG using standard methods (Keel *et al.*, 1992). The purified antibiotic DAPG inhibited the growth of *X. oryzae* pv. *oryzae* on peptone-sucrose agar (PSA) plates in laboratory assays. The extracted material was compared to that of an authentic sample in thin layer chromatography and high performance liquid chromatography. Its spectral properties (^{1}H NMR, IR spectra) and melting point were also compared with reference sample of DAPG.

Generation of Phl⁻ Mutant in a DAPG Producing *P. fluorescens* strain PTB 9

To make a critical assessment of the role of DAPG in bacterial blight suppression, Phl^- mutants were generated by transposon mutagenesis. Transposons (in donor strain) were mobilized into DAPG producing *P. fluorescens* PTB 9 by the method of Voisard *et al.* (1988) and Bangera and Thomashow (1999) described below:

The donor [S17- (1pir)/pUT•Km: : phl TR#1] was plated onto LB plate containing ampicillin (100µg/mL) and kanamycin (50µg/mL) and the recipient strain *P. fluorescens* PTB 9 was plated onto LB. Single colonies were picked and grown in liquid culture for 12 h with km (100 µg/mL) at 27°C. The donor *E. coli* was diluted 1: 100 in LB broth without antibiotics and the recipient was diluted1: 5 in LB broth. The recipient strain *P. fluorescens* PTB 9 was grown for 3 hours at 27°C in plain LB broth. Sterile 0.45 µm filter paper squares were placed onto LB plates and pre-warmed to 27°C. One mL of recipient and 1.5 mL of donor strains were transferred to an eppendorf tube and spinned down at 5000 rpm for 3 min (so pili are not disrupted). Cells were resuspended in 150 µL of LB broth by gently pipetting up and down and 25 µL of donor cells were spotted onto the filter, followed by 25 µL of the recipient. For controls, 25 µL of each strain used was spotted onto filters in separate plates. Plates were not turned over as that will cause the suspension to run off the filter squares. After incubation, overnight bacteria from the filter squares were resuspended in 1 mL of LB broth by vortexing and pipetting, if necessary. The cells were transferred to eppendorf tubes and spinned down. Most of the supernatant was aspirated leaving about 100 µL to resuspend the cells. The cells were resuspended in the ~100 µL remaining media and were plated on double selective media: one antibiotic to kill the donor strain and the second to select for transformants of the recipient strain. If the recipient lacks a resistance marker, it may be possible to use minimal medium because the *E. coli* donor is an auxotroph. The transconjugants were screened *in vitro* by dual plate assay against rice bacterial blight pathogen, *X.oryzae* pv. *oryzae* (to identify mutants with altered antibiotic activity). Strains that lacked the ability to inhibit *X. oryzae* pv. *oryzae* were selected and checked with PCR to confirm the transposon insertion.

Evaluation of *Pseudomonas fluorescens* Strain PTB 9 and its Phl$^-$ Mutants in a Greenhouse Experiment for the Biological Suppression of Bacterial Blight of cv. IR24 Rice

A greenhouse experiment was carried out in CAS in Botany, University of Madras, Chennai, to evaluate the biocontrol efficiency of wild type *Pseudomonas fluorescens* strain PTB 9 and its mutants (PTB 9a, PTB 9b, PTB 9c, PTB 9d and PTB 9e) with altered antibiotic activity, for suppression of bacterial blight. Rice seeds of a bacterial blight susceptible cv. IR24 were used in this study. Seeds were bacterized with the wild type and its mutant strains and then evaluated in the greenhouse (evaluation procedure as previously described). Seeds soaked in plain 0.1 per cent sterile cmc solution served as check. These plants were then clip-inoculated with a bacterial suspension of *Xoo* culture (OD 0.1 at 600 nm) on 45th day after sowing. Development of bacterial blight lesions was scored 14 days after the inoculation. The lesion lengths and leaf lengths were measured and the disease severity was determined in terms of reduction in lesion length.

Results and Discussion

Six *P. fluorescens* were short-listed from more than 637 plant-associated bacteria (Velusamy, 2003). These strains were afforded more than 50 per cent of bacterial blight (BB) suppression in cv.IR24 rice under nethouse and field experiments. Among these, maximum BB suppression (64.46 per cent) was afforded by strain PTB9 in field plots (listed in Table 1). There has been no previous report on the production of DAPG by plant-associated bacteria from India and therefore, this study assumes special importance as a major study on DAPG-production by Indian strains of *Pseudomonas* which also act as biocontrol agents for rice bacterial blight (Velusamy and Gnanamanickam, 2003a, b; Velusamy *et al.*, 2003).

Analysis of Mechanisms Mediating Suppression of *Xanthomonas oryzae* pv. *oryzae* by Bacterial Strains

The crude extract from *P. fluorescens* PTB 9 was assayed for the antibiosis to *X. oryzae* on PSA medium by the dual plate method. A 100 µL of crude extract (100 µg/mL) showed inhibition of pathogen, compared to no inhibition in control (100 µL methanol) Fig. 15.3.

Production of 2,4-diacetylphloroglucinol (DAPG) Antibiotic

Antibiotic 2,4-diacetylphloroglucinol was extracted from culture filtrate of the most efficient strain *Pseudomonas fluorescens* PTB 9 (confirmed by PCR as a DAPG-producer) (Fig. 15.2) by using a standard method. A sample of 40 µg of the antibiotic was obtained per mL of 72 h grown bacterial culture filtrate. The purified culture filtrate was pooled for biological activity.

Analytical Detection of 2,4-diacetylphloroglucinol

The purified DAPG extracted from *P. fluorescens* PTB 9 (showed antibacterial activity against *Xoo*) had matching R_f and retention time (RT) values as that of the authentic DAPG (result not shown). Its spectral (^{1}H NMR, IR) profiles plus its melting point of 165-167°C confirmed the production of DAPG (Fig. 15.4).

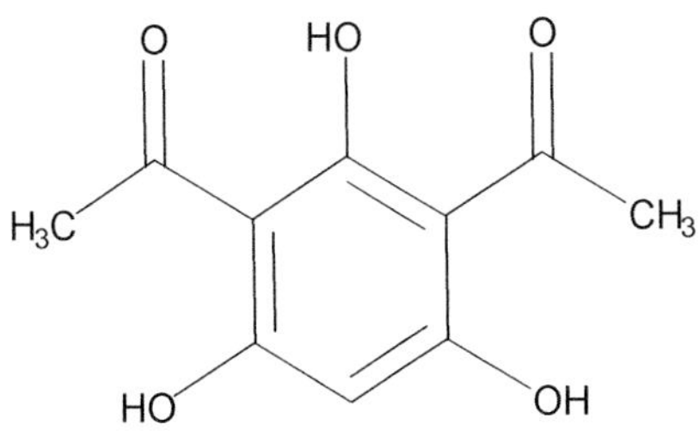

Fig. 15.4: Structure of 2,4-diacetylphloroglucinol

Genetic Analysis of Traits Involved in Biocontrol of Rice BB

Transpositional Mutagenesis

The *Tn5*-Km transposon (received as a gift from Dr. Linda Thomashow USDA-ARS, U.S.A) was delivered from *E. coli* S17-1/ pir (pUT•Km: : phl TR#1) to *Pseudomonas fluorescens* strain PTB 9. The transpositional mutants arose at a frequency of approximately 3×10^{-6} per recipient cell plated. A total of 1,125 Km^r transconjugants were analysed to identify mutants altered with DAPG activity. Fifty of the 1,125 transconjugants were screened in dual plate assays against the rice BB pathogen, *X. oryzae* pv. *oyzae* and were found altered in their antibacterial activity (aba). While 37 strains out of 50 still inhibited the pathogen, others did not inhibit or only partially inhibited the pathogen when compared with wild strain PTB 9. Fig. 15.5 shows some of the transconjugants showing inhibition of *X. oryzae* pv. *oryzae* while several others showing lack of inhibition.

PCR-Based Analysis of *Tn5*-Km Mutants of *Pseudomonas fluorescens* Strain PTB 9 Defective in DAPG Production

The site of *Tn5* insertion was determined in S17-1/ pir (pUT•Km: : phl TR#1) mutants of *Pseudomonas fluorescens* PTB 9 defective in DAPG production by PCR analysis.

Heat-lyzed DNA from the wild type strain and its Phl⁻ mutants were extracted by a standard method (Raaijmakers *et al.*, 1997). Then the heat-lyzed bacterial DNA was used to confirm transposon insertion by PCR analysis. The DNA of the wild type amplified a characteristic 745-bp fragment and other Phl⁻ mutants did not show such amplification (Fig. 15.6).

Performance of *P. fluorescens* PTB 9 and its DAPG-Deficient (Phl⁻) Mutants in BB Disease Suppression Under Greenhouse Condition

Wild type strain PTB 9 and five of its Phl⁻ mutants were evaluated in a greenhouse experiment for their ability to control the bacterial blight disease. BB lesions of more than 15-25 cm length appeared on untreated IR24 rice plants at 15 days after inoculation. Plants grown from seeds treated with Phl⁻ mutants of *Pseudomonas fluorescens* strain PTB 9 had longer BB lesions of 15-25 cm length while the plants grown from seeds treated with the wild type parent strain *P. fluorescens* PTB 9 had short lesions of 3-5 cm (Table 15.2). The wild type strain *P. fluorescens* PTB 9 afforded 59.52 per cent reduction of bacterial blight, while the Phl⁻ mutants (PTB 9a, PTB 9b, PTB 9c, PTB 9d and PTB 9e) gave 19.65, 17.11, 23.77, 20.81 and 18.11 per cent reductions of bacterial blight, respectively. Therefore, the protection from BB incidence afforded by the Phl⁻ mutants was much less than the level of protection shown by *P. fluorescens* PTB 9.

Table 15.2: Evaluation of *Pseudomonas fluorescens* Strain PTB 9 and its Phl⁻ mutants in a greenhouse experiment for suppression of bacterial blight (BB) in cv. IR24 rice

Name of Bacterial Strain		Mean BB Lesion Length (cm)[ac]	Difference in Lesion Length from Control (cm)	Percent Disease Suppression	LSD value 5%	LSD value 1%
Wild type (*P. fluorescens* PTB 9)		7.66	11.27[b]	59.52	5.8	6.6
Phl⁻ mutants	PTB 9a	15.21	3.72	19.65	5.8	6.6
	PTB 9b	15.69	3.24	17.11	5.8	6.6
	PTB 9c	14.43	4.50	23.77	5.8	6.6
	PTB 9d	14.99	3.94	20.81	5.8	6.6
	PTB 9e	15.50	3.43	18.11	5.8	6.6
Check		18.93	0.00	0.00	0.0	0.0

[a] Mean of three replications.

[b] Reduction in lesion length significant at 1 per cent by statistical analysis by LSD

[c] Mean of normalized lesion lengths (mean lesion length in control plants–mean lesion length in treated plants).

REFERENCES

Bangera, G.M. and Thomashow, L.S. (1996). Characterization of a genomic locus required for synthesis of the antibiotic 2,4-diacetylphloroglucinol by the biological control agent *Pseudomonas fluorescens* Q2-87. *Mol. Plant-Microbe Interact.*, **9**: 3155-3163.

Bangera, G.M. and Thomashow, L.S. (1999). Identification and characterization of a gene cluster for synthesis of the polyketide antibiotic 2,4-diacetylphloroglucinol from *Pseudomonas fluorescens* Q2-87. *J. Bacteriol.*, **191**: 3155-3163.

Broadbent, D., Mabelis, R.P. and Spenser. H. (1976). C-Acetylphloroglucinols from *Pseudomonas fluorescens*. *Phytochem*. 15: 1785 p.

Defago, G., Berling, C.H., Burger, U., Haas, D., Kahr, G., Keel, C., Voisard, C., Wirthner, P. and Wuthrich, B. (1990). Suppression of black root rot of tobacco and other diseases by strains of *Pseudomonas fluorescens*: potential applications and other mechanisms. pp. 93-108 In: *Biological control of soilborne plant pathogen*, (D. Hornby, R.J. Cook, Y. Henis, W.H. Ko, A.D. Rovira, B. Schippers and P.R. Scott eds.) CAB International, Oxon, U.K.

Dowling, D.N. and O'Gara, F. (1994). Metabolites of *Pseudomonas* involved in the biocontrol of plant disease. *Trends Biotechnol.*, **12**: 133-141.

Duffy, B.K. and Defago, G. (1997). Zinc improves biocontrol of *Fusarium* crown root rot of tomato by *Pseudomonas fluorescens* and represses the production of pathogen metabolites inhibitory to bacterial antibiotic biosynthesis. *Phytopathology*, **87**: 1250-1257.

Duffy, B.K. and Defago, G. (1999). Environmental factors modulating antibiotic and siderophore biosynthesis by *Pseudomonas fluorescens* biocontrol strains. *Appl. Environ. Microbiol.*, **65**: 2429-2438.

Fenton, A.M., Stephens, P.M., Crowley, J., O'Callaghan, M. and O'Gara, F. (1992). Exploitation of gene(s) involved in 2,4-diacetylphloroglucinol biosynthesis to confer a new biocontrol capability to a *Pseudomonas* strain. *Appl. Environ. Microbiol.*, **58**: 3873-3878.

Garagulya, A.D., Kiprianova, E.A. and Boiko. I.O. (1974). Antibiotic effect of bacteria from the genus *Pseudomonas* on phytopathogenic fungi. *Mikrobiol. Zh.* (Kiev), **36**: 197-202.

Keel, C., Schinder, U., Maurhofer, M., Voisard, C., Laviller, J., Burger, U., Wirthner, P., Haas, D. and Defago, G. (1992). Suppression of root diseases by *Pseudomonas fluorescens* CHAO: Importance of the bacterial secondary metabolite 2,4-diacetylphloroglucinol. *Mol. Plant-Microbe Interact.*, **5**: 4-13.

Keel, C., Weller, D.M., Natsch, A., Defago, G., Cook, R.J. and Thomashow, L.S. (1996). Conservation of the 2,4-diacetylphloroglucinol biosynthesis locus among fluorescent *Pseudomonas* strains from diverse geographic locations. *Appl. Environ. Microbiol.*, **62**: 552-563.

Levy, E., Gough, F.J., Berlin, K.D., Guiana, P.W. and Smith, T.J. (1992). Inhibition of *Septoria tritici* and other phytopathogenic fungi and bacteria by *Pseudomonas fluorescens* and its antibiotics. *Plant Pathol.*, **41**: 335-341.

Nowak-Thompson, B., Gould, S.J., Kraus, J. and Loper, J.E. (1994). Production of 2,4-diacetylphloroglucinol by the biocontrol agent *Pseudomonas fluorescens* Pf-5. *Can. J. Microbiol.*, **40**: 1064-1066.

Raaijmakers, J.M., Bonsall, R.F. and Weller, D.M. (1999). Effect of population density of *Pseudomonas fluorescens* on production of 2,4-diacetylphloroglucinol in the rhizosphere of wheat. *Phytopathology*, **89**: 470-475.

Raaijmakers, J.M., Weller, D.M. and Thomashow, L.S. (1997). Frequency of antibiotics-producing *Pseudomonas spp.* in natural environments. *Appl. Environ. Microbiol.*, **63**: 881-887.

Reddy T.K.K. and Borovkov, A.V. (1970). Antibiotic properties of 2,4-diacetylphloroglucinol produced by strain *Pseudomonas fluorescens* strain 26-0. *Antibiotiki* (Moscow), **15**: 19-21.

Reddy, T.K.K., Khudiakov, Y.P. and Borovkov, A.V. (1969). *Pseudomonas fluorescens* strain 26-0, producing phytotoxic substances. *Mikrobiologiya*, **38**: 909-913.

Tada, M., Takakuwa, T., Nagai, M. and Yoshii, T. (1990). Antiviral and antimicrobial activity of 2,4-diacetylphloroglucinol, 2-acylcyclohexane-1, 3-diones and 2-carboxamidedocyclohexane-1, 3-diones. *Agric. Biol. Chem.*, **54**: 3061-3063.

Thomashow, L.S. and Weller, D.M. (1995). Current concepts in the use of introduced bacteria for biological disease control: mechanisms and antifungal metabolites, pp. 187-235. In: *Plant-Microbe Interactions, Vol. 1.* (G. Stacey and N. Keen eds.), Chapman & Hall, New York, N.Y.

Velusamy, P. (2003). Biological control of bacterial blight of rice by plant-associated bacteria: molecular and genetic analyses of metabolites for their role in disease control. Ph.D. Thesis, University of Chennai, India.

Velusamy, P. and Gnanamanickam, S. S. (2003a). Plant-associated bacteria, 2, 4 diacetylphloroglucinol (DAPG) production and suppression of rice bacterial blight in India. *Curr. Sci.*, **85**: 1270-1273.

Velusamy, P. and Gnanamanickam, S. S. (2003b). *Plant-associated bacteria, 2,4-diacetylphloroglucinol (DAPG) production and suppression of rice bacterial blight in India.* National conference on 'Biological Control of Crop Diseases: Present Perspectives and Future Challenges' at Centre for Advanced Studies in Botany, University of Madras, Guindy Campus, Chennai, India. March 5-6, 2003.

Velusamy, P., Kavitha, S., Preeti Vasudevan and Gnanamanickam, S.S. (2003). Role of plant-associated bacteria and their antimicrobial products in the suppression of major rice diseases in India. In: *Proc. 6th Intl workshop on PGPR* at Calicut, India October 5-10, pp. 256-259.

Voisard, C., Rella, M. and Haas, D. (1988). Conjugative transfer of plasmid RP1 to soil isolates of *Pseudomonas fluorescens* is facilitated by certain large RP1 deletion. *FEMS Microbiol. Lett.*, **55**: 9-14.

16

Heart Rot and Root Rot Diseases of Trees: A Case Study from the Campus of Pune University

J.G. VAIDYA* & G.G. DESHPANDE

Department of Botany, University of Pune, Pune – 411 007, India

In fungi, rot is main disease occurring on live standing tree. In this process softening, discoloration, and often disintegration of succulent plant tissue occur as result of fungal or bacterial infection. Fungal rot has beneficial as well as damaging influences on forest stand. They are essential in the earth's carbon cycle by returning billions of tons of carbon each year to atmosphere. Wood destroying fungi also contribute to soil fertility and natural soil forming process. They add organic matter with highly desirable ion exchange properties of soil and contribute to cycling of essential mineral elements. These rots are detected by microscopic or by cultural diagnosis. These rots break down the cell wall components like lignin, hemicelluloses and cellulose (which assign the structure to the tree) into simple sugar. This result in the disintegration and ultimate dissolution of wood substance. When fungi degrading all components of wood like cellulose, hemicellulose, lignin, discoloration occurs, giving white colour, so it is called as white rot. In other type only cellulose is degraded leaving behind lignin, thus giving brown colour to the decay wood, hence called as brown rot, More than 10,000 species of fungi cause different types of rot. Rot diseases are caused by a wide variety of wood colonizing

* jitendra@unipune.ernet.in

fungi causing decay of trunk, large branches and roots. Wood rotting fungi releases some specific enzyme, which digest wood component like cellulose, hemicellulose, and lignin wood rot fungi will eventually infect and deteriorate wood beneath the bark.

There are various essential conditions for growth of wood rotting fungi.

Source of Infection

If infected wood is in contact with sound wood, infection gets transmitted from decayed wood to sound wood. Prior to infection, the fungus is present in soil with organic matter in the form of hyphal fragments, cords or rhizomorph. The infection may also occur through air borne spores.

Substrate

Cellulose, lignin and other cell components are suitable substrate for fungal growth. Soluble sugar, lipid, peptides and other primary metabolism play important role in establishment of mycelia from spore

Moisture

The moisture content of the wood controls the development of fungi on wood. All decaying fungi require moderate amount of water for its growth. If there is no sufficient moisture, fungi become dormant. Moisture of about 30 to 50 per cent is required by wood rotting fungi. Fungal spores do not germinate, if moisture is below than 25 to 30 per cent.

Oxygen

All decay fungi require air for its growth, many species die quickly in absence of oxygen. Fungi need oxygen for oxidation pf sugars, which is used for growth and as source of energy. The break down of carbohydrates in respiration process produces water and carbon dioxide. If there is no interchange of air, fungus will die from suffocation by carbon dioxide.

Temperature

The growth of wood rotting fungi is affected by temperature. The optimum temperature condition for growth of wood decay fungi ranges from 65°C to 95°C. All fungi show little or no growth at freezing temperature or slightly more. Growth is slowed down, if the temperature increases above 95°.

In addition, light usually has retarding influence and exposure to intense light as bright sunlight can kill mycelium of some species. The acidity and alkalinity of wood can also have an effect on its growth. Rot fungi grow on material that is slightly acidic, and very few can tolerate alkaline conditions. Treatment of wood with nitrogenous material stimulates growth of rot fungi.

Rots are further classified in to heart rots and root/butt rots.

Heart Rot

Decay in central core stem of live standing tree is called heart decay or heart rot. Wood that decay by heart rot fungi is called cull. Generally heart rot decay occurs between 15- 30 years age of tree. There are many fungi causing heart rot. These fungi generally belong to Hymenomyceteceae majority under the genera of *Phellinus.* Heart rot decay occurs in living tree as well as dead wooden logs. At infection court moisture, temperature, nutrients, microflora and natural extractive determines the establishment of decay fungi. High moisture prevents heart rot fungi to progress in sapwood. When wood of live standing tree is freshly exposed following injury, then the biological substances, especially phenolic substances protect the invading action of microorganisms. Discoloration and decay are compartmentalized within tissue, present at time of wounding (Shigo 1972). (This process is referred as CODIT, *i.e.* Compartmentalization of Decay in Trees.) Toxic natural extractive, which impart resistance, to decay during development of heartwood are deposited in cell wall. The decay resistance of outer heartwood becomes progressively less from base to upwards of tree.

Spores are formed on basidia and disseminated from pore surface of fruiting body. Approximately 10,000 spores are released over night by members of Hymenochaetaceae, generally with 4 to 5 pores/ square mm. It may be due to change in electrical charge of spore, which may cause repulsion from basidium (like charges repel) and expulsion of the spore from pore. It is possible that fungal spores are attracted in an electrostatic way to wound area as drift by host. All the growth requirements (moisture, temperature etc.) are necessary for the fungal spore to germinate and infect again the host tissue. The newly formed hyphae release enzyme and non- enzymatic deploymerizing agent for digestion of cell wall material. The fungus becomes dormant until a future time when tree is stressed or wounded. Hyphae move from one cell to another cell through pits, and then enter the cell wall, bore holes are formed by enzymatic digestion of secondary wall as the decay progress.

Root and Butt Rots

Root disease affect growth of forest trees. Root and Butt rots are wood decay disease and also overlap with stem decay. They enter into lower part of tree through a wound. It penetrates the lower part of tree or roots directly. It kills the cambial tissues while growing in inner wood. The fungus kills cambium at the root collar and the tree dies. When collar is girdled, the fungus decays the inner wood at base of stem so called as butt rot. Root rot fungi can be distinguished between two groups *i.e.* fungi causing typical root rot and do not attack heart wood as in *Ganoderma lucidum* other fungi that may cause rot and butt rot as in *Fomes annonus* (= *Heterobasidion annosum*).

Root and Butt rots are slowly affecting diseases. Fungus colonizes and continues to use colonized roots as a food base. From this, energy rich food base can be transferred to new host trees by root contact or by means of specialized structure like rhizomorph or by spore released from fruit bodies.

In susceptible host, pathogen may spread quickly into root and cause disease. In forest, healthy root is infected when it comes in contact with decayed root. Some fungi are able to colonize freshly cut stumps through basidiospores (Petch, 1921). This ability for colonization decreases with time as shown in *H. annonsum* (Rishbeth 1951). Rhizomorph can use energy from stump or killed tree to grow through soil along roots, infect nearby roots. This rhizomorphic growth can be upto several meters through soil. Mycelial mat can be seen under bark of trees, which are colonized.

Age of host influences the disease; older trees can tolerate infections while young get infected in short time. Healthy roots react strongly against spread of infection.

A Case Study

Ganoderma and *Phellinus* are root rot and heart rot diseases respectively which have studied from the campus of Pune university. Pune University Campus flora on whole is typically xerophytes or as a sub-tropical deciduous plants. Campus shows exotic woodland type vegetation due to of *Dalbergia melanoxylon* Guill and Perr. In past decade's the university has developed big plots of *Gliricidida, Leucaena, Caesalpinia* plantations. These diseases were found in older plantation like *Gliricidia, Dalbergia,* Subabool and other garden plant. Trees on which these diseases were found at least 60 years old. *Phellinus* occur throughout year but *Ganoderma* occur only during monsoon season.

External symptoms occur on the trees due to infection of *Ganoderma* and *Phellinus* diseases.

1. Formation of cavities. (Plate 16.1, Fig. b)
2. Detachments of bark and break down branches. (Plate 16.1, Fig. c)
3. Exposure of central bole. Fungus grows on exposed tissue. (Plate 16.1, Fig. b).
4. Tree bents to one side but still continue to live. (Plate 16.1, Fig. a).
5. Change in canopy.
6. Bark is cracked and peeled off.
7. Some branches are dead.
8. Fruit body formation along the ridges. (Plate 16.1, Fig. c).
9. Chlorosis of leaves.
10. Branches are shortened.

In this case study 15 species of *Ganoderma* and 17 species of *Phellinus* were found on live standing tree as well as on dead stump. (Tables 16.1 and 16.2.). In which *Phellinus poonesis* and *Ganoderma poonensis* are proposed as new species. *Ganoderma lucidum* var. microspore is also newly proposed variety.

Table 16.1: List of Species on Live Standing Tree

Fungal Species	*Host Plant*
Ganoderma curtisii	*Phoenix sylvertris*
Ganoderma lipsiense	*Casurina equisitifolia*
Ganoderma lucidum var. lucidum	*Citrus aurantium*
Ganoderma orbiformum	*Leucaena leucophala*
Ganoderma perzonatum	*Tamarindus indica*
Ganoderma praelongum	*Leucaena leucophala*
Ganoderma stipitatum	*Tamarindus indica*
Ganoderma testaceum	*Peltophorum petrocarpus*
Phellinus aureobrunneus	*Albizzia lebbeck*
Phellinus badius	*Acacia nilotica*
Phellinus coffeatoporus	*Acacia nilotica*
Phellinus crocatus	*Albizzia lebbeck*
Phellinus griseoporus	*Dalbergia melanoxylon*
Phellinus lloydii	*Atrocarpus integrifolia*
Phellinus merrillii	*Caesalpinia coriaria*
Phellinus minutiporus	*Albizzia lebbeck*
Phellinus rickii	*Swietenia mahogany*
Phellinus sublinteus	*Peltophorum petrocarpus*
Phellinus lamaensis	*Olea dioeca*
Phellinus pectinatus	*Dalbergia melanoxylon*
Phellinus orientalis	*Thespecia* sp.
Phellinus poonesis	*Brousssonetia papyrifera*

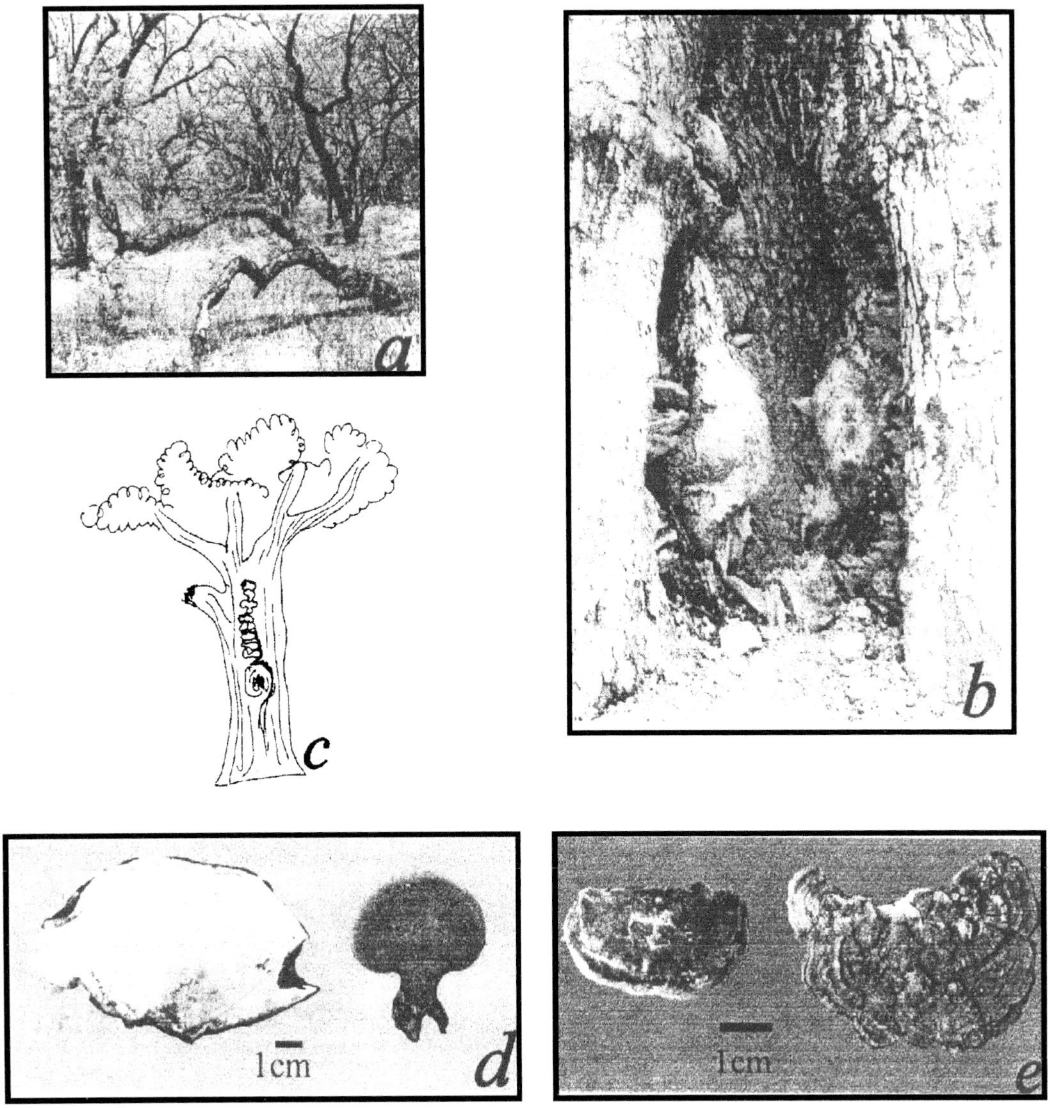

Plate 16.1: Fig. (a) Infected trees of *Acacia nilotica,* Fig. (b) Infected tree of *Peltophorum petrocarpus,* Fig. (c) Sketch of infected tree of *Swietenia mahogany*, Fig. (d) *Ganoderma orbiformum*: lower surface and upper surface, Fig. (e) *Phellinus linteus*: lower surface and upper surface

Table 16.2: List of Dead Stump

Fungal Species	Host Plant
Ganoderma applanatum	*Phoenix sylvestris*
Ganoderma lucidum var. capense	*Pogmia glabra*
Ganoderma philippi	*Albizzia lebbeck*
Ganoderma sessiliforme	*Pogamia glabra*
Ganoderma lucidum var. microspore	*Leucena leucephala*
Ganoderma poonensis	*Pithecolobium saman*
Phellinus adamantinus	*Eucalyptus globus*

Phellinus species were observed on four positions on host plant. (Table 16.3). Two species of *Phellinus; i.e. P. aurobrunneus, P. admantimus* were found on single plant of *Albizzia lebbek. P. merrillii* and *P. sublintus* are abundant in Pune university campus. *Phellinus rickii* was found on ridges of *Swietenia mahogany* (Plate 16.1, Fig.c). *Phellinus sublinteus* grow in cavity of *Peltophorum petrocarpus.* (Plate 16.1, Fig. b). Due to *Phellinus* infection various symptoms occur on studied host plant. (Table 16.5).

Table 16.3: Position of Fruiting Body of *Phellinus* on Host Tree

Host	On Center of Main Stem	Above Root	Side of Main Bole	Side on Branches
Acacia nilotica	*P. merrillii*	*P. badius*	*P. pappinus*	—
Albizzia lebbek	*P. aureobrunneus* *P. minutiporus*	*P. crocatus*	—	—
Artocarpus integrifolia	—	—	*P. merrillii* *P. liovdii*	—
Azadirachata indica	*P. sublinteus*	*P. llovdii*	—	*P. merrillii*
Broussonetia papyrifera	*P. poonesis*	*P. poonesis*	—	—
Caesalpinia coriaria	—	—	*P. merrillii*	—
Dalbergia melanoxylon	*P. lamensis*	—	—	*P. pectinatus*
Eucalyptus globus	*P. adamantinus*	—	—	—
Gliricidia sepium	*P. merrillii*	—	*P. merrillii*	*P. merrillii*
Mangifera indica	*P. linteus*	*P. linteus*	—	—
Peltophorum ferrugineum	*P. sublinteus*	—	*P. sublinteus*	—
Swietenia mahogany	—	*P. rickii*	—	—
Thespesia species	—	*P. orientalis*	—	—
Tamarindus indica	—	*P. adamantinus*	—	—

Fruiting bodies of *Ganoderma* occur on three positions on host plant. (Table 16.4.) Fruiting body in monsoon also varies from species to species like *Ganoderma stipitatum, Ganoderma multiplicatum* shows basidiocarp immediately after first rain. This may be due to reflex of fungus to the stress that are impose on continuously growing perennial mycelia in butt or root region. While majority of species occur during heavy rain probably for such species humidity and light are important for fruiting. Mycelial growth in decaying part gets activated during first rain and subsequent growth may occur during rainy season. Except species like *Ganoderma lucidum* var. microsporus, *Ganoderma lucidum* var. lucidum fruit very late in rainy season. It seems, this species require high moisture and pre decomposition of the substrate by other microflora. The basidiocarp of species of *Ganoderma* never occurs throughout the year.

Table 16.4: Position of Fruit Body in *Ganoderma* Species

Position of Fruiting	*Name of Ganoderma Species*
Butt rot	*G. applanatum, G. curtisii, G. lipsiense, G. lucidum* var. capense, *G. multiplicatum, G. orbiformum, G. perzonatum, G. testaceum, G. sessili forme*
Root rot	*G. lucidum* var. lucidum, *G. praelongum, G. stipitatum, G. lucidum* var. microsporus, *G. philippi*
Butt root rot	*G. poonesis*

Table 16.5: Symptoms Occurring Due to *Phellinus* Infection on Studied Tree Species

Host Name	*Bark Cracking*	*Cavity Formation*	*Branch Fallen*	*Tree Bending*	*Dead Tree*	*Canopy Open*
Acacia nilotica	+	+	–	+	–	–
Albizzia lebbeck	+	–	+	–	–	–
Artocarpus integrifolia	–	–	–	–	–	+
Broussonetia papyrifera	–	–	+	–	–	–
Dalbergia melanoxylon	+	–	–	+	–	–
Eucalyptus globus	–	–	–	–	+	–
Ficus bengalensis	–	+	+	–	–	+
Gliricidia sepium	+	+	+	–	–	–
Mangifera indica	–	–	–	–	+	–
Peltophorum ferrugineum	+	+	+	–	–	+
Swietenia mahogany	–	+	+	–	–	+
Tamarindus indica	–	+	+	–	–	+

+ : Present

– : Absent

In subabool plantation, *Ganoderma* damaged some of the trees and it is spread from tree to tree by root. In *Gliricidia* plantation due to *Phellinus* infection many tree are bending or damaged. These diseases are slowly destroying all plantations. After some year there will be no existence of any plantation in Pune University campus. So there is need of protection against these diseases. Many authors have suggested control measure.

REFERENCES

Annonymous (1980). *Timber decay and its control.* Technical Note No. 53. Prences Risborough Laboratory

Bakshi, B.K. (1976). *Forest Pathology: Principle and Practices in Forestry*, Published by the Controller of Publication, Delhi. pp. 55-72.

Boddy, L. and Rayner A.D.M. (1983). Origins of decay in living deciduous trees: the role of moisture content and are appraisal of the expanded concept of decay. School of Biological science, university of bath, Claverton Down. pp. 625-641.

Cartwright, K. St. G. and Findlay, W. P. K. (1958). *Decay of timber and it's prevention*. H.M. Stationery office, London, 294 pp.

Gottlieb, A.M. and Wright, J.E. (1999). Taxonomy of *Ganoderma* from southern South America: subgenus *Elfvingia. Mycol. Res.,* **103(10)**: 1289-1298.

Gottlieb, A.M. and Wright, J.E. (1999). Taxonomy of *Ganoderma* from southern South America: subgenus *Ganoderma. Mycol. Res.,* **103(6)**: 661-673.

Gottlieb, A.M. Saidman, B.O. and Wright, J.E. (1998). Isoenzymes of *Ganoderma* species from southern South America. *Mycol Res.,* **103**: 415-426.

Larsen, M.J. and Cobb-poulle (1990). *Phellinus* (Hymenochaetaceae). A survey of the world taxa. *Synopsis Fungorum 3, Fungiflora-Olso–Norway.*

Monoclavo and Ryvarden (1997). A nomenclatural study of the Ganodermataceae. Donk Synopsis Fungorum. *Fungi Flor:Olso Norway,* 1-114.

Ryvarden, L. (2000). Studies in neotropical polypores: A preliminary key to. neotropical species of *Ganoderma* with a laccate pileus. *Mycologia,* **92(1)**: 180-191.

Shigo, A.L. (1976). Compartmentalization of discolored and decayed wood in trees. *Beihefte zu Material und organismen,* **3**: 221-226.

Shortle, W.C. (1979). Mechanism of compartmentalization of decay in living trees. *Phytopathology,* **69**: 1147-1151.

Tainter, F.H. and Baker F.A. (1996). *Principle of Forest Pathology*. John Wiley and Sons, Inc. 743, 760-767.

Vaidya, J.G. (1995). *Biology of the Fungi*. First Edition, Satyajeet Prakashan, Pune 1-2, 340-344.

Vaidya, J.G. and Rabba, A.S. (1995). The genus *Phellinus* from Bhimashankar forest. *Biologia Indica,* **5**: 47-56.

Varadpande, D.G. (1973). The flora of Ganeshkhind, Pune. Pune University. *J. Sci and Tech.,* **44**: 97-132.

Wagener, W.W. and Davidson, R.W. (1954). Heart rot in living trees. *Bot Rev.,* **20**: 61-134.

http://www.nrc.ca/irc/cbd/cbd111e.html: "Decay of tree."

http://www.ag.uinc.edu/~vista/pdf_pubs/642.pdf: "Wood rots and decays"

http://www.caf.wvuedu/kearneysville/disease_descriptions/woodrot.html: "Wood rots"

http://www.minesotascience.com/rot.htm: "Fungi and wood rots"

http://www.fs.fed.us/r10/spf/fhp/sefhp/sertrot/html: "Heart rot live trees"

http://www.fs.fed.us/r2/fhm/bugcrud/decay.html: "Introduction to decay"

http://ohioline.osu.edu/hyg-fact/3000/3300.html: "Wood rot"

http://www.for.gov.bc.ca/revenue/manuals/scaling/ch4.html: "Common defects"

http://www.fore.canterburg.ac.n2/Hobson%20439/putb3.pdf: "Wood decay"

http://www.forestpathology.org/concepts.html: "Diseases concepts"

Hartig^{+} (1824), Etheridge^{+} (1970), Shigo^{+} (1967, 1984):

http://www.syllabus.syr.edu/ESF/pdmanion/forpath/wood.html: "Heart rot disease cycle, reducing Impacts of decay liabilities"

17

VA Mycorrhiza A New Biotechnological Tool as Biocontrol Agent: Indian Scenario

SUDHIR CHANDRA* & HARBANS KAUR KEHRI

Department of Botany, University of Allahabad, Allahabad – 211 002 (U.P.), India

Because of the advantage of their universal presence, minute size and astonishingly rich physiological qualities, microbes play the key role in maintaining the biological equilibrium on earth and in their absence, life on the planet will be extinct. Different forms of interrelationship between plants and microbes have developed over million of years. On the extreme end, destruction of one by other takes place. On the other hand, in the struggle for existence between the two efficient opponents various forms of symbiotic relationships have emerged. The possible beneficial role of a symbiotic association between plant roots and fungi was brought out for the first time in 1842 by Vittadini and the name "Mycorrhiza" was given to it by Frank in 1885. Since then our knowledge on mycorrhiza-plant relationship has grown somewhat with slow pace until about 1970 when there was a sudden upsurge of interest on a specialized type of endomycorrhiza, vesicular-arbuscular mycorrhiza (VA

* *Emeritus Professor*

Mycorrhiza). The prodigious research made during last three decades clearly established its widespread occurrence in various plant species and under different agroclimatic conditions covering broad ecological range including deserts, forests and mangroves. It was also established that this symbiotic association benefits the plants through enhanced nutrient uptake, biological control of root pathogens, synergistic interaction with nitrogen fixing microorganisms, hormone production and drought resistance. In view of its utility to plants, this biotool has new attracted the attention of microbiologists, agronomists, horticulturists and foresters at the global level. Microbiologists, plant pathologists, ecologists, biochemists and biotechnologists of the country have been involved in the researches on VA Mycorrhiza and have gathered valuable information on the exploitation of this biotool for the welfare of the plant communities.

Extensive studies have also been carried out to explore the possibility of using mycorrhizal model system in inducing suppressions of potential soil-borne pathogens. Its relevance becomes all the more important since most of the soil-borne diseases, by and large, are not easily controlled by conventional methods.

Most of the studies conducted on the interaction of VAM fungi and fungal pathogens suggest that mycorrhizae decrease the severity of disease, while a few studies suggest that there is either no effect or there is an increase in the severity of disease. It has been pointed out that in view of the fact that VAM fungi deter the severity of disease they may be used as biocontrol agents. Various mechanisms by which VAM could suppress the fungal pathogens have been suggested on the basis of experimental evidences. Morphological, anatomical, biochemical or physiological changes in the host due to mycorrhization may suppress the pathogen and its activity and impart resistance against it. It has also been suggested that better nutrition of the host due to mycorrhization curtails the damage caused by the pathogen resulting in a curtailment of disease expression. Although generalizations regarding interactions between VAM fungi and fungal pathogens is difficult, however, the possibility of biologically controlling the pathogens through VAM fungi looks promising. In view of this many Indian mycorrhizasts have attempted to explore the possibility of and controlling the fungal pathogens through mycorrhization. The following account gives a description of their reports.

Jalali and Thareja (1981) observed suppression of chickpea (*Cicer arietinum*) wilt caused by *Fusarium oxysporum* f. sp. *ciceri* following soil inoculation with vesicular-arbuscular mycorrhizal fungus-*Glomus* sp. The endophyte colonized plant roots extensively which resulted in reduction in wilt incidence from 80 per cent in non-mycorrhizal to 26.6 per cent in mycorrhizal plants. Inoculation with *Glomus* sp. also increased nodule number, total dry matter production and 'P' content in those plants not infected by the pathogen. This suggested that mycorrhizal colonization may provide effective protection against infection with *Fusarium* and perhaps other root pathogens.

Krishna and Bagyaraj (1983) observed that the mycorrhizal fungus *Glomus fasciculatum* reduces the severity of peanut (*Arachis hypogaea*) disease and the number of sclerotia produced by *Sclerotium rolfsii*. Root and shoot dry weight of the host and their 'P' content were highest in plants inoculated with mycorrhiza only and lowest in plants inoculated with pathogen only. Simultaneous addition of mycorrhizal inoculation and the pathogen reduced disease severity.

Hegde and Rai (1984) studied the influence of *Glomus fasciculatum* on damping-off of tomato. In greenhouse pot tests, inoculation with the mycorrhizal fungus *G. fasciculatum* either simultaneously with or 2 weeks before inoculation with *Pythium aphanidermatum* reduced damping-off and increased plant height, and shoot and root weights compared with plants inoculated with the damping off pathogen alone or 2 weeks before mycorrhizal inoculation. Mycorrhizal inoculation alone increased root but not shoot weight over that of uninoculated controls.

Parvathi *et al.* (1985) studied the influence of root-infecting fungi on the development of *Glomus mosseae* in groundnut. In pot, test plants infected with *G. mosseae* alone exhibited greater mycorrhiza formation than those inoculated with *G. mosseae* + *Rhizoctonia solani* or *G. mosseae* + *Fusarium solani.* However,, when the other fungi were introduced separately into the soil after inoculation with *G. mosseae* mycorrhizal development was not affected.

Bisht *et al.* (1985) suggested that vesicular-arbuscular mycorrhiza (*Gigaspora calospora*) exerted an inhibitory effect on the development of pigeonpea (*Cajanus cajan*) blight caused by *Phytophthora drechsleri* f. sp. *cajani.* In studies at ICRISAT (1986), presence of vesicular-arbuscular mycorrhiza reduced *Phytophthora* blight of pigeonpea in glasshouse tests. In pot experiments promising results were obtained with the VAM fungi, *Glomus fasciculatum* and *G. monosporum,* in significantly increasing 'P' uptake.

Chakravarty and Mishra (1986) observed a reduction in *Cassia tora* wilt caused by *Fusarium oxysporum* when roots inoculated seedlings were transplanted to pot containing soil inoculated with chlamydospores of *Glomus fasciculatum* or *G. tenue.* The mycorrhizal fungi stimulated seedling growth and reduced the *F. oxysporum* population in the rhizosphere.

Jalali (1986) showed that mycorrhizal inoculation of mungbean resulted in a measurable reduction in infection by the root pathogen *Rhizoctonia solani.* This effect was more pronounced when seed pelleting with sporocarps of the endophyte was done. This suppressive effect was directly correlated with the ability of the mycorrhizal symbiont to develop in the absence of available soil phosphate. The studies indicated that soil conditions favourable to mycorrhizal colonization were in many respects not conducive to the growth and establishment of the test pathogen.

Chaturvedi *et al.* (1987) studied the occurrence of VA mycorrhiza in *Ascochyta* blight resistant and susceptible cultivars of chickpea. They examined the roots and soil from the root zone of 2 varieties (ICC 8160 and ICC 8161) with resistance to *Ascochyta rabiei* and 2 (ICC 2664 and Annegeri) with susceptibility for VA mycorrhiza at the time of maximum disease development. The soil from the root zone of the resistant cultivars harboured more VA mycorrhizal spores than did that of the susceptible cultivars. However, the difference in spore number between resistant ICC 8161 and susceptible ICC-2664 was not significant. The roots of the resistant varieties always harboured more VA mycorrhizal spores than did those of the susceptible varieties.

Bali and Mukerji (1988) investigated the effect of VAM fungi on *Fusarium* wilt of cotton and jute. The investigations were performed in 4 sets of experiments taking mycorrhizae *G. macrocarpum* and pathogens in different combinations. Mycorrhizal plants showed improved plant dry weight and per cent mycorrhizal infection than the non-mycorrhizal plants. Mycorrhizae were beneficial in reducing disease severity in host plant.

Ramraj *et al.* (1988) studied the biocontrol of *Macrophomina* root rot of cowpea and *Fusarium* wilt of tomato by using VAM fungi. Inoculation of *Glomus erunicatum,* besides increasing the biomass and yield of cowpea (*Vigna unguiculata*), induced tolerance to *Macrophomina* root rot infection. The disease incidence was reduced to 16 per cent in the inoculated plants as against 33 per cent in untreatd control. In tomato, *G. mosseae* enhanced the growth and also reduced *Fusarium* wilt. The mycorrhizal plants recorded only 11 per cent disease as against 46 per cent in the non-mycorrhizal plants.

Reddy *et al.* (1988) studied the influence of mycorrhiza on chickpea *Fusarium* wilt. Simultaneous inoculation of *Cicer arietinum* with vesicular arbuscular mycorrhizal fungi (*Glomus mosseae, G. constrictum* and *G. monosporum*) and *F. oxysporum* f. sp. *ciceri* had no effect on wilt incidence in the susceptible JG 62 and resistant WR 315 genotypes. However, the wilt tolerant genotype, K 850 had 25 per cent wilt in soil with *Fusarium* + VAM, less than with *Fusarium* alone (35 per cent).

Kehri and Chandra (1989) studied the interaction between *Glomus fasciculatum* (Thaxter sensu. Gerd.) and Trappe and *Macrophomina phaseolina* causing dry root-rot of chickpea cv. T-3, under normal as well as strained water schedule. Addition of *G. fasciculatum* in the soil caused an increase in mycorrhizal colonization of the roots under both the water schedules. The increase in mycorrhizal colonization was always associated with a decrease in the population of pathogen in the roots and an improvement in the growth and yield in chickpea.

Jalali *et al.* (1990) studied the interaction between vesicular-arbuscular mycorrhizal endophyte and *Macrophomina phaseolina* in mungbean. It was reported that mycorrhizal inoculation significantly restricted the spread of the pathogen in host root tissues. Disease incidence was reduced from 77.9 per cent in pathogen inoculated to 13.3 per cent in mycorrhiza + pathogen inoculated plants. Dual inoculation with mycorrhizal endophyte and the pathogen also resulted in increased total dry matter production, nitrogen, phosphorus and potassium contents of such plants as compared to pathogen-inoculated plants.

Sundaresan *et al.* (1993) reported that pre-establishment of the vesicular-arbuscular mycorrhizal fungus in cowpea (*Vigna unguiculata*) roots reduced colonization by *Fusarium oxysporum* and severity of the disease as measured by a reduction in vascular discolouration. They observed induction and accumulation of phytoalexins in cowpea roots infected with mycorrhizal fungus. Production of phytoalexins was always higher in mycorrhizal plants than in non-mycorrhizal ones and there appeared to be direct correlation between phytoalexin concentration and the degree of mycorrhizal association. They suggested that the production of phytoalexin compounds in mycorrhizal plants could be one of the mechanisms imparting tolerance to disease.

Jayaraman and Kumar (1995) on the other hand, observed VAM fungi-pathogen interaction in two cultivars of chickpea. In cv. BG 256, VAM fungus *Glomus fasciculatum* enhanced the activity of pathogenic fungus *Solerotium rolfsii* that causes collar rot disease. However, in cv. JG 62 the VAM fungus not only nullified adverse effects of the pathogen but also improved plant growth compared with control.

Chandra *et al.* (1995) evaluated the efficacy of a number of VAM fungi *viz., Glomus fasciculatum, G. mosseae, G. aggregatum, G. claroideum, G. macrocarpum, G. mosseae, G. aggregatum, G. claroideum, G. macrocarpum* and *G. multicaule* recovered from agricultural fields raising leguminous crops in checking the dry root rot of greengram, blackgram and cowpea due to *Macrophomina phaseolina.* The interaction between VAM fungi and pathogen was examined by inoculating the crops with selected VAM fungi alone as well as in combination with selected isolate of pathogen in sterilized as well as unsterilized soil. The rate of mortality, mycorrhizal status, changes in the population of pathogen and yield were recorded. When the inoculum of VAM fungi was added along with that of pathogen, a reduction in the rate of mortality and population of pathogen was recorded in all the three crops. This change was coupled with an increase in the mycorrhization in the roots as well as yield. This finding clearly indicated that VAM fungi were instrumental in reducing the damage caused to the crops due to the pathogen. The study under pot conditions was repeated under microplot conditions and more or less similar findings were recorded.

Singh and Singh (1995) studied the effect of *Glomus aggregatum* inoculation and P-amendments on *Fusarium oxysporum* f. sp. *ciceri* causing chickpea wilt. 80 and 20 per cent plants were found healthy and showing yellowing respectively at both 0 and 40 mg P/kg soil after 45 days of sowing in the presence of mycorrhizae and wilt pathogen. All plants, inoculated with *F. oxysporum* f. sp. *ciceri* alone, were found to have wilted after 45 days of sowing at all three levels of P. On the other hand, none of the plants showed wilting symptom in the soil inoculated with VAM fungus. The chickpea plants showed significant resistance against the wilt pathogen in presence of *G. aggregatum*. The 20, 20 and 60 per cent plants showed yellowing symptom (partially infected) at 0, 40, 80 mg P/kg soil after 45 days sowing. The observations revealed that the presence of mycorrhizal fungi significantly delayed infection, as well as reduced its percentage. The root colonization with *G. aggregatum* protected chickpea plants up to 80 per cent, against 100 per cent wilting due to *F. oxysporum* f. sp. *ciceri* in absence of mycorrhizal fungus at variable levels of P. The wilting process in mycorrhizal chickpea plant was found to be significantly slow as compared to non-mycorrhizal plants.

Sivaprasad *et al.* (1995) studied the vesicular-arbuscular mycorrhizal colonization relation to foot-rot disease intensity in black pepper. A study was undertaken to analyze the relationship, if any, between VAM colonization and foot-rot disease intensity in black pepper. Three major black pepper-growing tracts of Kerala, namely, Idukki, Wynad, and Cannanore districts, were selected for the study. Pepper plants from existing plantations with varying foot-rot disease intensities were selected randomly from different locations. The percentage of vines dead per standard was considered as the index of disease intensity. The observations made during the pre-monsoon period gave a definite negative relationship between disease intensity and VAM colonization percentage and spore count in the rhizosphere soil. The roots were discoloured and the percentage of VAM colonization decreased with increase in disease intensity. For example, sample 14 with 90 per cent dead vines recorded five per cent VAM colonization in the root and a spore count of 15 in rhizosphere soil, as against 51 per cent colonization and 33 spores recorded in a healthy plant (sample 16). The healthy plants, irrespective of locations, consistently showed higher VAM colonization and spore count compared to the diseased plants. Analysis of the 38 pepper root and rhizosphere samples collected during the period before and after the monsoon indicated a general trend of decrease in foot-rot disease intensity of black pepper with increase in VA mycorrhizal colonization.

Sivaprasad *et al.* (1999) studied the foot rot of black pepper and rhizome rot of ginger as influenced by vesicular arbuscular mycorrhizal fungi (VAMF). Effect of inoculation with different VAMF on foot rot disease of black pepper and rhizome rot of ginger was tested under field conditions. Black pepper cuttings inoculated with *Glomus monosporum* at nursery level and transplanted to the field significantly reduced the foot rot incidence from 100 per cent infection and 83 per cent mortality observed in control to 33 per cent. Similarly, in ginger the rhizome rot infection of 78 per cent and the mortality of 47 per cent was brought down to 32 and 17 per cent by inoculation with *G. mosseae*. Ability of VAMF to stimulate plant growth and impart resistance against pathogen was found as independent traits of VAMF.

Reddy and Hindumathi (1999) conducted experiments in irrigated and non-irrigated soils during kharif and rabi with different varieties of sorghum by using mycorrhizal fungi since sorghum roots appear to be good hosts of mycorrhizae. In all cases charcoal rot was declined to a great extent in mycorrhizal inoculated plots. No disease was observed in the varieties Rs–29 and E–36–1 inoculated with mycorrhizae and pathogen. The disease incidence was reduced upto 80 per cent even in the variety, CSV–8R which is considered as highly susceptible to charcoal rot. The reduction in disease severity by competition of sites, though the main reason is by strengthening of the host plant so that the damage caused by the pathogen was offset by improved plant growth.

Chandra *et al.* (2000) tried to manage the soil-borne diseases of two important horticultural crops *viz.*, guava and papaya. An experiment was designed to determine the efficiency of the VAM isolates recovered from various orchards of Allahabad *viz.*, *Glomus aggregatum, G. claroideum, G. fasciculatum, G. mosseae* in improving the growth and biomass of both the crops. VAM inoculants were provided to the crops under pot conditions in sterilized/unsterilized soil. Out of five VAM inoculants, *G. aggregatum* was found to show highest potential in colonizing the roots of guava and papaya and improve their performance. It was also shown to suppress the indigenous population of pathogens in the roots. On this basis, it was selected and its potential to act as biopesticide in controlling the diseases of guava and papaya was evaluated. Dual inoculation experiment was designed to examine the effect of selected VAM and pathogen inoculants on the performance of guava and papaya under pot conditions in sterilized and unsterilized soil. The crops were provided with combination of VAM (*G. aggregatum*) and pathogen [*F. oxysporum* (SA) for guava, *F. semitectum* (NG) for papaya] inoculants. The VAM inoculants *G. aggregatum* was found to be effective in suppressing the root infection in both the crops not only in sterilized but also in unsterilized soil. It was found to be equally effective in counteracting the negative effects of pathogen on the performance of both guava and papaya in terms of growth of roots and shoots.

Conclusion

Although a number of Indian crops suffer from severe fungal diseases, only few have been investigated with reference to their response to mycorrhizal fungi as biocontrol agents. A perusal of published reports shows that majority of mycorrhizasts have concentrated their attention to some of the important leguminous crops including chickpea, mungbean, greengram, blackgram, pigeonpea, groundnut and cowpea. Wilts and root-rots have been reported to severely affect the productivity in these crops. It is heartening to note that they have been successfully controlled through the use of mycorrhizal fungi. Few other crops *e.g.* sorghum, guava, papaya, ginger and black pepper have also been investigated and have been shown to give a positive response to mycorrhizal fungi as biocontrol agent for some of the diseases caused to them. It may be mentioned that the published data related to potentiality of mycorrhizal fungi as biocontrol agent is of limited value because the observations are based in nearly all the cases on pot experiments. For practical application of the results of these studies, microplot and field trials should be carried out. If possible, multilocational trials should also be made in one lot, keeping all conditions similar.

Only fragmentary information is available on the mechanisms involved in the suppression of the severity of soil-borne diseases by VAM endophytes. Concerted efforts should be made to generate meaningful information on the biocontrol efficiency of mycorrhizal symbionts and mechanisms involved in such controls. In depth studies on non-nutritional influences of mycorrhizal endophytes on the pathogens may yield a more conclusive picture.

Abiotic factors such as drought, salinity and stress are expected to affect the interactions between VAM symbiont and pathogens. Efforts should be made to collect information with reference to this and effective correlation models should be developed integrating managemerlt of abiotic factors and pathogens through VAM symbiont.

Attempt should be made to develop integrated management systems including effective VAM for suppressing the soil-borne diseases. This appears to be a practical approach especially for fungal disease complexes.

REFERENCES

Bali, M. and Mukerji, K.G. (1988). Interaction between VA Mycorrhizal fungi and root microflora of jute. In: *Plant roots and their environment*: *Proceedings of an ISSR Symposiumm* (Eds. B.L. McMichael and H. Persson). Amsterdam, Netherlands, Elsevier Science Publishers, B.V., 396-401.

Bisht, V.S., Krishna, K.R. and Nene, Y.L. (1985). Interaction between vesicular-arbuscular mycorrhizal fungi and *Phytophthora drechsleri* f. sp. *cajani*. *International Pigeonpea Newsletter*, **4**: 63-64.

Chakravarty, P. and Mishra, R.R. (1986). Influence of endotrophic mycorrhizae on the fusarial wilt of *Cassia tora* L. *Jour. Phytopathol.*, **115(2)**: 130-133.

Chandra, S., Kehri, H.K. and Maheshwari, S. (1995). Macrophomina dry root-rot of leguminous crops and its management through VAM biopesticides. In: *Detection of plant pathogens and their management.* (Eds. J.P. Verma, Anupam Varma and Dinesh Kumar), Angkor Publishers (P) Ltd., New Delhi, 357-364.

Chandra, S., Kehri. H.K. and Dwivedi. R. (2000). Managing soil-borne diseases of horticultural crops through VAM fungi. In: *Glimpses in Plant Sciences.* (Eds. K.R. Aneja, M.U. Charaya, A. Aggarwal, D.K. Hans and S.A. Khan). Pragati Prakashan, Meerut, 82-93.

Chaturvedi, C., Singh, R. and Sharma, A.K. (1987). Occurrence of vesicular-arbuscular (VA) mycorrhiza in *Ascochyta* blight resistant and susceptible cultivars of chickpea. *Current Science*, **56(16)**: 831-831.

Hegde, S.V. and Rai, P.V. (1984). Influence of *Glomus fasciculatum* on damping-off of tomato. *Current Science*, **53(11)**: 588-589.

Jalali, B.L. (1986). Vesicular-arbuscular mycorrhiza: Current Status. In: *Vistas in Plant Pathology*, (Eds. Anupam Varma and J.P. Verma), Malhotra Publishing House, New Delhi, 437-450.

Jalali, B.L., Chhabra, M.L. and Singh. R.P. (1990). Interaction between vesicular-arbuscular mycorrhizal endophyte and *Macrophomina phaseolina* in mungbean. *Indian Phytopathology*, **43**: 527-530.

Jalali, B.L. and Thareja, M.L. (1981). Suppression of *Fusarium* wilt of chickpea in vesicular-arbuscular mycorrhiza inoculated soils. *International Chickpea Newsletter*, **4**: 21-22.

Jayaraman, J. and Kumar, D. (1995). VAM fungi-pathogen-fungicide interactions in gram. *Indian Phytopathology*, **48**: 294-299.

Kehri, H.K. and Chandra, S. (1989). Interaction of *Glomus fasciculatum* and *Macrophomina phaseolina* in chickpea. *Journal of Indian Botanical Society*, **68**: 363-365.

Krishna, K.R. and Bagyaraj, D.J. (1983). Interaction between *Glomus fasciculatum* and *Sclerotium rolfsii* in peanut. *Canadian Journal of Botany*, **67(9)**: 2349-2357.

Parvathi, K., Venkateswarlu, K. and Rao, A.S. (1985). Influence of root infecting fungi on the development of *Glomus mosseae* in groundnut. *Current Science*, **54(19)**: 1006-1007.

Ramaraj, B., Sharmungam, N. and Reddy, C.D. (1988). Biocontrol of *Macrophomina* root rot of cowpea and *Fusarium* wilt of tomato by using VAM fungi. *Mycorrhizae for Green Asia*, (Eds. A. Mahadevan, N. Raman and K. Natrajan), Madras University, Madras, 350-251.

Reddy, B.N. and Hindumathi, A. (1999). Biological control of charcoal rot of sorghum by using mycorrhiza. Abstract *Proc. National Conference on Mycorrhiza,* Barkatulla University, Bhopal: 62.

Reddy, M.V., Rao, J.N. and Krishna, K.R. (1988). Influence of mycorrhiza on chickpea *Fusarium* wilt. *International Chickpea Newsletter,* **19**: 16.

Singh, R.S. and Singh, D. (1995). Effect of *Glomus aggregatum* inoculation and P-amendments on *Fusarium oxysporum* f. sp. *ciceri* causing chickpea wilt. In: *Mycorrhizae: Biofertilizers for the Future,* (Eds. Alok Adholeya and Sujan Singh), TERI, New Delhi, 119-123.

Sivaprasad, P., Joseph, P.J. and Sulochana, K.K. (1999). Foot rot of black pepper and rhizome rot of ginger as influenced by arbuscular mycorrhizal fungi (AMF) in the field. Abstract: *National Conference on Mycorrhiza,* Barkatulla University, Bhopal, 61.

Sunderesan, P., Raja, Naa, V. and Gunasekaran, P. (1993). *J. Biosciences,* **18(2)**: 291-301. cf. Dynamics of plant-pathogen–mycorrhiza interactions in legume crops (H.C. Chand and B.L. Jalali). In: *Trends in Microbial Exploitation,* (Eds. Bharat Rai, R.S. Upadhyaya and N.K. Dubey), B.H.U., Varanasi, 71-77.

18

Microbial Management of Plant Diseases: An Overview

R.C. RAJAK, A K. PANDEY, A.K. SINGH & ROHIT SHARMA

Department of Biological Science, R.D. University, Jabalpur – 482 001 (M.P.), India

Introduction

Substantial progress of a nation is entirely dependent upon availability of food to human and animal populations. The predicted growth of world population from 5.7 billion in 1993 to approximately 8 billion by 2020, present a major global challenge to meet the necessary increase in food production. Mahatma Gandhi at Naokhali in 1946 said, that "to the hungry, God is bread; the God, bread should prevail in every home and hut of the country". To recognize the importance of food production, Late Mrs. Indira Gandhi released a special stamp titled "the wheat revolution" in 1968. Despite vast knowledge and advanced technology, agricultural production has not been up to mark.

One of the major reasons for this incompetence is loss of agricultural productivity due to diseases caused by plant pathogens, which are responsible for yield reductions of almost 20 per cent in the principal food and cash crops worldwide (Oerke *et al.*, 1994; Rohilla *et al.*, 2001). Losses may be more severe when highly susceptible varieties are widely grown (Ragsdale *et al.*, 1993). Significant yield improvements are potentially achievable in stable food crops through different inputs, including effective management of losses due to diseases.

Green revolution in India during 1960s no doubt brought about self sufficiency in food, however, excessive use of plant protection chemicals and inorganic fertilizers for maximizing crop yields and change in traditional cultivation practices, resulted in deterioration of chemical, biological and physical health of cultivated lands (Nautiyal, 1999). The uses of naturally occurring microbes as biofertilizers, bioprotector, bioremediator, biopesticides without damaging the agro-ecosystems are the new and propagative ways towards sustainable agriculture.

The four major problems encountered with conventional chemical pesticides are pest resistance, toxic residues, secondary pest and pest resurgence. Truly satisfactory solutions to pest problems will require a shift to understanding and promoting naturally occurring microbial agents and other inherent strengths as components of total agricultural ecosystems and designing our cropping systems in a way that these natural forces keep the pests within the acceptable bounds (Smith *et al.*, 1992).

Biological control has played a major role in disease management since the beginning of agriculture. Farmers learn disease management practices through careful observation and by trial and error that takes advantage of natural biological control, even though they may not recognize its contributions to the well being of their crops and livestock's. The following points emphasize the need for the biological control of (especially soil borne) plant pathogens (Satyaprasad & Manoharachary, 1997):

1. Development of resistant strains of fungi to fungicides.
2. Indiscriminate treatment of soil with fungicides upsetting biological balance leading to severe out break of diseases.
3. Residual effects of pesticides and their resistance to bio-deterioration.
4. Increased/decreased levels of disease due to environmental variation and genotypic response to agro-climatic conditions.
5. Non-pathogenic fungi becoming pathogenic and disease of less importance becoming major diseases.
6. Management of soil-borne disease by traditional methods has become laborious, expensive and non-viable hence biological control.

Therefore, the objective of this article is to highlight the status of microbial control of plant pathogens.

Biological Control: Concept

Biological control has been defined by workers in different ways. Garret (1956) gave credit to Sanford (Sanford, 1926) as the pioneer to start microbial control of plant pathogens by working with *Streptomyces scabies* causing potato scab and *Bacillus subtallis* as microbial control agents. Some of the common definitions are:

Snyder (1960) defined Biological control as "an interruption of host-parasitic relationships through biological means by imparting resistance to host or by modifying the cultural practices of the crop so as to reduce the infection". Baker and Cook (1974) defined it as "the reduction of inoculum density or disease producing activities of a pathogen in its active or dormant stage".

Garrett (1956) defined it as " any condition under which, or practices whereby survival or activity of a pathogen is reduced through the agency of any other living organism (except man himself) with the result that there is a reduction in the incidence of disease cause by the pathogen".

Para-biological control would be defined as pest species or the pest resources (in whole or in part) for the reduction of negative effects or pest species density to levels at or below their economic thresholds". However the lack of economic (or aesthetics) thresholds for many pest species requires the more practical definition for both biological and para-biological control.

"Direct and purposeful manipulation of natural enemies, pest competitors (in whole or in part), or the resource required by these orgnisms for the control of pest species or their negative impact". National Academy of Sciences defined biological control as " the use of natural and modified organism, genes or gene product to reduce the effects of undesirable organisms (pest) and to favour desirable organism such as crops, trees, animals, and beneficial insects and microorganism". Several reports dealing with various aspects of biological control have been reviewed from time to time (Baker, 1968; Baker & Cook, 1974; Nigam, 1997; Baker & Dickman, 1992; Chet, 1989; Mukerji & Garg 1988a.b; Agarwal, 1991; Mukerji *et al.*, 1992; Cook, 1993; Papavizas 1992; Sharma & Sankaran, 1988; Aggrawal & Mehrotra, 1992; Loper & Lindow, 1993; Lumsden & Vaughn, 1993; Reichelderfer, 1981; Satyaprasad & Manoharachary, 1997; Blakeman, & Fokkema, 1982; Fravel, *et al.*, 1986).

Sharma and Paul (1998) outlined the characteristics of an ideal biocontrol agents as follows:

1. It should survive for prolonged periods in soil either in active or passive form.
2. There must be high probability of its contacting the pathogen either directly or indirectly by diffusion of metabolites.
3. Its multiplication in laboratory should be simple and inexpensive.
4. It should be amenable to a simple, efficient and inexpensive process of packing, distribution and application.
5. If possible it should be specific to the target organism.
6. It should not be a health hazard in its preparation, distribution and application.
7. It should be active under the required environmental conditions.
8. It should be efficient and economical.

Mechanism

Efforts made by the scientists involved in plant disease management have reached to a point where two broader strategies have been clearly defined. These are as follows:

Act on Pathogens

Being natural component of ecosystem, bio-control agents interact directly and when grow in association, adversely affect the pathogens (Baker and Cook, 1974). This phenomena is popularly known as 'Antagonism' which operates through the following ways:

Competition

Competition occurs when there is a "simultaneous demand upon the same resources by two populations". Competition occurs between microorganisms when space or nutrients are limited. The term has been defined by Clarke (1965) as," the acute demand in access of the immediate supply of material or condition on the part of two or more organisms". Bio-control practices in which microbial antagonists are used often involve competitive interaction between antagonists and pathogen for nutrients *viz.*, nitrogen, carbon and iron (Baker, 1968; Baker *et al.*, 1986 b). Iron is an essential nutrient for virtually all forms of life (Winkelmann, 1991). The most successful and thoroughly investigated example is Siderophores production. Siderophores are extracellular secondary metabolites belongs to either pyroviridins or pseudobactin produced by certain bacteria (*Aerobacter aerogens, Arthrobacter pascens, Pseudomonas cepacia, P. fluorescens*), actinomyces (Streptomyces spp.), fungi (*Penicillium* sp.) etc. Siderophores production in various groups of microorganisms have been thoroughly investigated by many workers (Neiland, 1995; Trick *et al.*, 1983). Siderophores production by fungal strains have been extensively reviewed by Renshaw *et al.* (2002). They have the ability to bind with iron. When the amount of iron available in the environment is limited, these microorganisms have competitive advantage over pathogens. The production of siderophores have been linked to their suppressive ability (Altman *et al.*, 1970; Loper & Buyer 1991). A direct correlation has been observed *in vitro* between siderophore synthesis and their capacity to inhibit germination of chlamydospores of *Fusarium oxysporum* (Elad & Baker, 1985). *Pseudomonas fluorescence* has been reported to interact and cause lysis of hyphae of *Gaeumannomyces graminis* var. *tritici* responsible for Take All Disease of Wheat. The bacterium produces pseudobactin, which is involved in promotion of plant growth and disease control (Burr *et al.*, 1978). Various aspects of use of siderophores in biological control of plant diseases have been extensively reviewed (Papavizas 1985; Nelson & Thies 1985, 1986; Weller 1988; Baker and Dickman 1992; Barton & Hemming, 1993; Kessel *et al.*, 2002). There are several siderophores, which have been tested and are successfully in use (Table 18.1).

Table 18.1: Siderophores Produced by Fungi

Sl.No.	Fungi	Siderophore
(a)	**Ascomycota**	
1.	*Coniothyrium* sp.	Siderophore
2.	*Alternaria* sp.	Ferricrocin, Corpgen, Neocoprogen I, Neocoprogen II. N-dimethylcoprogen, Hydroxyisoneocoprogen.
3.	*Curvularia* sp.	Ferricrocin, Corpgen, Neocoprogen I, Neocoprogen II.
4.	*Leptosphaerulina* sp.	Corpgen
5.	*Stemphyllium* sp.	Corpgen B, Dimerum acid.
6.	*Aspergillus* sp.	Ferricrocin, Ferrichrome, Ferrichrome C, Asperchromes, Cis-fusarinine, Fusarinine A, Fusarinine C, Triacetylfusarinine C.
7.	*Paecilomyces* sp.	Ferrirubin.
8.	*Penicillium* sp.	Ferrichrome, Ferrirhodin, Ferrirubin.
9.	*Cylindrocarpon* sp.	Dimerum acid, Neocoprogen.
10.	*Fusarium* sp.	Malonichrome, Cis-fusarinine, Fusarinine C, Corpgen B, Dimerum acid, N-dimethylcoprogen.
11.	*Gibberella* sp.	Fusarinine B, Fusarinine C.
12.	*Nectria* sp.	Fusarinine B, Fusarinine C, Corpgen B.
13.	*Trichoderma* sp.	Ferricrocin, Fusarinine B, Fusarinine C, Corpgen B, Corpgen, Palmitoylcoprogen.
14.	*Hymenoscyphus* sp.	Ferricrocin, Ferrichrome, Ferrichrome C
15.	*Botrytis* sp.	Ferrirhodin.
16.	*Microsporum* sp.	Ferricrocin, Ferrichrome C.
17.	*Trichophyton* sp.	Ferricrocin, Ferrichrome, Ferrichrome C.
18.	*Oidiodendron* sp.	Ferricrocin, Ferrichrome C.
19.	*Histoplasma* sp.	Ferrichrome B, Coprogen B, Dimerum acid.
20.	*Onygena* sp.	Neocoprogen II.

Contd...

Table 18.1–Contd...

Sl.No.	*Fungi*	*Siderophore*
21.	*Sporothrix* sp.	Hydroxamate
22.	*Ascodesmis* sp.	Coprogen
23.	*Wilcoxina* sp.	Ferricrocin
24.	*Colletotrichum* sp.	Ferricrocin
25.	*Protomyces* sp.	Ferrichrome
26.	*Neurospora* sp.	Ferricrocin, Ferrichrome, Ferrichrome C, Neurosporin, Coprogen, N-dimethlisoneocoprogen II.
27.	*Dipdascopsis* sp.	Ferrichrome, Ferrichrome C.
28.	*Zygozyma* sp.	Ferrichrome, Ferrichrome C.
29.	*Candida* sp.	Hydroxamate, Phenolate
(b)	**Basidiomycota**	
1.	*Amanita* sp.	Hydroxamate
2.	*Agaricus* sp.	Ferrichrome, Des ferrirhodin, Fusarinine C
3.	*Boletus* sp.	Ferricrocin
4.	*Suillus* sp.	Hydroxamate
5.	*Lentinus* sp.	Siderophore
6.	*Coriolus* sp.	Siderophore
7.	*Gloeophyllum* sp.	Phenolate
8.	*Postia* sp.	Phenolate
9.	*Pisolithus* sp.	Hydroxamate
10.	*Phanerochaete* sp.	Siderophore
11.	*Phlebia* sp.	Siderophore
12.	*Peniophora* sp.	Siderophore
13.	*Sistotrema* sp.	Siderophore
14.	*Cryptococcus* sp.	Ferrichrome C
15.	*Graphiola* sp.	Ferrichrome
16.	*Aessosporon* sp.	Hydroxamate
17.	*Leucosporidium* sp.	Rhodoturulic acid
18.	*Rhodosporidium* sp.	Rhodoturulic acid
19.	*Sporidiobolus* sp.	Rhodoturulic acid
20.	*Tilletiaria* sp.	Ferrichrome
21.	*Neovossia* sp.	Ferrichrome, Ferrichrome C, Tetraglycylferrichrome
22.	*Cintractia* sp.	Ferrichrome
23.	*Farysia* sp.	Ferrichrome
24.	*Microbotryum* sp.	Rhodoturulic acid
25.	*Moesziomyces* sp.	Ferrichrome
26.	*Scizonella* sp.	Ferrichrome, Ferrichrome A

Contd...

Table 18.1–Contd...

Sl.No.	*Fungi*	*Siderophore*
27.	*Sorosporium* sp.	Ferrichrome
28.	*Sphacelotheca* sp.	Ferrichrome A, Ferrichrome C.
29.	*Sporisorium* sp.	Ferrichrome
30.	*Ustilago* sp.	Ferrichrome, Ferrichrome A.
(c)	**Zygomycetes**	
1.	*Basidiobolus* sp.	Rhizoferrin
2.	*Glomus* sp.	Hydroxamate
3.	*Cunninghamella* sp.	Rhizoferrin, Ferrichrysin
4.	*Mortierella* sp.	Rhizoferrin
5.	*Absidia* sp.	Hydroxamate
6.	*Rhizopus* sp.	Rhizoferrin, Hydroxamate
7.	*Mycotypha* sp.	Rhizoferrin
8.	*Phycomyces* sp.	Rhizoferrin
9.	*Chaetostylum* sp.	Rhizoferrin
10.	*Cokeromyces* sp.	Rhizoferrin

Source: Renshaw *et. al.* (2002); Hawksworth *et. al.*, (1995).

Antibiosis

It is also called as Amensalism. In reference to bio-control, antibiosis is a situation where the metabolites secreted by a non-pathogenic microbes inhibits/kills the pathogen, but the producer agent remain unaffected. Importance of antibiosis in the suppression of plant diseases has been well realized by many workers including Weindling (1934), Fravel (1988), Weindling and Emerson (1936). Secondary metabolites produced by several strains of *Gliocladium virens*, *Trichoderma viride*, *T. harzianum*, a virulent strain of *Agrobacterium radiobacter* has shown significant biopesticidal potential against many pathogens (Bell *et al.*, 1982; Dennis & Webster, 1971 a.b; Wright, 1956).

Trichoderma has been recognized as exceptionally good model, because it is ubiquitous, easy to isolate and culture, grow rapidly on many substrates, affect a wide range of plant pathogen, is rarely pathogenic to higher plants, act as mycoparasite, competes well for food and site, produces antibiotics and has an enzyme system capable of attacking a wide range of plant pathogens (Lewis & Papavizas, 1987 a, b; Mukhopadhayay, 1987; Hermosa *et al.*, 2000; Well, 1988; Jariwala *et al., 1997*). *Trichodermin* a potential fungicidal antibiotic has been reported by Godfredsen & Vangedal (1965). Wright (1956) reported gliotoxin from *T.viride*. Dennis and Webster (1971 a, b) reported a volatile antibiotic from *Trichoderma* sp. with coconut odour, which inhibited *Rhizoctonia solani*. It was later on identified as alkylpyrones by Calydon & Allan (1987). Some microbial metabolites successfully employed for disease management are listed in Table 18.2.

Table 18.2: Antibiotics of Microbial Origin

Sl.No.	*Producer Organism*	*Antibiotic*
(a)	**Actinomyces**	
1.	*Actinomadura madurae*	Simaomicin
2.	*A. hibisca*	Pradimicins A, B & C
3.	*Actinomadura* sp.	Sch 40873
4.	*Actinomadura* sp.	Macrolactams
5.	*Actinomadura* sp.	Macrolactumdisaccharide
6.	*Actinoplanes* sp.	Sch 42137
7.	*Micromonospora* sp.	Rustimicin
8.	*Streptomyces hygroscopicus*	Validamycin
9.	*S. coelicolor*	Pigmented antibiotic
10.	*S. hachijoensis*	Trichomycin A
11.	*Streptomyces* sp. HK-803	Phosphazomycin C
12.	*S. roseviridis*	Enactin congeners
13.	*S. cyanoalbus*	Antifungal & Antibacterial
14.	*Sreptomyces* sp. RK-16	Phosmidosine
15.	*Sreptomyces* sp. UK-10	HA-2-91
	Sreptomyces sp.*	Phoslactomycin E, Gualamycin, Jietacins, Photoxazolines B, C.
16.	*Sreptomyces* sp.	A1 21
17.	*S. plantensis* A-136	TAN-9500 A
18.	*Sreptomyces* sp. SD-581	AB023 (Polyene)
19.	*S. violaceoniger*	Tubercidin
20.	*S. hygroscopicus*	Shurimycin A & B
21.	*Streptoverticillium fervens*	FR-900848
(b)	**Bacteria**	
1.	*Bacillus subtillis NB 22***	Iturin
2.	*B. subtilis*	Volatile antifungal
	*B. cerues**	Zwittermicin
3.	*B. subtilis* F-29-3	Fengycin
4.	*Erwinia herbicola* AIII	Herbicolin A & B
5.	*Gluconobacter oxydans* F-2702	F-2702
6.	*Lysobacter gumosus*	Catacandis
7.	*Pseudomonas syringe**	Pseudomycins
	Agrobacterium radiobacte Strain K84*	Agrocin
8.	*Serratia plymuthica*	CB-25-1

Contd...

Table 18.2: Antibiotics of Microbial Origin

Sl.No.	*Producer Organism*	*Antibiotic*
(c)	**Fungi**	
1.	*Aspergillus fumigatus*	Synerazol
2.	*A. sydowii*	Deoxymulundo-candin
3.	*Crepidotus fulvotomentosus*	Strobilurin E
4.	*Flagellascypha pilatii*	Pilatin
5.	*Monilinia fructicola*	Chloromonilicin
	*Penicillium rubrum***	Palitanin
6.	*Penicillium vermiculatum*	2-Methylsorbic Acid
7.	*Peniophora polygenia*	Volatile and Non-volatile Antifungal
8.	*Pycnidiophora dispersa*	Lanomycin and Glucolanomycin
9.	*Sparassis crispa*	SC I & SC II
	Gliocladium virens *	Asperfuran, Gliovirin, Gliotoxin.
	Tichoderma viride *	Viridin, Gliotocin.
	T. hamatum *	Sesquiterpene, Trichoderma, 1, 2, 3
	T. harzianum *	Sesquiterpene, Trichoderma, 1, 2, 3
10.	*Tolypocladium inflatu*	Ergokonin

Commercially Available; * Fungicidal ** Bactericidal;

There are several other reports where antibiotics have been successfully employed in plant disease management. Blasticidins produced by *Streptomyces griseocharomogens* was used successfully for the first time in the control of Rice Blast Diseases (Kono *et al.*, 1968, Takeuchi *et al.*, 1985). Significant control of several fungal diseases was obtained by application of very small doses of cycloheximide (Massie *et al.*, 1968; Klingstrom & Lundeberg, 1978). Early blight of tomato, *Botrytis* disease of Lettue, Brown Spots a Rice, Apple Blossom and *Alternaria* Blight of Rape and Mustard were successfully controlled by Griseofulvin (Brian *et al.*, 1951; Hussain & Thakur 1962). It is a most effective and less phytotoxic antibiotics, however, manufacturing cost is of a major constrain in its wide application (Misato & Yamaguchi, 1977). Blast of Rice Disease has been successfully controlled by the application of Kasugamycin, a water-soluble antibiotic obtained from *S. kasugamycin* (Misato & Yamaguchi, 1977). Development of resistance in pathogenic strain is the major drawback of its application. Polyoxins produced by *S.cacaoi* var. *asoensis* showed significant control efficacy against sheath blight of rice pathogen and diseases incited by *Alternaria* sp., in fruit orchards (Sasaki *et al.*, 1968; Paul, 1998). Validamycin produced by *S. hygroseopicus* has been found to be effective against *Rhizoctonia* sp., incites web blight, damping off, seed decay, root rot and scurf of several crops and southern blight of vegetation as well as sheath blight of rice plant (Misato & Yamaguchi, 1977). Mycobocillin and Versicolin were found to be highly effective against *Pyricularia oryzae* and *Colletotrichum gloeosporioides* respectively (Nandi *et al.*, 1975). Disease suppressive role of secondary metabolites of fluorescent pseduomonads have also been well realized in several publications (Thomashow & Weller, 1988; Howell & Stipanovic 1979, 1980; Ganamanickam & Mew, 1992; Gananamanickam *et al.*, 1998). Several such compounds as Phearines (Thomashow & Weller, 1988), Pyoluterorin (Howell & Stripanovic, 1979), 2,4-diacetyl-phdoroglucinol (Shanahan, 1992), Pyrrolnitrin (Howell & Stripanovic, 1979) have been isolated from

Pseduomonads. Some of these compounds have broad-spectrum activity against several bacterial and fungal pathogen (Updhyay & Mishra, 1998; Howell & Stripanovic, 1979; Thomashow & Weller, 1988). Banerjee (2000) recorded very high antagonistic potential of *Trichoderma viride* #85, *Aspergillus terrus* # 61, *A. niger* # 63; *Ascotricha chartarum* # 59 and *Chaetomium globosum* against *Fusarium moniliformae* # 75 responsible for severe seedlings blight of *Trigonella foenum-graecum*(methi). These isolates were originally recovered from sewage-irrigated soil. Different antibiotics are in use for the management of plant diseases, Table 18.2.

Role of HCN production by fluorescent pseudomonads in plant disease management have also been tried in few instances (Howie & Suslow, 1991; Schippers, 1988; Brisbane & Rovira, 1988; Defago *et al.*, 1990; Voisard *et al.*, 1989; Janisewicz & Roitman, 1988). Baker *et al.* (1986; 1987) reported detrimental effects of HCN on plant growth.

Hyperparatism

It often refers to, parasitism of one microorganism by others. Myco-parasitism is one of the significant components of this phenomenon, which is defined as " Parasitism of one fungus by another". It occurs when one fungus exists in intimate association with another from which it derives some or all of its nutrients while conferring no benefit in return (Lewis *et al.*, 1989). In other words, it is the attack of a secondary parasite on a primary parasite. On the basis of mode of parasitism classified myco-parasites into two broad groups":

1. *Necrotrophic or destructive*: where the myco-parasite destroys the host cell slightly before or after invasion and utilized nutrients from the dying or dead host cell.
2. *Biotrophic or Balanced*: where myco-parasite is able to obtain nutrient from living cells with little or no apparent harm to its host. Three types of biotrophic myco-parasitism have been recognized:
   (a) The contact parasite may merely contact the host cell without penetration, obtaining nutrients through a 'buffer cell'
   (b) The internal parasite may penetrate the host hyphae and grow from cell to cell.
   (c) The haustorial parasite penetrates the host cells with an infection peg from appressorium and produce distinct haustoria within the host hyphae.

In general when a myco-parasite come into contact with a pathogen, hyphal may subsequently coil around the host and hook like structure *i.e.* apressoria are formed. At this stage, recognition phenomena involving aggutilinin(lectin) from the host binds to carbohydrate residues on the cell wall of myco-parasite. Thus it produce hydrolytic enzymes (B-1, 3, gluconase, cellulose and chitinase and different combinations of enzymes that are appropriate to the cell wall substrate (Baker & Dickman, 1992; Ridout *et al.*, 1988; Sharma *et al.*, 1998).

The hyphae of *Trichoderma* has been shown to parasitise many pathogenic fungi *viz.*, *Rhizoctonia*, *Sclerotium*, *Sclerotinia*, *Helminthosporium*, *Fusarium*, *Armillaria*, *Colletotrichum*, *Verticillium*, *Venturia*, *Endothia*, *Pythium*, *Phytophthora*, *Rhizopus*, *Diaorthe*, *Fusicladium*, *Botrytis*, *piricularia*, *Macrophomina etc.* (Beagle-Ristaino & Papavizas, 1985; Chet *et al.*, 1981; Haque *et al.*, 1992, Sharma *et al.*, 1998).

Mechanism and potential of mycoparasitism in biological control have been extensively reviewed in many publications (Kranz 1973; Kuhlman 1980; Mathur & Mukherji, 1981; Sharma & Heather, 1981 a b; Kranz, 1981; Tsundea *et al.*, 1981; Chet *et al.*, 1981; Papavizas 1985; Sundheim 1986; Lumsden, 1981; Well 1988; Haque *et al.*, 1992; Dayal, 1989; Wolfehechel & Lensen 1992).

It is widespread and very common phenomena in various groups of fungi. However, the major genera belongs to this category are *Piptocephalis, Syncephalis, Dimargaris, Dispira, Tieghemiomyces, Comiothyrium minitans* (Manocha, 1988; Dube and Podile, 1988).

Sharma & Paul (1998) stated, "Practical application of myco-parasitism in disease management has yet to be exploited due to some physical and nutritional constraint. However, it has significant potential for eradication of pathogen and remains an attractive strategy in biological control.

Act on the Host

There are instances where non-pathogenic strains of fungi induced significant resistance in host against pathogens. For instance, many virulent strains of fungi possess the ability to breech intact host barriers and invade tissues, thereby potentially inducing resistance. Freeman and Rodrequez (1993) have reported that a single gene manipulation in the cucurbit anthracnose fungus " *Colletotrichum magna*. Convert the fungus into a non pathogenic endophytic mutualist" with ability to protect cucurbits against the pathogenic wild type. They have proposed "the use of such mutant as biocontrol agent". A transgenic strains of *Gliocladium virens* containing *Neurospora crassa* gene for *Bacillus tubilin*, a Benomyl resistant gene, have very high biocontrol potential against *Rhizoctonia solani* (Mischke, 1993). Villajuan *et al.* (1996) demonstrated that damping of disease of cucumber caused by *R. solani* could be controlled by using non-pathogenic bi-nucleate *Rhizoctonia*. There are several formulations with biocontrol agents which are effective against pathogens, some of them are listed in Table 18.3.

In addition to fungal and bacterial strains, attenuated strains of viruses have also reported as potential biocontrol agents for many plant diseases. Significant reduction in pathogenicity of many fungal pathogens *viz., Endothea parasitica, Gaeumannomyces graminis tritici, Helminthosporium victoriae* and *Rhizoctonia solani* due to ds RNA have been reported. Grente and Sauret (1969), have successfully controlled Chestnut blight caused by *E. parasitica* by hypovirulent strain of the same fungus carrying ds RNA. Tomato and Citrus plants were protected against virulent strains of TMV and Citrus Tristeza virus respectively by using mild strains. Mild strains of potato spindle tuber viroid (PSTV) protected potato plants against severe strains of viroid. There are certain limitations in the use of virus as biocontrol agents. Firstly, there is a possibility of hybrid formation between mild and virulent genomes. Secondly, uniform distribution of mild strains may not be possible in all cases and thirdly, correct selection of stable, mild strains of virus for commercial use (Purkayastha, 1998).

Application Techniques

Delivery System

In order to achieve significant level of control of plant diseases, optimization of suitable delivery systems is one of the most important steps. These are coordinated to be the major limitations, however, several techniques have been employed for the delivery of biocontrol agents. Biocontrol agents have been applied in liquid broths (Keer, 1980; Marois *et al.*, 1982) as root dippings (Yuen *et al.*, 1983), Seed treatments (Agarwal & Singh, 2003; Windels, 1981; Harman *et al.*, 1980; Kehri & Chandra, 1991; Agarwal & Tripathi, 1999), in organic matter (Akhtar *et al.*, 1982; Sivan & Chet, 1986; Upadhyay & Mukhopadhyay, 1983, 1986), Soil and in vermiculate or in clays such as pyrax (Fravel *et al.*, 1983; Maul *et al.*, 1980). Some inert granules like diatomceous earth (Beckman & Rodriguez-Kabana 1975) or lignite (Jones *et al.*, 1984) have also used to deliver the organisms. To achieve success in the management of aerial diseases, BCA has been directly applied by forcing the inoculum suspensions thorough a tube leading to the cutting blade where it comes in contact with the cuttings *i.e.* control agent delivered by a puring action. Forcing impregnated hardwood dowels into timber with a nail gun and firing metal tipped birth dowels into the trunk with a rifle are the other methods used to delivered BCA

(Sharma, 1993). Various aspects of delivery systems have been extensively reviewed in many publications (Agarwal & Singh, 2003; Naik & Sen, 1988).

Table 19.3: Commercial Microbial Biocides Used in Agriculture and Forestry

Sl.No.	*Trade Name of the Biocide, used for Pathogen*	*Biocontrol Agent*	*Target Pathogen*
1.	BINAB-T	*Trichoderma* sp.	*Chondrostereum purpureum/ Endothia parasitica/Verticillium malthousei*
2.	Dagger G	*Pseudomonas fluorescens*	*Pythium ultimum/Rhizoctonia solani.*
3.	Galltrol	*Agrobacterium radiobacter*	*Agrobacterium tumifaciens*
4.	Mycostop	*Streptomyces* sp.	*Alternaria brassicola*
5.	Norbac 84-C	*Agrobacterium radiobacter*	*Agrobacterium tumifaciens*
6.	P.G. Suspension	*Phlebia gigantea*	*Heterobasidion annosum*
7.	Polygandron	*Pythium oligandurm*	*Pythium ultimum*
8.	Gliogard	*Gliocladium virens*	*Pythium/Phytophthora*
9.	F-Stop	*Trichoderma harzianum*	*Rhizoctonia solani/Sclerotium* spp.
10.	Aspire	*Candida oleophila*	*Monilinia fructicola* and several other fungi
11.	Kodiak	*Bacillus subtilis*	Several root pathogens
12.	Bio-Save 10LP, 110	*Pseudomonas syringe* (two strains)	*Monilinia fructicola* and several other fungi
13.	Phalada 102G	*Gliocladium virens*	*Fusarium/Phytophthora/Rhizoctonia/Pythium*
14.	Phalada 104 PF	*Pseudomonas fluorescens*	Air-borne fungal and Bacterial Pathogens.
15.	Phalada 105 TH	*Trichoderma harzianum*	*Fusarium/Phytophthora/Rhizoctonia/Pythium*
16.	Phalada106 TV	*Trichoderma viridae*	Air, water, soil-borne pathogens
17.	Bioderma	*Trichoderma* spp.	Fungal, Bacterial plant pathogens
18.	Actinovate	*Streptomyces lydicus*	Soil-borne pathogens
19.	AQ 10 Biofungicide	*Ampelomyces quisqualis* M-10	Powdery Mildew

Source: Sharma 1998, Sharma & Paul, 1998.

Conclusion

There are several commercial and regulatory challenges involved in utilizing biocontrol agents, *viz.* Viable market size, high performance and consistency, persistence of effect, safety, stability, product differentiation, indigenous microbes saprophyte, minimal capital costs cheap and easy to produce and using applied conventional technology (Powell *et al.*, 1990; Jones, *et al.*, 1984).

The mystery of microbes should be unraveled and their potentials are applied as biocontrol agents. By using microbial technology we will be in harmony with natures checks and balance as a consequence of which sustainable and profitable agriculture can be enjoyed which is beneficial to all eco-friends including human beings.

REFERENCES

Agarwal, Abha & Tripathi, H. S. (1999). Biological and chemical control of *Botrytis* grey mould of chickpea. *J. Myc. Pl. Pathol.*, **29**: 52-56.

Agarwal, G. P. (1991). Biological Plant Protection: Recent Development. Proc. 78th Ind. Sc. Cong.: Part II. Presidential address, pp. 1-20.

Agarwal, P. & Singh, T. (2003). Biocontrol agents in the management of plant pathogens. In: *Plant Protection: A Biocontrol Approach*, (Ed. Pravin Chandra Trivedi), pp. 180-193.

Aggarwal, Ashok & Mehrotra, R. S. (1997). Biological Control of Plant Pathogens. In: *New approaches in Microbial Ecology*, (Eds. J. P. Tiwari, G. Saxena, N. Mittal, I. Tiwari and B. P. Chamola), Aditya Books Pvt. Ltd. New Delhi, pp. 411-426.

Akhtar, C.M., Amin, C.M., & Salim, C.M. (1982). Investigation on the Biological control of *Rhizoctonia solani* and *Fusarium oxysporum* f sp *lini* Causing root rot and wilt disease in cotton and Linseed crops respectively Faislabad, Pakistan, Ayub Agri Res Inst., 63p.

Altman, J., Toussoun, R.V., Bega, T.A., & Nelson, P.E. (1970). *Root Diseases and Soil Borne Pathogens*. University of California Press, Berkeley.

Baker, R. (1968). Mechanisms of biological control of soil borne pathogens. *Ann. Rev. Phytopathol.*, **6**: 263-294.

Baker, K.E. & Cook, R.J. (1974). *Biological Control of Plant Pathogens.* Amer. Phytopathol. Soc., St. Paul, Minnesota.

Baker, P.A.H.M., Lamers, J.G., Bakker, A.W., Marugg, J.D., Weisbeek, P.J. & Schippers, B. (1986 b). The role of siderophores in potato tuber yield increase by *Pseudomonas putida* in short rotation of potato. *Neth. J. Plant Pathol.*, **92**: 256-294.

Baker, P.A.H.W., Baker, A.W., Marugg, J.D., Weisbeek, K.P.J & Schippers, B. (1987). Bioassay for studying the role of *Pseudomonas* spp. In short potato rotations. *Soil Biol. Biochem.*, **19**: 451-457.

Baker, R. & Dickman, M. B. (1992). Biological control with fungi. In: *Soil Microbial Ecology: Applications in Agricultural and Environment Management*, (Ed. Blaine, F.). Metting Jr., pp. 275-305.

Baker, R., Elad, T. & Sneh, B. (1986 a). Physical, biological and host factors in iron competition in soils. In: *Iron, Siderophores and Plant Diseases*, (Ed. Suinburne, T. R.). Plenum Press, New York, pp. 77-84.

Banerjee, Ratna. (2000). Management of Fusarial Blight of Methi (*Trigonella foenum-graecum*) through some fungi inhabiting sewage irrigated soil. A thesis submitted for PhD. In Microbiology to R.D. University.

Barton, L.L. & Hemming, B.C. (1993). *Iron chelation in plants and soil microorganisms*. Academic Press, New York.

Beagle-Ristaino. J.E. & Papavizas, G.C. (1985). Biological control of Rhizoctonia stem canker and black scurf of potato. *Phytopathol.*, **75**: 560-564.

Beckman, P.A., & Rodriguez-Kabana, R. (1975). A system for the growth and delivery of biological control agents to the soil. *Phytopathol.*, **65**: 819-821.

Bell, D. K., Wels, H. D. & Markham, C. R. (1982). In vitro antagonism, of *Trichoderma* species against six fungal plant pathogens. *Phytopathol.*, **72**: 379-381.

Blakeman, J. P. & Fokkema, N. J. (1982). Potential of Biological Control of Plant Disease on Phylloplane. *Annu. Rev. Phytopathol.*, **20**: 167-192.

Brian, P. W., Wright, J. M., Stubbs, J. & Way, A. M. (1951). Uptake of antibiotic metabolites of soil microorganisms by plants. *Nature*, **167**: 347-349.

Brisbane, P. G. & Rovira, A. D. (1988). Mechanism of inhibition of *Gaeumannomyces graminis* var. *tritici* by fluorescent pseudomoans. *Plant Pathology,* **37**: 104-111.

Burr, T. J., Schroth, M. N. & Suslow, T. (1978). Increased potato yields by treatment of seedpines with specific strains of *Pseudomonas fluorescens* and *P. putida. Phytopathol.,* **68**: 1377 -1383.

Chet, I. (1989). Mycoparasitism-Recognition, Physiology and Ecology. In: *New Directions in Biological Control,* (Eds., Baker, R. and Dunn, P. C.). R. Liss, New York, pp. 725-733.

Chet, I., Harman, G. E. and Baker, R. (1981). *Trichoderma hamatum:* Its hyphal interaction with *Rhizoctonia solani* and *Pythium* spp. Microb. Eco., **17**: 29-38.

Clarke, F.E. (1965). The concept of competition in microbial ecology. In: *Ecology of soil borne pathogens,* (Eds. Baker, K.F. & Synder, W.C.), California Uni. Press, Berkley, pp. 339-345.

Claydon, N. & Allan, M. (1987). Antifungal alkyl pyrons of *Trichoderma harzianum. Trans. Br. Mycol. Soc.,* **88**: 503-513.

Cook, R. J. (1993). The role of Biological Control in Pest Management in the 21st Century. In: *Pest Management: Biologically Based Technologies,* (Eds. R. D. Lumsden and J. L. Vaughn.), Washington, D. C. pp. 10-20.

Dayal, R. (1989). Biological Control of Plant Diseases by Mycoparasitism. In: *Plant Microbe Interactions,* (Ed. K. S. Bilgrami), Narendra Publishing House, Delhi, pp. 199-210.

Defago , G., Berling, C. H., Burger, U., Haas, D., Kahr, C., Keel, C., Voisard, C., Wirthner, P. & Wuthrich, B. (1990). Suppression of black root of Tobacco and other root diseases by strains of *Pseudomonas fluorescens:* potential applications & mechanisms. In: *Biological Control of Soil borne Plant Pathogens,* (Eds. Hornby, D, Cook, R. J., and Y. Heins), CAB International, Wallingford, Oxford, U.K., pp. 93-108.

Dennis, C. & Webster, J. (1971, a). Antagonistic properties of species groups of *Trichoderma* I. Production of volatile antibiotics. *Trans. Brit. Mycol. Soc.,* **157**: 25-30.

Dennis, C. & Webster, J. (1971, b). Antagonistic properties of species groups of *Trichoderma* II. Production of volatile antibiotics. *Trans. Brit. Mycol. Soc.,* **157**: 41-50.

Dube, H. C. & Podile, A. R. (1988). Biological control of fungal diseases of plants. In: *Perspectives in Mycology & Plant Pathology,* (Eds. V.P. Agnihotri, A.K. Sarbhoy & Dinesh Kumar), Malhotra Publishing Company, New Delhi, pp. 496-509.

Elad, Y. & Baker, R. (1985). The role of competition for iron carbon in suppression of chlamydospore germination of *Fusarium* spp. by *Bendomonas* spp. *Phytopathol.,* **75**: 1053-1059.

Fravel, D. R. (1988). Role of antibiosis in biocontrol of Plant Diseases. *Annu. Rev. Phytopathol.,* **26**: 75-91.

Fravel, D. R., Davis, J. R. & Sorensen, L. H. (1986). Biological control of *Verticillium* wilt of potato in the field by *Talaromyces flavus. Phytopathol.,* **76**: 863-867.

Fravel, D. R., Papavizas, G. C. & Marois, J. J. (1983). Survival of ascospores and conidia of *Talaromyces flavus* in field soil and pyrax. *Phytopathol.,* **73**: 821-826.

Freeman, S. & Rodriquez, R. J. (1993). Genetic conversion of a fungal plant Pathogen to a nonpathogenic, endophytic mutualist. *Science,* **260**: 75-78.

Garrett, S.D. (1956). *Biology of Root infecting fungi.* Cambridge Univ. Press.

Gnanamanickam, S.S. & Mew, T.W. (1992). Interaction between *Sarocladium oryzae* and stem attacking fungal pathogens of rice. In: *Crop Diseases: Innovative Techniques and Management*, (Eds. Sivaprakasa, M.K. & Seetharaman, K.), Kalayani Publishers, New Delhi, pp. 255-265.

Gnanamanickam, S. S., Valasubramanian, R., Thara, K. V. & Chatterjee, A. K. (1998). Microbial Antagonists for Rice Disease Control: Molecular Approaches. In: *Microbes: For Health, Wealth and Sustainable Environment*, (Eds. Ajit Varma), MPH, New Delhi, pp. 371-388.

Godtfredsen, W.O. & Vangedal, S. (1965). Tichodermin, a new sesqueterpine antibiotic. *Acta Chemica Scand.*, **91**: 1088.

Grente, J. & Sauret, S. (1969). L'hypovirulence exclusive est-elle controled par des determinants cytoplasmiques. *C.R. Acad. Sci. Paris*, **268**:3173-3176.

Haque, S., Elter Hamul & Gaffar. A. (1992). Efficacy of *Trichoderma* spp. and *Rhyzobium meliloti* in the control of root rot of fenugreck. *Pak. J. Bot.*, **24**: 217-221.

Harman, G. E., Chet, I. & Baker, R. (1980). *Trichoderma hamatum* effects on seed and seedling disease induced in radish and pea by *Pythium* sp. or *Rhizoctonia solani*. *Phytopathol.*, **70**: 1167-1172.

Hawksworth, D. L., Kirk, P. M., Sutton, B. C. & Pegler, D. N. (1995). *Ainsworth & Bisby's Dictionary of the Fungi*, 8th edn, CAB International, Wallingford.

Hermosa, M R., Grondona, I., Iturriaga, E. A., Diaz-Minguez, J. M., Castro, C., Monte, E. & Garcia-Acha, I. (2000). Molecular characterization and Identification of Biocontrol Isolates of *Trichodema* spp. *Appl. and Env. Microbiol.*, pp. 1890-1898.

Howell, C. R. & Stripanovic, R.D. (1979). Control of *Rhizoctonia solani* on cotton seedlings with *Pseudomonas fluorescens* and with an antibiotic produced by the bacterium. *Phytopathol.*, **69**: 480-482.

Howell, C.R. & Stripanovic, R.D. (1980). Suppression of Phytium ultimum induced damping off on cotton seedlings by *Pseudomonas fluorescene* and its antibiotic pyoluteorin. *Phytopathol.*, **70**: 712-715.

Howie, W. J. & Suslow, T. V. (1991). Role of antibiotic biosynthesis in inhibition of *Pythium ultimum* in cotton spermosphere and rhizosphere by *Pseudomonas fluorescence*, pp. 393-399.

Hussain, A. & Thakur, R. N. (1962). Control of *Alternaria* blight of rape and mustard by Griseofulvin. *Plant Dis. Reptr.*, **46**: 672-673.

Janisewicz, W. J. & Roitman, J. (1988). Biological control of blue mold and gray mold on apple and pear with *Pseudomonas cepacia*. *Phytopathol.*, **78**: 1697-1700.

Jariwala, S., Chandra, C., Upadhyay, R. S. & Bharat, Rai. (1997). Role of *Trichoderma* in Biological Control of Plant Disease. In: *New approaches in Microbial Ecology*, (Eds. J. P. Tiwari, G. Saxena, N. Mittal, I. Tiwari and B. P. Chamola), Aditya Books Pvt. Ltd., New Delhi, pp. 474-502.

Jones, R.W., Pettit, R. E. & Taber, R.A. (1984). Lignite and Stillage: carrier and substrate for application of fungal biocontrol agents to soil. *Phytophpathology*, **74**: 1167-1170.

Keer, A, (1980). Biological control of crown gall through production of agrocin 84. *Plant Disesase*, **64**: 25-30.

Kehri, H.K. & Chandra, Sudhir. (1991). 'Antagonism of Tricoderma viride to Macrophomina phaseolina and its application in the control of dry root rot of mung'. *Indian Phytopath.*, **44(1)**: 60-63.

Kessel, G. J. T., Haas, B. H. De., Werf, Woke van Der & Jurgen, Kohl. (2002). Competitive substrate colonization by *Botrytis cinevea* and *Ulocladium atrum* in relation to Biological control of *B. cinevea* in Cyclamen. *Mycol. Res.*, **106(6)**: 716-728.

Klingstrom, A. & Lundeberg, G. (1978). Control of *Lophodermium* and *Phacidium* needle cast and *Sleroderris* canker in *Pinus sylvestris. Environ. J. Forest. Pathol.*, **8**: 20-25.

Kono, Y., Tkeuchi, S. & Yonehara, H. (1968). Studies on blasticidin A. *J. Antibiot.*, **21**: 433-438.

Kranz, J. A. (1973). Host list of the rust parasite *Eudarluca caricis. Nova Hedwigia.*, **24**: 169-180.

Kranz, J. A. (1981). Hyperparasitism of biotrophic fungi. In: *Microbial Ecology of Phylloplane*, (Ed. J. P. Blakeman), Academic Press, London, pp. 327-352.

Kuhlman, E. G. (1980). Hypovirulence and hyperparasitism. In: *Plant Disease*, Vol. 5. (Ed. J. G. Horsfall and E. B. Cowling), Academic Press, New York, pp. 363-380.

Lewis, J. A. & Papavizas, G. C. (1987 a.). Application of *Trichoderma* and *Gliocladium* in alginate pellets for control of *Rhizoctonia* damping off. *Plant Pathol.*, **36**: 438-446.

Lewis, J. A. & Papavizas, G. C. (1987 b.). Reduction of inoculum of *Rhizoctonia solani* in soil by germlings of *Trichoderma hamatum. Soil Biol. Biochem.*, **19**: 195-201.

Lewis, J. A., Papavizas, G. C. & Lumsden, R. D. (1989). A new biocontrol formulation for *Trichoderma* and *Gliocladium. Phytopathol.*, **79**: 1160-1166.

Loper, J.E. & Buyer, J.S. (1991). Siderphores in microbial interactions on plant surfaces. *Molec. Plant-Microbe Interact.*, **4**: 5-13.

Loper, J.E. & Lindow, S. E. (1993). Role of competition and antibiosis in suppression of plant diseases by bacterial biocontrol agents. In: *Pest Management: Biologically based Technologies*, (Lumsden & Vaughn eds), ACS, Washington, D.C., pp. 144-165.

Lumsden, R. D. (1981). Ecology of mycoparasitism. In: *Fungal Community. Its Organization and Role in the Ecosystem*, (Eds. Wicklow, D. T. and Carrol, G. C.), Marcel Dekker, New York, pp. 295-297.

Lumsden, R. D. & Vaughn, J. L. (1993). *Pest Management: Biologically Based Technologies*. ACS, Washington, D.C.

Manocha, M.S. (1988). Biotrophic mycoparasitism: A model system for investigations in host parasite interactions. In: *Perspectives in mycology and plant pathology*, (Eds. V.P. Agnihotri, A.K. Sarbhoy & Dinesh Kumar). Malhotra Publishing House, New Delhi, pp. 462-495.

Marois, J. J., Johnston, S. A., Dunn, M. T. & Papavizas, G. C. (1982). Biological control of *Verticillium* wilt of eggplant in field. *Plant Dis.*, **66**: 1166-1168.

Massie, L. B., Cole, H. & Duick, J. (1968). Pathogen variation in relation to disease severity and control of *Sclerotinia* dollar spot of turfgrass by fungicides. *Phytopathol.*, **58**:1616-1619.

Mathur, M. & Mukherji, K. G. (1981). Antagonistic behaviour of *Claosporium spongiosum* against *Phyllactinia dalbergia sissoo. Angew. Bot.*, **55**:78-81.

Maul, S. B., Lemke, P. A, Gerner, W. L. & Yoder, J. B. (1980). Method and apparatus for sterile cultivation of cells on solid substrate. *U.S. Patent*, **4**: 204-364.

Misato, T. Ko, K. & Yamaguchi, T. (1977). Use of antibiotics in Agriculture. *Adv. Appl. Microbiol.*, **21**: 53-88.

Mischke, S. (1993). Improvement of the Biocontrol Fungus, *Gliocladium virens,* by Genetic Manipulation. In: *Pest Management: Biologically Based Technologies,* (Eds. R.D. Lumsden and J.L. Vaughn.), Washington, D.C., pp. 355-356.

Mukerjii, K. G. & Garg, K. L. (1988 a). *Biocontrol of Plant Diseases,* Vol. I. CRC Press, Florida, USA.

Mukerjii, K. G. & Garg, K. L. (1988 b). *Biocontrol of Plant Diseases,* Vol. II. CRC Press, Florida, USA.

Mukerji, K. G., Tewari, J. P., Arora, D. K. & Saxena, G. (1992). *Recent Developments in Biocontrol of Plant Diseases.* Aditya Book Pvt. Ltd., New Delhi, pp. 195.

Mukhopadhyay, A. N. (1987). Biological Control of Soil-borne plant pathogens by *Trichoderma* spp. *Indian J. Mycol. Pl. Pathol.,* **17(1)**: 1-9.

Naik, M. K., & Sen, B. (1992). Biocontrol of Plant diseases caused by *Fusarium* species. In: *Recent development in Biocontrol of Plant Diseases,* (Eds. K. G. Mukerjii, J.P. Tiwari, D.K. Arora, G. Saxena), Aditya Books Pvt. Ltd., New Delhi, pp. 38-51.

Nandi, J., De, B. K. & Bose, S. K. (1975). Evaluation of mycobacillin and versicolin as agricultural fungicides: I Antifungal spectrum and phytotoxicity. *J. Antibiot.,* **28**: 988-992.

Nautiyal, C. S. (1999). Bioinoculants for sustainable Agriculture: Recent status and constraints. In: *Microbial Biotechnology for Sustainable Development and Productivity,* (Eds. R.C. Rajak), Scientific Publishers, Jodhpur. India.

Neiland, J. B. (1995). Siderphores: structure and function of microbial iron transport compounds. *Journal of Biological Chemistry,* **270**: 26723-26726.

Nelson, E. E. & Thies, W. G. (1985). Colonization of *Phellinus weirii* infested stumps by *Trichoderma viride* I. Effect of isolated and inoculum base. *Eur. J. For. Path.,* **15**: 425-431.

Nelson, E. E. & Thies, W. G. (1986). Colonization of *Phellinus weirii* infested stumps by *Trichoderma viride* II. Effect of isolated and inoculum base. *Eur. J. For. Path.,* **16**: 56-60.

Nigam, Nalini. (1997). Fungi in Biocontrol. In: *New Approaches in Microbial Ecology,* (Eds. J.P. Tiwari, G. Saxena, N. Mittal, I. Tiwari and B.P. Chamola), Aditya Books Pvt. Ltd., New Delhi, pp. 428-457.

Oerke, E. C., Dehne, H. W., Schonbeck, F. & Webber, A. (1994). *Crop Protection and Crop Production.* Amsterdam: Elsevier.

Papavizas, G. C. (1985). *Trichoderma* and *Gliocladium*: Biology, Ecology and potential for biocontrol. Ann. Rev. *Phytopathol.,* **23**: 23-54.

Papavizas, G. C. (1992). *Biological Control of Selected Soil - borne Plant Pathogens with Gliocladium and Trichoderma.* ISBN. 0-306-442-223-230.

Paul, A. K. (1998). Antifungal antibiotics - Present status and future Prospects. In: *Microbes for health, wealth and sustainable environment,* (Eds. Ajit Verma), Malhotra Publishing House, New Delhi, pp. 347-369.

Powell, K.A., Faull, J.L. & Renwick, A. (1990). The commercial and regulatory challenge. In: *Biological Control of Soil borne Plant Pathogens,* (Eds. D. Hornby), CAB, International, Wallingford, U.K., pp. 45-64.

Purkayastha, R. P. (1998). Crop protection: advanced biological strategies. In: *Trends in microbial exploitation,* (Eds. Rai, B., Upadhyay, R. S. & Dubey, N. K.), ISCNR, Deptt. of Botany, BHU, Varanasi, U.P., pp. 89-96.

Ragsdale, N. N. Henry, M. J. & Sisler, H. D. (1993). Minimizing non-target effects of fungicides. In: *Pest Control with Enhanced Environment Safety* (Eds. Duke, S.O. Menn, J.J. and Plimmer, J.R. , pp. 332-341. Washington, ACS, 357 pp.

Reichelderfer, K. H. (1981). Economic feasibility of biological control of crop pests. In: *Biological Control in Crop Production*, (Eds. Papavizas, G.C.). Allenbeld, Totowa, N.J., pp 403-417.

Renshaw, J. C. Robson, G. D., Trinci, A. P. J. Wiebe, M. G., Livens, F. R., Collison, D. & Taylor, R. J. (2002). Fungal Siderophores: Structures, functions and applications. *Mycol. Res.*, **106(10)**: 1123-1142.

Ridout, C. J., Coley-Smith, J. R. & Lynch, J. M. (1988). Fraction of extracellular enzyme from a mycoparasite strain of *Trichoderma harziznum. Enzyme Microb. Technol.*, **10(3)**: 180-187.

Rohilla, R., Singh, R. L. Singh, U. S. Singh, R., Duveiller, E. & Singh, H. B. (2001). Recent Advances in the Management of Plant Diseases Using Chemicals. *Indian J. Plant Pathology*, **19(1&2)**: 1-23.

Sanford, G. B. (1926). Some factors affecting the pathogenicity of *Actinomycetes scabies. Phytophpathol.*, **16**: 525-547.

Sanford, G. B. & Broadfoot, W. C. (1931). A note on the biological control of root rots of cereals. Studies of the effects of other soil inhibiting microorganism on the virulence of *Ophiobolus graminis* Sacc. *Sci. Agric.*, **11**: 512-528.

Sasaki, S., Ohta, N., Eguchi, J., Furukawa, T., Akashiba, T., Tsuchiyama, T. & Suzuki, S. (1968). Polyoxins, antifungal antibiotics. VIII. Mechanism of action on sheath blight of rice plants. *Nippon Shokubutsu Byori Gakkaiho.*, **34**: 272-279.

Satyaprasad, K., & Manoharachary, C. (1997). Biological control of soil borne disease-an overview. In: *New approaches in Microbial Ecology*, (Ed. J.P. Tiwari *et al.*), pp. 503-515.

Schippers,B. (1988). Biological control of pathogen with rhizobacteria. *Phil. Trans Rpy. Soc. Lond.*, **318**: 283-293.

Shanahan, P., O, Sullivan, D. J., Simpson, P., Glennon, J.D. & O, Gara. (1992). Isolation of 2, 4-diacetylphloroglucinol from a fluorescent Pseudomonad and investigation of Physiological parameters influencing its production. *Appl. Environ. Microbiol.*, **58**: 352-358.

Sharma, J. K. & Heather, W. A. (1981 a). Antagonism by three species of *Cladosporium* to three of *Melampsora larici populina* Klelo. *Aust. J. For. Res.*, **11**: 283-287.

Sharma, J. K. & Heather, W. A. (1981 b). Hyperparasitism of *Melampsora larici-populina* by *Cladosporium herbarum* and *C. teniussimum. Ind. Phytopath.*, **34**: 395-397.

Sharma, J. K. & Sankaran, K. V. (1988). In: *Biocontrol of Plant Diseases*, Vol.II, (Eds. K. G. Mukerji & K. L. Garg), C.R.C. Press Inc., Boca Raton, Florida, pp. 1-24.

Sharma, Mamta, Mittal, Neelima & Mukherji, K.G. (1998). Fungi: Tool for Plant Disease Management. In: *Microbes: For health, wealth & sustainable environment*, (Ed. Verma, A.), Malhotra Publishing House, New Delhi, pp. 185-239.

Sharma, P.D. (1993). In: *Fungal ecology and Biotechnology*, (Eds. B. Rai, D.K. Arora, N.K.Dubey & P.D. Sharma), Rashtogi Publications, Merrut, pp. 1-33.

Sharma, P.D. (1998). Environment friendly biocide strategy for crop protection. In: *Trends in Microbial exploitation*, (Eds. Bharat Rai *et al.*), Int. Soc. Con. Nat. Resources, Varanasi.

Sharma, P. D. and Paul, P. K. (1998). Recent tactics of Biological Control. In: *New Trends in Microbial Ecology*, (Eds. Bharat Rai and M.S. Dkhar), Dept. of Botany, NEHU, Shillong and ISCNR, Dept. of Botany, BHU, Varanasi, pp. 272-286.

Sivan, A., and Chet, I. (1986). Biological control of *Fusarium* species in cotton, wheat and muskmelon by *Tricoderma harzianum*., *J. Phytopath*., **116**: 39-47.

Smith, T.A., Baker, J.H.A. & Owen, W.J. (1992). Insensitivity of Septoria tritici and Ustilago maydis to inhibitors ornithine decarboxylase. *Mycol. Res.*, **96**: 395-400.

Sundheim, L. (1986). Use of hyperparasites in biological control of biotrophicplant pathogens. In: *Microbiology of the Phyllosphere*, (Eds. N.J. Fokkeman and J. van den Heuvel), Cambridge Univ. Press, Cambridge, pp. 333-347.

Snyder, W. C. (1960). Antagonism as a plant disease control principle. In: *Biological and Chemical Control of Plants and Animal Pests*, Washington, D.C. USA, American Association for the Advancement of Science, 1960.

Takaysu, T., Nakayama, H., Shimaju, A., Furthata, k., Ikeda, K., Furihata, K., Seto, H. & Otake, N. (1985). Rustimicin, a new macrolide antibiotic active against wheat stem rust fungus. *J. Antibiotit.*, **38**: 1806-1809.

Tahvonen, R. & Avikainen, H. (1987). The Biological control of seed-borne *Alternaria brassicola* of cruciferous plants with a powdery preparation of *Streptomyces* sp. *J. Agric. Sci. Finland*, **59**: 199-208.

Takeuchi, S., Hirayama, K., Ueda, K., Sakai, H. & Yonchara, H. (1985). Blasticidin S, a new antibiotic. *J. Antibiot.*, **11A**: 1-5.

Thomashow, L. S. & Weller, D. M. (1988). Role of a phenazine antibiotic from *Pseudomonas fluorescens* in biological control of *Gaeumannomyces graminis* var. *tritici*. *J. Bacteriol.*, **170**: 3499-3508.

Trick, C.G., Anderson, R.J., Gillam, A., & Harrison, P.J., (1983). Prorocentrin: an extracellular siderophore produced by the marine dinoflagellate *Prorocentrum minimum*. *Science*, **291**: 306-308.

Tsundea, A. & Hiratsuka, Y. (1981). Biological Control of pine stem rusts by mycoparasites. *Proc. Japan Acad.*, **57**: 337-341.

Upadhyay, J.P. & Mukhopadhyay, A. N. (1983). Effect of non-volatile and volatile antibiotics of *Tricoderma harzianum* on the growth of Sclerotium rolfsii Indian. *J. Mycol and Pl. Pathol.*, **13**: 232-233.

Upadhyay, J.P. & Mukhopadhyay, A.N. (1986). Biological control of *Sclerotium rolfsii*, *Tricoderma harzianum* in sugarbeet. *Trop. Pest Manage.*, **32**: 215-220.

Upadhyay, R. S. & Mishra, Sandhya (1998). Exploitation of fluorescent *Pseudomonas* for plant growth and biocontrol of soil- borne plant pathogens. In: *Trends in Microbial Exploitation*, (Eds. Bharat Rai, R.S. Upadhyay and N.K. Dubey), pp. 97-112.

Villajuan, A., Kageyama, K. & Hyakumachi, M. (1996). Biocontrol of *Rhizoctonia* damping-off of cucumber by non-pathogenic binucleate *Rhizoctonia*. *Eur. J. Plant Pathology.*, **102**: 227-235.

Voisard, C. Keel, C., Hass. & Defago, G. (1989). Cyanide production by *Pseudomonas fluorescens* helps suppress black root rot of tobacco under gnotobiotic conditions". *EMBO J.*, **8**: 351-358.

Weindling, R. (1934). Studies on a lethal principle effective in the parasitic action of *Trichoderma lignorum* on *Rhizoctonia solani* and other soil fungi. *Phytopathol.*, **24**: 1153-1179.

Weindling, R. & Emerson, D. H. (1936). The isolation of a toxic substance from the culture filtrate of *Trichoderma, Phytopathol.*, **26**: 1068-1074.

Well, H.D. (1988). Tricoderma as a biological agent. In: *Biocontrol of Plant Disease,* Vol. 1, (Eds. Mukerji, K.G. and Garg, K.L.). CRS Press. Inc., Florida, pp. 71-82.

Weller, D. M. (1988). Biological control of soil borne plant pathogens in the rhizoshere with bacteria. *Ann. Rev.. Phytopahol.*, **26**: 379-407.

Windels, C. E. (1981). Growth of *Pencillium oxalicum* as a biological seed treatment on pea seed in soil. *Phytopathol.*, **71**: 929-933

Winkelmann, G. (1991). Importance of siderophores in fungal growth, sporulation and spore germination. In: *Frontiers in Mycology*, (Ed. D.L. Hawksworth) , CAB International, Wallingford, pp. 49-65.

Wolfehechel, H & Hensen, D.F. (1992). Use of *Tricoderma harzianum* and *Gliocladium virens* for the biological control of post-emergence damping off and root rot of cucumbers caused by *Phythium ultimum J. Phytopathol.* (Berl), **136(3)**: 221-230.

Wright, J. M. (1956). The production of antibiotics in soil. III. Production of Gliotoxin in wheat straw burried in soil. *Ann. Appl. Biol.*, **44**: 461-466.

Yuen, G.Y., Mc Cain, A.H. & Schroth, M.N. (1983). The relation of soil type to suppression of Fusarium wilt of carnations. *Acta Hort.*, **141**: 95-102.

19

Waste Management: An Environmental Biotechnology Way

ONKAR J. CHAKRE

Bioresource Information Division, Wealth of India Project, National Institute of Science Communication & Natural Resources, CSIR, Dr. K.S. Krishnan Marg, New Delhi – 110 012, India

Waste... to the clinician, a vector in disease; to the ecologist, a disruption; to the conversationalist, a wasted resource; to the producer, harassment; to the processor, a source of profit; to the slum dwellers, a source of income; to the environmentalist, a source of pollution; and to the biotechnologist, a source of utilization and management for a sustainable future. They originate in the body, home, the factory and the field. A recent World Health Organization publication states: "Man cannot live in dignity in the midst of his own waste". However, many circumstances combine to make it necessary for man to live in such a state.

Environmental biotechnology in relation to waste involves management of accumulating hazardous waste materials and disposing them at source, ensuring better quality environment with safe or free from risks to human health for sustainable development.

Waste can be broadly classified into solid and liquid wastes. Domestic activities such as personal hygiene, preparation of food, washing clothes, household refuges or garbage, mature compost, human excreta, dead animals, straw, discarded tyres, batteries, waste papers, packing materials such as drums, cans, cartons, squeezy bottles, plastic bags or upholstery; medical wastes include unused syringes, removed plaster of Paris, colostomy bags, babies disposal nappies, incontinence pads, etc, are some of the examples of solid waste. Food and vegetable processing industry produces water wastes, which consists of hydrocarbons and small amounts of proteins and fats, dissolved compounds, pesticides, herbicides, and cleaning chemicals; fermentation industry include brewing, distilling, and wine manufacture. These industries produce liquid waste with high Biochemical Oxygen Demand (BOD) and Chemical Oxygen Demand (COD). The waste water contains tannins, phenols and organic acids. Dairy industries contribute substantially to the pollution of surface water and soil. It may contain proteins, salts, fatty substances, lactose, phosphorus, nitrogen and various cleaning chemicals. Meat, poultry and fish industries produce various kinds of blood, fats, and residues from the intestine, paunch grass and manure, nitrogen, and a high BOD. The executive director of The United Nations Environment Programme (UNEP), Klaus Toepfer, said, "In India more than two million people are expected to die because of the incomplete burning of biomass." The open fire used by the majority of the ordinary Indians to cook their foods also contributes to this.

Environmental biotechnology is different from other areas of technology, which have been in use to maintain wastes as a valuable resource and restore environmental quality since the term existed. In agricultural biotechnology, DDT (2, 2, 2–dichloro diphenyl trichloroethylene), dieldrin, parathion and other pesticides are routinely spread to eradicate mosquitoes and other microorganisms, which cause plant diseases. The pesticide residues are brought into aquatic eco-systems through surface water run-off and thus pollute the water. Since the demand for pesticides use is increasing, there is a need to develop simple, quick, cost-effective and sensitive assays to detect pesticide residues or pollutants in the aquatic environment. Under these circumstances, immuno assays of various types have been developed for monitoring harmful chemicals. The enzyme-linked immunosorbant assay (ELISA) is one of the analytical methods has proved valuable serological tool used for quick analysis/ screening of not only bacteria, viruses and other fungal pathogens, which cause plant diseases, but also determine pesticide residues in foods, water, air, soil, etc. ELISA uses labeled specific antibodies and enzymes to measure pollutant levels.

Environmental biotechnology has successfully developed genetically engineered biopesticides in crop protection. The transgenic plants that biosynthesize the *Bacillus thuringiensis* (Bt) insecticidal proteins *in situ* has been proved highly effective against the cotton bollworm and caterpillar insects that attack lettuce, cabbage, spinach, broccoli and other vegetable crops. More recently, *Pseudomonas* strains enzymatically generate hydrogen cyanide which suppresses the black root rot disease of tomato, which is caused by the fungus *Thielaviopsis basicola*. These transgenic bio-pesticides are currently available for sale in US market as commercial products, which are likely to displace a large number of highly toxic *xenobiotic* chemicals used in crop protection today.

Animal biotechnology is aiming towards higher quality of cattle populations from fewer species, specially important in regions where wastes from intensive animal rearing is leading to serious soil and water pollution. In this regard, anaerobic digesters have been used to process manure in countries like Germany, Denmark, Sudan, China, India and Zaire. Thousands of dairy farms have been constructed with waste storage lagoons, which hold millions of liters of liquefied waste. In a digester, anaerobic bacteria feast on the sludge, releasing about 1600 m^3 of biogas per day. The biogas contains about 55-60 per cent methane, 40 per cent CO_2, other gases, and a small amount of H_2S, which is used as lubricating engine oil. The digested slurry is spread on the land, which works as farmyard manure. The dairy separates out solids for compost and pumps the liquid in lagoons. It is not only spread on crops, but also used to feed the cattle. Recycling animal waste can keep water cleaner and generate electricity. Moreover, this waste is free from obnoxious odour, which otherwise may lead to souring eyes and breathing problems. Since the early seventies, the Government of India has been supporting the installation of gobar gas plants (anaerobic digesters) in rural cow-owning households and small-income farmers as an alternative to wood and kerosene for cooking, heat, electricity and other purposes. However, due to the economic constrains and relatively poor know-how of the farmers, the idea is still nascent.

Another area of environmental protection is the treatment or cleaning of industrial wastewater by microbes. A new biotech process to treat industrial waste- waters, especially those from coal and petrochemical industries has been developed by the German Company. The microbial species employed in the processes are *Pseudomonas*, *Candida*, *Cryptococcus*, and *aurobacter*. This process has a three-fold higher purification capacity as compared to conventional methods. The presence of nitrogen and phosphorus compounds in wastewater streams released into the environment leads to many problems. Nitrogenous compounds can be removed by applying aerobic nitrification coupled with anaerobic de-nitrification. This process was found to be suitable for the treatment of high ammonia industrial waste-water. Industrial waste-water treatment systems can be enriched with phosphate-accumulating microorganisms to produce a phosphate-free waste stream. The metabolic pathway for phosphate removal using nuclear magnetic resonance (NMR) as a noninvasive, nondestructive tool can also be used. High-rate aerobic waste water treatment processes that occupy less land and produce less sludge have given impetus by recent and proposed EU directives on sludge treatment and disposal. The membrane bioreactor is another example of these high–rate processes which may provide complete retention. Bio-filters, bio-films, biocatalysts and biomaterials, biosensors, flocks (suspended aggregated microorganisms), biodegradation and bioremediation, etc, have been proposed remediation strategies for polluted industrial sites, waters and atmospheres.

There is increasing interest in water reuse because of the growing problems of water supply and increasing cost of disposal, recycling and reclamation. Further cleanup of water treatment effluent is required before reuse through granular activated carbon filtration systems. The resulting water supply and toxicity must be monitored. Despite these initiatives, public acceptance of water reuse is thought to be low, unless the water has passed a natural step *i.e.* by being returned to a river.

Medical waste is generally defined as any solid waste that is generated in the diagnosis, treatment, or immunization of human beings or animals in research. David Nelson, President and CEO of Environ search, an Environmental Research Institute at Utah, USA, claimed that hospital wastes contain 25-50 per cent mostly halogenated plastic materials, are disposed in single chamber, low temperature incinerators. According to the conservative estimate only 15-20 per cent of hospital wastes in India claimed to be hazardous. Unfortunately, no biotechnology based disposal methods are available in the country to take proper care of the highly infectious biodegradable components and halogenated plastics discharged by most of our ill managed hospitals and nursing homes.

When things get more severe one can either resort to conventional methods, such as prevention, and reduction, reduce employment of degradable materials, recycling, incineration, pyrolysis and landfill or to modern innovative methods such as composting, biodegradability and bioremediation. Composting can be used as a method to stabilize and decrease sewage sludges, industrial and municipal wastes. Recently, composting has been applied to treatment of hazardous waste such as explosives and petroleum waste. Although the installation of incineration technology may substantially reduce the volume of waste, it may generate highly toxic residues if handled improperly.

Bioremediation is a novel approach of naturally stimulating microorganisms to rapidly degrade hazards organic contaminants to environmentally safe levels in soils, substrate materials, water, sludge and residues. Bioremediation involves "Bio-augmentation", wherein the *naturally occurring or genetically engineered* microorganisms are inoculated in the contaminated soil. Environmental use of microorganisms for the remediation of hazardous waste is a new emerging approach to treat the most drastic mixed waste problem. These contain a mixture of organic wastes, heavy metals and high energy radio nuclides. In this context, an extremely radiation-resistant bacterium, *Deinococcus radiodurans*, is being genetically engineered with biodegradation genes to render it suitable for the treatment of mixed wastes. The broad-specificity *Pseudomonas* enzyme, toluene dioxygenase, has been cloned, expressed and shown to be active in *D. radiodurans*, even in the presence of high fluxes of ionizing radiation. Ananda Chakraborty's invention of 'Super bug', to clean oil spills, won the case in US Supreme Court in 1980 and established the right to patent Genetically Altered Microorganisms (GAMs) for industrial, medical, agricultural and environmental applications. To provide the bacteria with sufficient O_2, the process of "Bioventing" is adopted wherein oxygen is pumped into contaminated soil. Bioremediation is an important soil and ground- water remediation strategy because it harnesses naturally occurring biological process; destroys or immobilizes contaminants rather than transfers them from one environmental medium to another; and conserves financial resources. There are lots of successful stories that portray the importance of bioremediation process. In 1975, a massive leak from a military fuel storage facility in Hanahan, South Carolina resulted in the release of about 80, 000 gallons of kerosene. It was then found that the naturally occurring bacteria present in the site were using the contaminant as their food and this was the first step that led the scientists to work on bioremediation on contaminated soil. After following the biotreatbility analysis in 1992 it was found that the contamination in the area was reduced by 70 per cent, and the groundwater that once contaminated 5000 ppb of toluene was left with nothing. Today, the US follows the bioremediation process for 90 per cent of their contaminated sites. Bioremediation strategies using Polymerase Chain Reaction (PCR) techniques are also being addressed to induce some soil organisms capable of degrading hydrocarbons in contaminated soils in land fill sites at rates 3.6 fold higher than the naturally occurring species.

The design of bioremediation system involves the following factors; substrate bioavailability, organic substrate, temperature, p*H*, moisture content, inorganic nutrients, toxins, and CO_2 concentration. Land forming, composting, intrinsic bioremediation and slurry bioreactor are some of the currently accepted methods, which are simple, rapid, inexpensive and low exposure risks procedures used for bioremediation contaminated soil and water.

Hazardous waste is dumped in a more or less uncontrolled manner. More industrialized nations and also some developing countries have come to realize how costly such ignorance may be. For instance, a total of US $8.6 billion was allotted for clean-up operations in the period from 1986-1990. In the Netherlands, the costs involved for clean-up activities have been estimated at US $5.6 billion while; the cost for handling hazardous waste dumpsites in West Germany will amount to at least US $10 billion. The only reliable way to really solve the waste problem is to reduce hazardous waste at source, *i.e.* to prevent waste from accumulating.

The waste analysis is one of the important tools in the reduction of waste formation. Environmental biotechnologists suggest techniques of monitoring, and controlling the waste during the manufacturing process. The management of the industry should take care of disposing their waste or properly recycled for the product manufacturing. If the above techniques are followed, definitely one can reduce the wastages up to some extent and may save the human environment. Environmental biotechnology is an emerging field of science and is fast expanding. Unfortunately, global economies and environmental biotechnology never come anywhere close to reality. Hence, eternal vigilance has to be priced for effective management of wastes for a cleaner and healthier environment; without which, things will not "simply work out" through environmental biotechnology. As far as waste is concerned, let all of us come up with a narrow mind with a motto, "OUT OF SIGHT OUT OF MIND". Therefore, the message is: environmental biotechnology solutions to waste management are simply not good enough; our ancient wisdom was always environmental and only industrial biotechnology holds these promise of providing the final solution. "Using the natural way for waste management can no longer remain a vision; it must become a reality.

REFERENCES

Anand, V.L. (2001). *Biohorizon*, Conf, IIT, New Delhi, Abstr. No, EB 01.

Bramucci, M.G. and Nagarajan, V. (2000). *Biotechnology*, **18:** 501-506.

Ghose, T.K. (2001). *Indian Chem Engr, Section B*, **43(2)**: 118-122.

Goldstein, H. (2002). *IEEE Spectrum*, **39(1)**: 72-77.

Indian Sci Abstr. (2002). July, **38 (14)**, Abstr No 014486.

Thassitou, P.K. and Arvanitoyannis (2001). *Trends Food Sci. and Technol.*, **12**: 185-196.

Udo, E. Simonis and E.U. von Weizsacke (1990). *Economics*. Edited by Institute for Scientific Co-operation, Tubingen, FRG, **42**: 39-53.

Wackett, Lawrence P. (2000). *Biotechnology*, **18**: 19-21.

Walters, Mary J. and Robinson, Gary K. (1997). *Biotechnology*, **15**: 280-285.

Http://www.lisdl.iisd.ca

Http://www.bee.cornell.edu./sustain/fuelcell/index.htm

Http://www.milkproduction.com

20

Problems in Management of Apple Scab in Kashmir: A Case Study

B.L. PUTTOO

Emeritus Scientist (ICAR), Division of Plant Pathology, SKUAST (J) Udehywalla, Jammu – 180 002 (J&K), India

In India, apple scab caused by *Venturia inaequalis* has been a disease of National Importance since 1974 (Anon., 1974), when entire apple plantation in Jammu and Kashmir state was under severe attack and in subsequent years it spread to adjoining apple belts and pockets of Himachal Pradesh. Uttaranchal, and far away apple trees of Arunachal Pradesh. Manipur, Sikkim and Tami Nadu. The disease is now endemic in these areas.

Spread of pathogen is through air borne conidia and wood infection. All the commercially grown varieties at present are susceptible involving emerging leaves, floral buds, fruit-lets. developing fruits, apical twigs by producing mousy gray scab lesions [Figs. 1(a, b)]. Under severe infection condition the leaves turn yellow and fall down prematurely. Fruitlet infection lead to fruit deformation and fall, while fruit infection during development phase produce scabby fruits of little commercial value. The fungus perpetuates in autumn fallen leaves on ground as a saprophytic phase and releases ascospores in early spring that become air borne and become the primary source of infection on emerging foliage and floral parts. Primary scab lesions on new growth produce a numerous asexual conidia that give rise to secondary infection through many cycles/generations in a season depending up on weather conditions.

Apple being a cash crop, calls for thorough management to ward-off scab invasion. Both cultural methods and fungicide application approaches are recommended. Cultural methods include collection and destruction of fallen leaves from orchard floor, deep ploughing of the orchard land and pruning of wood scab. Serial application of fungicides on the emerging foliage and fruits as a prophylactive measure and periodic sprays after the scab appearance as curative or eradicative measure till pre-harvest stage are suggested. Initially the operational schedule comprised:

1. Collection and burning of fallen leaves and fruits from orchard area after harvest.
2. Spraying available fungicides at following phenological stages of host:
   (a) Green tip,
   (b) Pink bud,
   (c) Petal fall,
   (d) Fruit-let,
   (e) Three weeks after fruit-let,
   (f) Three weeks before harvest.

Knowledge about the scab disease under local conditions of Kashmir valley and availability of fungicides at country level was quite inadequate during early seventies when scab appeared in orchards of Kashmir in epidemic state. To meet the sudden challenge posed by the severity of apple scab in orchard areas, lot of concern at fruit growers level and administrative levels were expressed which forced a situation to tackle the problem on the experiences gathered out side the country and also utilization of any available fungicide to fight the challenge posed. (Malik *et al.*, 1976). Fortunately, only, mancozeb based fungicide (Dithane M45) was immediately available and repeated applications of this fungicide during on-season at 3 per cent concentration did prove useful in reducing the scab incidence in treated trees. By and by, in subsequent seasons the fungicide became popular with the orchardists, who became used to its usage even at much higher concentrations up to even 5 per cent for 8 to 9 seasons exclusively. Initially the pinch of higher input cost of the fungicide was not felt due to the subsidy provided by the Government on the cost but the ineffectiveness of the fungicide at recommended concentration was adequately observed. The efficacy of recommended concentrations was verified under local orchard conditions (Dar *et al.*, 1981; Puttoo and Basuchaudhary, 1984) and the data confirmed the ineffectiveness of Dithane M45 at 3 per cent concentration in Kashmir. Emergence of field tolerant strains of scab pathogen in apple orchards that had a history of exclusive and continuous usage during past eight to nine years with Dithane M45 was established (Basuchaudhary and Puttoo, 1984). Similar approach followed by some growers with carbendazim based systemic fungicide Bavistin 50 WP that quickly followed its introduction in apple plantations in 1976 also lead to the emergence of resistant strains in the pathogen" making disease management difficult and making room for other fungicides having other modes of action and properties in a serial application schedule. (Qasba and Shah, 1987). Puttoo (1984) carried out studies on the adaptation of *V. inaequalis* to various fungicides in Jammu and Kashmir. Puttoo and Basuchaudhary (1985) reported the *in vitro* adaptability potential of scab pathogen to different fungicides. Fungal isolates exhibited differential response when adapted for 2-12 generations. It was of higher order with carbendazim, and of medium order with unspecifically protectants and low order with sterol inhibitors. Base-line sensitivity in two wild isolates of *V. inaequalis* against 21 different fungicides formulations available in India was established (Puttoo and Basuchaudhary, 1986). Qasba, *et al.* (1988) studied the genetic analysis of carbendazim resistance in *V. inaequalis*. Four levels of tolerance were identified and crosses between the isolates revealed segregation into parental phenotypes with no recombinant, indicating the mutation as the cause of resistance. Other fungicides belonging to protectant and systemic groups were also available for field trials that later got entry in spray schedule (Puttoo, *et al.*, 1976). As a result of repeated field trials conducted in different agro-climatic apple growing zones of Kashmir the following fungicides were suggested to be included in the schedule:

Unspecifically Acting Fugicides

Mancozeb 75 WP @ 0.3 per cent, Ziram liquid 27 w/w @ 0.4 per cent, Ziram 80 WP @ 0.2 per cent, Zineb 75 WP @ 0.2 per cent, Captan 50 WP @ 0.25 per cent, Captafol 80 WP @ 0.2 per cent, Dithionan 75 WP @ 0.075.

Specifically Acting Fungicides

Dodine 65 WP @ 0.06 per cent, Benomyl 50 WP @ 0.03 per cent, Carbendazim 50 WP @ 0.03 per cent, Thiophenate methyl 50 WP @ 0.03 per cent, Trizoles @ 0.12 to 0.05 per cent

Single appliction technique that involves the application of higher dose of Foltaf 80 WP, Delan 75 WP or Syllit 65 WP (1.5 per cent) at advance green tip stage to ward off primary scab infection followed by three cover sprayes with other fungicides during secondary scab development phase was also evaluated and suggested. Usefulness of Urea application @ 5 per cent to trees, one to two weeks prior to general leaf-fall in autumn was also found effective in high altitude and Karewa (upland) orchard areas of valley and suggested so as to inhibit the formation and development of ascospores in fallen leaves as a result of microbial antagonism in orchard floor to suppress the primary inoculum.

An integrated management approach for apple scab thus worked as under:

After Harvest

1. Spray Urea @ 5 per cent, 2 weeks before general leaf fall in high altitude and upland orchard areas.
2. Prune the scab lesion bearing apical twigs during dormancy and burn them.
3. Collect the fallen leaves and scabby fruits from orchard floor and adjoining fields and destroy them. This operation calls for community approach in contiguous belts.

Spring Season

1. Deep plough the orchard floor so that the left over leaf litter is pulverized with the soil.
2. Conduct first spray at green tip stage with any fungicides having persistence and redistribution property.
3. Apply second spray at pink bud stage with any protectant fungicides.
4. Spray any systemic fungicides at 80 per cent petal fall stage.
5. Spray any sterol bio-synthesis inhibitor fungicide at fruit-let stage.

Summer Season

1. Apply any other protectant or systemic fungicide not used in previous sprays 3 weeks after the fourth spray.
2. Repeat fungicidal sprays with fungicides also known to improve the fruit colour just three weeks before harvest.

Single Application Technique (SAT)

This approach of scab management is aimed to provide the fungicide cover to the emerging foliage and fruits through the property of persistence, redistribution, tenacity specific to captafol, dodine, chlorothalonil, based fungicides when applied at higher doses at advance pink bud stage as first spring spray. Fruit-let and summer season sprays are applied as in normal schedule. It is helpful in managing scab with four sprays and results in saving application costs of two sprays.

Full Bloom Spray

Orchard areas where inoculum pressure is quite heavy and highly susceptible cultivars are dominant full bloom spray with systemic fungicides prove useful in addition to other measures.

Integrated Management Schedule of Apple Scab Based on Forecasting System

Advances made in epidemiological research of apple scab over a period of years in orchard areas of Kashmir valley, Himachal Pradesh, Uttaranchal and other apple growing states of India have established the relevance of disease monitoring and forecasting in managing it with reduced number of operations and sprays and usage of fungicides on the basis of their inherent eradicative, curative, antisporelant and protective properties. These results are based on the coordinated efforts initiated by ICAR (Gupta and Thakur, 1992).

Need for conducting first spray is calculated on the maturity and discharge of ascospores that cause primary scab and infection periods for subsequent sprays. The strategy includes the methods of reducing primary inoculum by orchard sanitation, quick decomposition of leaf litter by spraying Urea (5 per cent), carbendazim (0.05 per cent) or dodine (0.1 per cent) in autumn. Usage of fungicides in alteration, belonging to ergosterol inhibiting group for their curative and eradicative nature is recommended during on-season.

Detailed spray schedule has been worked as:

Phenological Stage of Host	*Chemical/Fungicide*[1]	*Quantity/100 L water*
Green tip	Chlorothalonil	400 g
	Dodine	100 g
	Mancozeb	400 g
	Captan	300 g
	Dithianon	75 g
Pink bud	Hexaconazole	30 ml
	Fenarimol	40 ml
	Bitertinol	50 g
	Myclobutanil	40 g
	Mancozeb+Sulphur	300 + 200 g
Petal fall	Hexaconazole	30 ml
	Carbendazim	50 g
	Thiophenate methyl	50 g

Contd...

Phenological Stage of Host	*Chemical/Fungicide*[1]	*Quantity/100 L water*
Fruit-let	Dodine	75 g
	Bitertinol	50 g
	Fenarimol	40 g
	Myclobutanil	40 g
	Mancozeb	300 g
	Captan	300 g
Three weeks after fourth spray	Carbendazim + Mancozeb	25 g + 250 g
	Thiophenate methyl + Mancozeb	50 g + 250 g
	Captan	300 g
	Dithianon	50 g
Three weeks before Harvest	Mancozeb	300 g
	Captan	300 g
Post-harvest pre-leaf fall	Urea	5 kg
	Carbendazim	50 g
	Dodine	100 g

[1] Any fungicide in a stage.

Other Factors that Indirectly Result in Erratic Disease Control

1. Use of muddy water which flows through canals, streams or water channels around the orchards immediately does not permit proper suspension formation thereby reducing effectively of a fungicide.
2. Untimely sprays and improper coverage of the tree canopy also results in less disease control. Partial/selective application to the bearing trees/limbs only of a tree leads to build up of inoculum in the orchard and make management difficult.
3. Under dosing or overdosing of fungicide per unit volume of spray solution/suspension not only lead to erratic disease control but also pose serious problems of the emergence of tolerant fungal strains, pollution to orchard micro-climate. Overdosing enhances the quantities of fungicide requirements.
4. Use of expired and sub-standard fungicides and formulations.

Orchard Management

1. Planting field resistant commercially acceptable cultivars in new areas or in between old trees as replants. Grafting or budding germplasm having scab resistance upon old/unproductive trees in a phased manner.
2. Encourage collection and destruction of fallen leaves and fruits, adoption of recommended schedule on community basis so as to achieve timely blanket coverage especially in contiguous orchard belts.

3. Ensure alternation of different fungicides of different modes of action in schedule to mitigate the phenomenon of resistance/tolerance in the pathogen.

REFERENCES

Anonymous. (1974). *ICAR Group Discussion "Apple'Scab"*, Krishi Bhavan, New Delhi, March 5, 1974.

Basuchaudhary, K.C, Puttoo, B.L. (1984). Fungicide resistant strains of *Venturia inaequalis* in Kashmir: A Prediction. *Proc. British Crop Protec. Conf.–Pests and Diseases*, **6A(12)**: 509-513.

Dat, G.N., Puttoo, B.L., Ghani, M.Y. (1981). Ineffectiveness of Dithane M45 against *Venturia inaequalis* causing apple scab in Kashmir. *3rd Int. Symp. on Plant Pathology*, Indian Phytopathological Society, 14-18 Dec., New Delhi, p. 98 (Abstr.).

Gupta, V.K., Thakur, V.S. (1992). *Integrated management of apple scab disease.* Information Bulletin ICAR and Dept. of Mycol. and Pl. Path., UHF, Nauni, pp. 23.

Malik, A.G., Din, S.G., Kaul, M.L., Puttoo, B.L. (1976). Fighting apple scab epiphytotic in Kashmir. *Pesticides*, **10**: 44-46.

Puttoo, B.L. (1984). *Studies on the adaptation of Venturia inaequalis (Cke.) Wint. causing, apple scab disease to various fungicides in Jammu and Kashmir.* Ph.D. Thesis, BHU, Varanasi, UP., pp 108.

Puttoo, B.L., Basuchaudhary, K.C. (1984). Field performance of some commercial formulations of fungicides for the control of apple scab in Kashmir. *Indian Phytopath.*, **37**: 385 (Abstr.).

Puttoo, B.L., Basuchaudhary, K.C. (1985). *In vitro* adaptability potential of *Venturia inaequalis* to different fungicides. *Indian Phytopath.*, **38**: 608 (Abstr.).

Puttoo, B.L., Basuchaudhary, K.C. (1986). Management of apple scab pathogen resistant to fungicides. In: *Vistas in Plant Pathology*, (Eds. Varma, A., Verma, J.P.) Malhotra Publishing House, New Delhi, India.

Puttoo, B.L., Basuchaudhary, K.C. (1986a). Base-line sensitivity in two wild isolates of *Tenturia inaequalis* against different fungicides. *ISPP Chemical Control News Letter*, **7**: 14-15.

Puttoo, B.L., Din, S.G., Kaul, M.L., Fazili, M.A., Teng, S.N. (1976). Comparative efficacy of systemic and non-systemic fungicides in controlling apple scab (*Venturia inaeqllalis*). *Pesticides*, **10**: 51-52.

Qasba, G.N., Shah, A.M. (1987). Resistance of *Venturia inaequalis* (Cke.) Wint. to benzimidazole *fungicides. Int. J. Trop. Pl. Dis.*, **5**: 189-198.

Qasba, G.N., Shah, A.M., Dalai, M.A. (1988). Genetic analysis of carbendazim resistance in *Venturia inaequalis*. *Indian J. Expt. Biol.*, **26**: 592-593.

21

Impact of Biotechnology on Crop Improvement with Special Reference to Biotic and Abiotic Stresses

M.N. KHARE[1] & M. SHRIMALI[2]

[1] *Ex-Dean and Professor Emeritus, Jawaharlal Nehru Agricultural University, Jabalpur – 482 004 (M.P.), India*

[2] *Principal Scientist, Rajasthan Agricultural University, Agricultural Research Station, Durgapura, Jaipur – 302 018 (Raj.), India*

ABSTRACT

The agricultural production has crossed 200 mt. but to feed the ever increasing population which has crossed one billion in 2001, it is essential to increase the crop productivity and to check losses due to diseases and insect pests which are around 30 per cent. Since the structure of DNA and genes and their functions are known, the genes for desired triats are identified, isolated, transferred to the plant cell and from such transformed plant cells plants are regenerated. This is achieved by physical methods, through bacteria and viruses. Transgenic plants have been evolved with increased yields, nutritional benefits, slow ripening of fruits and vegetables with enhanced shelf life, resistant to diseases, insect pests, herbicides, increased quantity of pharmaceuticals and vaccines, greater tolerance to salinity, alkalinity, cold etc. Besides related and unrelated plants, genes have been used from animals and even humans. Biosafety cautions are still needed in the consumption of transgenic crop varieties.

The population of the country was 238 m in 1901, 351 m in 1951 and has crossed 1000 m in 2001 and is still increasing 1.8 per cent yearly. Out of the total geographical area of 329 m ha about 183 m ha is the gross cropped area. About 2/3 cropped area is rainfed. The crops suffer due to biotic and abiotic stresses. Efforts have been made and success achieved in crop improvement to get high yielding varieties and hybrids; newer and better crop production technology; plant protection measures–integrated pest management; post harvest technology; high quality seed for planting etc. but further advancements are needed in all directions to increase the productivity of crops. Biotechnology offers great hope in this direction. Genetic engineering, tissue culture, embryo transfer, cloning, genomics are some of the techniques to modify a plant as per need and planned manner. Hence, new varieties with desirable triats from various sources can be produced in shorter time than conventional breeding. Hence the emerging field of biotechnology is directly related to agricultural development leading to the ultimate upliftment of rural population. It has highly influenced the chemical and drug industry as well.

Available biodiversity needs to be explored to use genes responsible for desirable triats for production of transgenics. International Treaty on Plant Genetic Resources for Food and Agriculture has been made in November 2001 among FAO Member countries. This will enable fair and equitable sharing of benefits derived from their use for sustainable agriculture (Anishetty, 2004). Rao *et al.* (2004) have referred use of biotechnology in improvement of agricultural crops as 'green biotech' and that covering pharmaceuticals and biomedical products as 'red biotech'. Strengthening of approach for broad spectrum disease resistance genes appears possible through plant genomics.

Improved Yield

Yield is a complex genetic triat. Molecular breeding is precision breeding employed to evolve high yielding types with desired triats. Biotechnological innovations are able to increase agricultural production at lower input by increased yields with elimination or reduction in the use of pesticides and herbicides. The dwarfing technique has increased the productivity due to dwarf strong plants with increased reproductive phase. Varieties resistant to biotic and abiotic factors have been developed saving the losses by such calamities.

Genetic Modifications

Genetic modifications are made in crops for bringing improvement for different purposes. The structure and functions of genes are analysed and is now possible to determine the complete nucleotide sequence of all the genes in an organism and genetic linkage maps are prepared by restriction fragment length polymorphisms (RFLP) markers. A small mustard type *Arabidopsis thaliana* is considered a model dicot plant and rice for monocot in plant genomics. The functional genomics has been successfully studied to understand the gene control on its expression (Gour, 2002). The first disease resistance gene HMI was cloned in 1992 from maize (Johal and Briggs, 1992). Such genes are transferred to susceptible but high yielding varieties making the plant resistant.

Transgenic Plants and Seeds

Transgenic plants and seeds contain a gene or genes artificially inserted but not through pollination. Plants having transgenes are called genetically modified like Bt Cotton. In case of terminator technology the produced seeds are sterile, other components are intact like oil, protein etc. The last introduced gene produces toxins which kill the embryo. In all, three genes are involved in the process. For genetic modifications for crop improvement it is possible to identify the genes responsible for specific triats, better understanding has developed about the working of cells and organisms at the molecular, biochemical, physiological levels and techniques have been evolved to transfer genes from one plant species to another and from unrelated organisms, animals and even humans to plants. The transferred gene is called transgene and the plant developed by such transfer is called transgenic plant or transformed plant (Ghosh, 2000).

Many transgenic crop varieties have been evolved with increased nutrients, delayed fruit ripening, longer shelf life, better flavour, aroma, free from harmful contents, pharmaceuticals, vaccines, more nutritious feed for cattle etc. The maximum acreage under transgenic crop varieties under commercial production is in USA (74.8m ha) followed by Argentina (24.7m ha), Canada (7.4m ha), China (1.2m ha), South Africa (0.5m ha) and Australia (0.4m ha). More than 50 engineered transgenic crop varieties have been cleared for cultivation in USA and about a 100 more are under field tests (Namboodri, 2003).

Golden or yellow rice a transgenic rice has been developed by introducing two genes from daffodils and one from a bacterium into rice, it has more beta-carotene, a precursor to vitamin A. Another transgenic rice has been produced with two to four times increased iron level. Genes involved in the production of an iron binding protein and an enzyme which facilitates iron availability in the human diet were introduced. Transgenic soybeans have been developed with high expression of anticancer proteins which are naturally present in the seed. Oleic acid gene when introduced, resulted in increased unsaturated fat in oil. In tomato transgenic varieties rich in lycopene content, delayed ripening, improved flavour, increased shelf life have been developed. In Canola enhanced vitamin content and balance of fatty acids has been achieved. In Amaranthus lysine gene has been introduced for supply of essential amino acids. In coffee beans and tea leaves different genes are involved in the production of caffeine. These genes are turned off to develop plants to yield naturally decaffeinated products with full flavour and aroma. At present organic solvents are used to remove caffeine. In tobacco genetically engineered nicotine free varieties have been developed in which leaves do not synthesise nicotine (Chawla, 2003).

Tolerance to Biotic and Abiotic Stresses

Vidhyasekaran (2002) has reviewed the work done on molecular breeding for disease resistance. Disease resistance genes and defence genes can be identified, cloned and transferred to other cultivars. The resistance genes do not code for any toxic chemicals helpful in disease resistance, it is done by defence genes which are also cloned and transferred to a susceptible cultivar. Numerous *R* genes have been cloned and characterised, some are listed by Paul Khurana *et al.* (2002). Transgenic plants resistant to diseases have been developed by transferring genes from bacteria, fungi, viruses, insects, unrelated plants, chicken, crab, frog, fish and even humans.

Investigations on the biochemical nature of and the signals involved in plant reaction to pathogen, the genes involved in resistance reactions are identified and introduced into crop plants to protect them against diseases. Plants can be protected from diseases with transgenes derived from the pathogens themselves. Plant viral genes can protect plants from infection by the virus from which transgene is derived. Peptides and proteins with antimicrobial properties when produced by plants have the potential to strengthen plant resistance to fungal and bacterial pathogens. Fungi, bacteria, animals and humans all contain genes exceeding antimicrobial compounds which are identified, cloned and engineered into plants to resist against diseases. Based on the structural features *R* proteins are classified into five groups (Paul Khurana *et al.*, 2002). Although plants have mechanisms to protect themselves against plant pathogen attack, they do not have 'immune system'. Plants can be engineered to express an antibody against a protein crucial for pathogenesis resulting in a level of immunity or resistance to the pathogen. Multiple resistance to more than one pathogen is also incorporated in varieties as in rice against sheath blight and bacterial leaf blight pathogens.

Resistance to Insect Pests

Transgenic plants developed by transferring genes from *Bacillus thuringiensis* (Bt) are resistant to insect pests as in cotton against bollworm and budworm; in maize against maize borers; in potato against potato beetle etc. Bt contains crystalline Cry protein. The protein breaks down in the insect gut, releases a toxin called delta-endotoxin which binds with the intestine and forms pores in the lining, causing ion imbalance, paralysis of digestive system and death of the insect. Bt cottons have potential of giving higher yields and in case of Bt potato insecticide application cost is reduced with 40 per cent gain. In rice Cryl Ab, Cryl AC genes have conferred resistance against stem borers and leaf folders, and GNA (*Galanthus agglutinin*) gene against sucking pests (Mishra, 2004). Permission for commercial cultivation of Bt cotton in the country was given by Government of India on March 26, 2002. Bt crops reduce cultivation costs, environment contamination, pesticide residues in the produce, threat to nontarget organisms and health risks of field workers (Rao *et al.*, 2004).

Resistance to Diseases

In papaya resistance to papaya ring spot virus has been introduced through UH- Sun up and UH-rainbow. In sunflower resistance to Sclerotinia stem rot and Argentina looper has been introduced. Grapes with a silk worm gene have been developed which are resistant to Pierce's disease, an important bacterial disease. Lactoferrin an iron binding glycoprotein with antibacterial properties has been used to, have resistance against fire blight in pear. Transgenic varieties have been developed in a number of crops against different pathogens as in tomato (*Phytophthora infestans*); potato (*P. infestans, Verticillium dahliae, Ralstonia solanacearum, Erwinia carotovora*, PVY, PLRV, PVX); tobacco (*Rhizoctonia solani, Phytophthora parasitica, Cercospra nicotianae, Pseudomonas syringae* pv. *tabaci, P. solanacearum*, TMV, CMV, TNV); rice (*R. solani*, RYMV); *Brassica napus* (*R. solani, Sclerotinia sclerotiorum, Phoma lignorum*); apple (*Erwinia amylovora*) (Paul Khurana, *et al.*, 2002).

Transgenic tobacco plants producing chloroperoxidase encoded by *Pseudomonas pyrrocircia* exhibited higher resistance against anthracnose caused by *Colletotrichum destructivum* (Rajasekaran *et al.*, 2000). Lactoferrin an iron binding glycoprotein with antibacterial properties has been used to have resistance against fire blight in pear.

Genes from animals have been used to obtain resistance to diseases in transgenic plants. The hen egg lysozyme gene (chly gene) has been used to have resistance in potato against black leg disease caused by *Erwinia carotovora* pv. *alroseptica* (Serrano *et al.*, 2000). A gene encoding a lytic peptide (tachyplesin) from horse shoe crab has been introduced in potato for resistance to tuber rot due to *E. carotovora* (Allefs *et al.*, 1996). A gene encoding magainin III peptide cloned from African Clawed frog has been used for increased resistance to *Xanthomonas campestris* pv. *campestris* in cauliflower (Braun *et al.*, 2000). Antibacterial peptides have been isolated from frogs (Vidhyasekaran, 2002). Human lysozyme degrades fungal and bacterial cell walls. The gene was introduced in carrot for strong resistance against *Erysiphe heraclei* and *Alternaria dauci* and in tobacco against *Pseudomonas syringae* pv. *tabaci* (Nakajima *et al.*, 1997).

Insects possess antimicrobial proteins which have been used in developing plants resistant to diseases. Cercopin an antibacterial peptide has been derived from silk moth *Hyalophora cercopia*. Several stable analogs of cercopin B like SB-37, Shiva-1 and MB-39 have been synthesized (Vidhyasekaran, 2000). Transgenic tobacco plants with Shiva-1 with higher resistance to *P. solanacearum*, transgenic apple and pear plants with SB-37 having resistance to *E. amylovora* have been developed (Jaynes *et al.*, 1993; Norelli *et al.*, 1999; Reynoird *et al.*, 1999). Transgenic cauliflower plants with Shiva protein exhibited resistance to *X. campestris* pv. *campestris* (Braun *et al.*, 2000). A synthetic gene encoding an N-terminus-modified-melittin cationic peptide chimera when introduced to two potato cultivars resulted in transgenic potato resistant to *P. cactorum, Fusarium solani* and *E. carotovora* (Osusky *et al.*, 2000). Sarcotoxin an antimicrobial peptide from *Sarcophaga peregrina* (sarcotoxin IA) when introduced in tobacco plants exhibited resistance to *E. carotovora, P. syringae* pv. *tabaci, R. solani* and *Pythium aphanidermatum* (Mitsuhara *et al.*, 2000; Ohshima *et al.*, 1999).

Virus diseases can be controlled by inducing the plants to produce antibodies to viruses (Vidhyasekaran, 2000). Pathogens produce toxins and toxin insensitive plants exhibit enhanced resistance to pathogens. Such plants have been generated by cloning the responsible gene and transferring to develop transgenics (Vidhyasekaran, 1993). Paul Khurana *et al.* (2002) have listed some examples of transgenic plants in crops developed with resistance to fungal, bacterial and viral diseases.

Abiotic Stresses

Abiotic constraints to crop productivity relate to water, temperature, frost, nutrients etc. The genetically modified plants escape the damage. Substantial progress has been made in identifying genes to various abiotic stresses but according to Chopra and Sinha (1998) due to the dynamic nature of environment and plant, it may take time in achieving success through biotechnology in field crops. A salt tolerance gene from mangroves (*Avicennia marina*) has been identified, cloned and transferred to other plants. The transgenics were tolerant to higher salt concentration. Gut D gene from *Escherichia coli* is used to generate salt tolerant transgenic maize. Transgenic tomato has been developed which proper in the presence of high sodium salts (Chopra, 2002). Cold tolerance is being introduced in strawberry by transferring a gene from fish from Antarctica.

Herbicide Tolerance

Varieties in various crops have been developed with trangenes to provide tolerance to herbicides. Such herbicides are used which break down in soil quickly and leave no residue. In cotton, soybean and corn glyophosate resistance genes have been introduced for herbicide resistance.

Trees

Forest trees like poplar, aspen, spruce are transformed with various genes to provide resistance to insect pests, tolerance to herbicides and with more commercial component. In some varieties lignin content has been reduced to make wood pulp with ease.

Pharmaceutical and Vaccines from Transgenic Plants

Plants have been evolved by incorporating genes into edible plants to be used for commercial production of products like pharmaceuticals, health food, edible vaccines, insulin, edible oils harmless to heart etc. (Smith, 1996). Vaccines against gastro-intestinal tract diseases have been produced in potato, banana etc. An anticancer antibody is evolved in rice and wheat grains which is useful in diagnosis and therapy. Muskmelon plants with rabies vaccine genes have been developed by the scientists at University of Agricultural Sciences, Bangalore (Rao *et al.*, 2004). Tobacco variety that synthesises human haemoglobin and one which cleans the environment of dynamite residues have been developed (Rao *et al.*, 2004).

Methodology

Advancements have been made in the techniques related to gene transfer leading to genetic transformation and transgenic plants. The essential steps are:

1. Selection of single or group of genes with desirable characters for introduction in plants;
2. Identification and isolation of such genes;
3. Transferring the isolated gene/genes into the plant cell and
4. regeneration of complete plants from transformed cells or tissues. The gene transfer methods are of two types:

Vector Mediated

Agrobacterium tumefaciens a soil bacterium has a segment of DNA that is transferred from the bacterium, to plant cells. A strain of the bacterium is created which has Ti plasmid (T: tumour, i: inducing plasmid) with the gene to be transferred to the plant. A plasmid is a ring of DNA floating freely in the bacterial cytopalsm outside the main chromosoma, an extrachromosomal hereditary determinant. The bacterium is cultured with pieces of plant tissues/cells so that DNA transfer may occur. The transformed cells are used to regenerate plants whereas the nontransformed cells get killed by antibiotics. *A. rhizogenes* is also used for the purpose. Viruses are also used for gene transfer through vectors with a broad host range (Hayes *et al.*, 1988).

Vectorless or Direct DNA Transfer

A foreign DNA is transferred into plant genome physically by electroporation, particle bombardment, macroinjection and microinjection (Shillito *et al.*, 1985, Celis, 1984; Klein *et al.*, 1987; Wang *et al.*, 1988). Chemicals which stimulate nucleic acid uptake into protoplasts are long chain polycations or charged carbohydrate polymers (Deshayes *et al.*, 1985). Usually polyethylene glycol (PEG 6000) in combination with calcium phosphate precipitation of vector DNA is used to stimulate DNA uptake into protoplasts (Bakshi, 2000). The transformation is also conducted through liposome, silicon carbide fibre, ultrasound etc. (Chawla, 2003).

A number of DNA markers linked to resistance genes have been identified and some are listed by Paul Khurana (2002) for diseases of tomato, potato, lettuce, soybean, bean, pea, barley and wheat. For genetically complex plant disease resistance governed by polygenes, quantitative trait loci (QTL) mapping based on DNA markers is effective as for resistance to rice blast, late blight of potato, bacterial wilt of tomato etc. (Young, 1996; Paul Khurana *et al.*, 2002).

Success has been achieved in engineering transgenes into chloroplasts, instead of nucleus. It gives better gene oxpression and there is less danger of gene flow from the transgenics to non target cultivated or wild plant species (Chopra, 2002).

Risks

The transgenics with introduced alien genes may be linked to a greater unpredictability of their behaviour in natural environment and the consumers. They may become carriers of the transgenes received from other species which have secondary mobility and integrate into human cells. Food products from such genetically engineered plants may cause health hazards to humans and animals as allergens may be introduced in foods. In transgenics, fortified genes with resistance to such antibiotics when used as medicine do not work. The glyophosate causes the production of phytoestrogens which can provoke severe reproductive disturbances. Transgenic soybean can also cause allergy problems. The other problems are the fear of increase in toxins naturally occurring in parent line; inclusion of new allergens along with desired triat; inadvertent introduction of hormones into the food supply; possible transfer of resistance from genetically engineered organisms to pathogenic ones; possibility of higher level of Bt toxin in plants; development of super weeds and weeds resistant to herbicides; effect on existing biodiversity; genetic contamination etc. (Rao, 2001; Ghosh, 2000).

Considering the risks involved in transgenic crop varieties biosafety regulations have been made. According to WTO it is necessary to have proper labelling of transgenic food products in international shipment; to make information on transgenic crop varieties available; to make an international framework for regulating international trade in transgenic crops. Three committees have been formed:

1. Institutional Bio-safety Committee (IBSC)
2. The Review Committee on Genetic Manipulation (RCGM)
3. Genetic Engineering Approval Committee (GEAC).

The GEAC gives approval for field trials and commercial release or genetically engineered products. In India the Department of Biotechnology (DBT) has prepared guidelines for biosafety assessment.

Conclusion

It is essential to improve crop productivity and quality of the produce. Varieties and hybrids have been evolved by conventional breeding for higher yields with resistance to important insect pests and diseases. Recently, the progress made in biotechnology by understanding the plant genomics, proper genes could be identified, isolated, transferred to the plant cell, sand a complete plant could be regenerated from such transformed cells. DNA markers have been identified and introduction of desired characters into specific varieties could make it possible to have types resistant to important insect pests and diseases. It is essential to work in a way to combat the biosafety issues so that the transgenic crop varieties may be used freely.

REFERENCES

Allefs, S.J.H.M., E.R. de Jong, D.E.A. Florack, C. Hoogendoom and W.J. Steikema (1996). Erwinia soft rot resistance of potato cultivars expressing antimicrobial peptide tachyplesin. *Mo. Breed.*, **2**: 97-105.

Anishetty, M. (2004). Biotechnology–Genetic conservation vital. *The Hindu Survey of Indian Agriculture*, p. 143.

Bakshi, I.S. (2000). Gene transfer technology–an overview. In: *Glimpses in Botany*, (Eds. K.G. Mukerji, B.P. Chamola and A.K. Sharma), APH Publishing Corporation, New Delhi, pp. 213-222.

Braun, R.H., J.K. Reader and M.C. Christy (2000). Evaluation of cauliflower transgenic for resistance to *Xanthomonas campestris* pv. *campestris*. *Acta Horticulturae*, **539**: 137-143.

Celis, J.E. (1984). Microinjection of somatic cells with micropipettes: comparison with other transfer techniques. *Biochem. J.*, pp. 233-281.

Chawla, H.S. (2003). Introduction to Plant Biotechnology, 2nd Ed. Oxford and IBH Publishing Co. Pvt. Ltd., New Delhi, p. 538.

Chopra, Renu Khanna and S.K. Sinha (1998). Prospects of success of biotechnological approaches for improving tolerance to drought stress in crop plants. *Current Science*, **74(1)**: 25-34.

Chopra, V.L. (2002). Biotechnology in crop productivity: improving the genetic produce. *The Hindu Survey of Indian Agriculture*, pp. 227-229.

Deshayes, A., L. Herrera-estrella and M. Caboche (1985). Liposome mediated transformation of tobacco mesophyll protoplast by an *E. coli* plasmid. *EMBO J.*, **4**: 2731.

Ghosh, P.K. (2000). Transgenic seeds in Indian agriculture. In: *Proc. 1999 National Seminar on Seed Science and Technology*, (Eds. H.S. Shetty and H.S. Praksh), Department of Applied Botany and Seed Pathology, University of Mysore, Mysore, pp. 25-38.

Gour, P.M. (2001). Recent advances in genomics and bioinformatics. In: *Recent Trend in Seed Production Management*, Department of Plant Breeding and Genetics, JNKVV, Jabalpur, M.P. pp. 255-258.

Hayes, R.J., I.T.D. Petty, R.H.A. Coutts and K.W. Buck (1988). Gene amplification and expression in plants by a replicating geminivirus vector. *Nature*, **334**: 179.

Jaynes, J.M., P. Nagpala, L.B. Destefano, J.H. Kim, T. Denny and S. Cetiner (1993). Expression of a cercopin-B lytic peptide analog in transgenic tobacco confers enhanced resistance to bacterial wilt caused by *Pseudomonas solanacearum*. *Plant Science*, **89**: 43-53.

Johal, G.S. and S.P. Briggs (1992). Reductase activity encoded by HMI disease resistance gene in maize. *Science*, **266**: 789-793.

Klein, T.M., E.D. Wolf, R. Wu and J.C. Sanford (1987). High velocity microprojectiles for delivering nucleic acids into living cells. *Nature*, **327**: 70.

Mishra, B. (2004). Rice: Exploring new opportunities. *The Hindu survey of Indian Agriculture*, pp. 29-31.

Mitsuhara, I., H. Matsufuru, M. Ohshima, H. Kaku, Y. Nakajima, N. Murai, S. Natori and Y. Oheshi (2000). Induced expression of sarcotoxin IA enhanced host resistance against both bacterial and fungal pathogens in transgenic tobacco. *Mol. Plant Microbe Interact.*, **13**: 860-868.

Nakajima, H., T. Muranaka, F. Ishige, K. Akutsu and K. Oeda (1997). Fungal and bacterial resistance in transgenic plants expressing humanlysozyme. *Plant Cell Rep.*, **16**: 674-679.

Namboodri, A.N. (2003). Genetic engineering: Present status and some future trends. *Manorama Year Book 2003*. Malayale Manorama Bangalore, pp. 26-43.

Norelli, J.L., J.A.Z. Mills, M.T. Momol and H. Aldwinckle (1999). Effect of cercopin like transgenes on fire blight resistance of apple. *Acta Horticulturae*, **489**: 273-278.

Ohshima, M., I. Mitsuhara, M. Okamoto, S. Sawano, S. Natori and Y. Ohashi (1999). Enhanced resistance to bacterial disease of transgenic tobacco plants over expressing sarcotoxin IA, a bacterial peptide of insect. *J. Biochem.*, **125**: 431-435.

Osusky, M., G.Q. Zhow, L. Osuska, R.E. Hancock, W.W. Kay and S. Misra (2000). Transgenic plants expressing cationic peptide chimeras exhibit broad spectrum resistance to phytopathogens. *Nature Biotechnology*, **18**: 1162-1166.

Paul Khurana, S.M., D. Pattanayak, A.K. Chakrabarti and S.P. Naik (2002). Molecular approaches to plant disease management. *Annu. Rev. Phytopathol.*, **1**: 83-121.

Rajasekaran, K., J.W. Cary, T.J. Jacks, K.D. Stromberg and T.E. Clevel (2000). Inhibition of fungal growth *in planta* and *in vitro* by transgenic tobacco expressing a bacterial nonhemechloroperoxidase gene. *Plant Cell Reports*, **19**: 333-338.

Rao, C.K., C.S. Prakash and A. Shantharam (2004). Agricultural Biochemistry: Exceting value addition. *The Hindu Survey of Indian Agriculture,* pp. 175-178.

Rao, V.H. (2001). Biotechnology, Biodiversity, Conservation and WTO regime. In: *Souvenir, National Agricultural Fair 2001*. Acharya N.G. Ranga Agricultural University, Hyderabad, pp. 109-126.

Reynoird, J.P., F. Mourgues, Norelli, H.S. Adwinkle, M.N. Brisset and E. Chevrean (1999). First evidence for improved resistance to fire blight in transgenic pear expressing the attacin E gene from *Hyalophora cercopia. Plant Science,* **149**: 23-31.

Serrano, C., P.A. Johnson, H. Torres, M. Gebaner, M. Gutierrez, M. Moreno, X. Jordana, A. Venagas, J. Kalazich and L. Holuigue (2000). Expression of the chicken lysozyme gene in potato enhances resistance to infection by *Erwinia carotovora* sub sp. *atroseptica. American J. Potato Res.,* **77**: 191-199.

Shillito, R.D., M.W. Saul, J. Paszowski, M. Muller and I. Portrykus (1985). High efficiency direct gene transfer to plants. *Biotechnology,* **3**: 1099.

Smith, J.E. (1996). *Biotechnology*. Cambridge University Press, Cambridge. U.K.

Vidhyasekaran, P. (1993). *Principles of Plant Pathology*. BBS Publishers, New Delhi, pp. 166.

Vidhyasekaran, P. (2002). Molecular breeding for disease resistance. *Annu. Rev. Phytopathol.,* **1**: 51-81.

Wang, Y.C., T.M. Clein, M. Fromm, J. Cao, J.C. Stanford and R. Wu (1988). Transient expression of forein gene in rice, wheat and soybean cell following particle bombardment. *Plant. Mol. Biol.* **11**: 433.

Young, N.D. (1996). QTL mapping and quantitative disease resistance in plants. *Annu. Rev. Phytopathol.,* **34**: 479-501.

22

Biotechnology in the Management of Pearl Millet Downy Mildew

R.P. THAKUR & C.T. HASH

International Crops Research Institute for the Semi-Arid Tropics (ICRISAT), Patancheru – 502 324 (Andhra Pradesh), India
Email: r.thakur@cgiar.org

Introduction

Pearl millet [*Pennisetum glaucum* (L.) R. Br.] is one of the staple cereals of the semi-arid tropical production systems of Asia and Africa. It is cultivated as a rainfed crop on marginal soils with low nutrient and low water holding capacity by resource-limited farmers. In India, pearl millet is grown on about 9.5 million ha annually, mostly in the states of Rajasthan, Maharashtra, Gujarat, Haryana, Uttar Pradesh, Madhya Pradesh, Karnataka, and Tamil Nadu, both for grain and fodder. During the past 30 years, single-cross F_1 hybrids have become more popular than the local landraces and improved open-pollinated varieties. Currently, more than 70 hybrids, mostly from private seed companies and some from public organizations are being grown in India. These hybrids occupy about 60 per cent of the major pearl millet area of Gujarat, Maharashtra, Haryana, Uttar Pradesh, Rajasthan, and Tamil Nadu, and have contributed significantly towards enhanced productivity (875 kg ha^{-1}) and production (8.35 million tonne) despite gradual decline in total cultivated area in the country (Bhatnagar, 2003).

Of more than 100 diseases that affect pearl millet, downy mildew [*Sclerospora graminicola* (Sacc.) Schroet.], smut [(*Moesziomyces penicillariae* (Bref.) Vánky = syn. *Tolyposporium penicillariae* Bref.)], ergot (*Claviceps fusiformis* Loveless), rust (*Puccinia substriata* Ell. & Barth. *var. indica* Ramachar & Cumm.), and leaf blast [*Pyricularia grisea* (Cooke) Sacc.] are economically important (Singh *et al.*, 1993b; Hash *et al.*, 1997). Among these, downy mildew is the most widespread and is highly destructive. Downy mildew is present in all major pearl millet production regions of Asia and Africa. Research at the International Crops Research Institute for the Semi-Arid Tropics (ICRISAT) for the past three decades has contributed significantly in effective management of this disease. Systematic studies on biology, epidemiology, and pathogen variability have led to the development of effective disease management practices. Effective management of this disease has been achieved through chemical seed treatment and host plant resistance. The role of biotechnology has been critical in studies of pathogenic and genetic variability, identification of genetic markers for resistance genes, and transferring resistance genes into agronomic elite inbred lines through marker-assisted backcross breeding. In this chapter, we discuss briefly the disease, the pathogen, disease management through fungicides, and disease management through host plant resistance following both conventional and molecular breeding approaches.

The Disease

Prior to 1970, downy mildew of pearl millet was of sporadic occurrence and remained only of academic interest. However, shortly after the onset of the hybrid era in India during the late 1960s, it became a major production constraint for pearl millet in India. Since then it has been a major focus of pearl millet improvement research at ICRISAT and in the All India Coordinated Pearl Millet Improvement Project (AICPMIP) of the Indian Council of Agricultural Research (ICAR), and of private seed companies. The disease symptoms appear both on foliage and panicles. Infected seedlings have yellowish chlorotic foliage which under favorable weather conditions produce masses of sporangia mainly on the undersides of the leaves. The disease is systemic and thus the entire plant gets infected, with symptom appearing on panicles as green ears wherein florets are converted into leafy structures. In such malformed panicles very few, if any, grains are formed (Fig. 22.1).

Fig. 22.1: Foliar (Left) and Green Ear (Right) Symptoms of Downy Mildew on Pearl Millet

Young seedlings up to two-weeks old are more vulnerable to infection than older seedlings. However, young growing tissues, and side and nodal tillers also get infected. Grain yield losses due to downy mildew may not exceed 20 per cent per annum worldwide. However, the disease can assume epidemic levels when a single genetically uniform pearl millet hybrid is repeatedly and extensively grown in a downy mildew endemic area (Andrews, 1987; Singh *et al.*, 1987b; Thakur, 1998; Hash and Witcombe, 2002). As the pathogen population adapts rapidly to this uniform host, inoculum levels build and under favorable environmental conditions an epidemic then occurs. In India, downy mildew epidemics have been reported on popular single-cross hybrids, such as HB 3, NHB 3, BJ 104, MBH 110, MLBH 104, and ICMH 451 resulting in severe yield losses (Singh, 1995; Thakur, 1999). The widespread use of genetically uniform host cultivars, particularly single-cross F_1 hybrids, has led to the evolution of host-specific virulence in populations of the pathogen (Thakur *et al.*, 1992). In contrast to single-cross F_1 hybrids, disease losses have been very low on heterogeneous landraces and improved open-pollinated cultivars (unless a highly susceptible landrace is grown in near a susceptible hybrid or sown in a field having high levels of soil-borne inoculum). Thus, heterogeneous cultivars can be retained in cultivation much longer than genetically uniform single-cross hybrids. However, they do not provide the yield advantages of hybrids and so are not very popular with farmers in India.

The Pathogen

The pearl millet downy mildew is caused by an obligate Oomycete *Sclerospora graminicola* (Sacc.) Schröet. The fungus reproduces both sexually and asexually, and is largely heterothallic with known sexual mating types (Michelmore *et al.*, 1982; Ball *et al.*, 1986). Physiological specialization in the pathogen was first reported in 1981 (Shetty and Ahmed, 1981). Oospores, the sexual spores formed in necrotic leaf tissues of infected plants, are thick-walled structures that survive for 10 years or more in the soil and play a crucial role in survival, spread and as primary inoculum source of the pathogen (Singh, 1995). Sporangia, the asexual spores formed on chlorotic foliar tissues, play significant role in disease spread (Singh and Williams, 1980). A single sporangium on maturity produces several motile zoospores that require a thin film of water on the leaf surface to swim and germinate to produce infection hyphae. The roles of spornagia and oospores in downy mildew disease cycle are shown in Figure 22.2.

Genetic variability in plant pathogens is important for survival of their species and biotypes. The primary objective of many studies of variation in pathogenicity or virulence is to enable the exploitation of host plant genetic resistance for disease management (Crute, 1981). When selecting breeding material for disease resistance it is important to know which pathotype to use in the screening process, how the resistance is expressed and inherited, and whether the resistance is likely to be adequate and durable. It is also important to determine whether the pathogen can change in response to the selection pressure resulting from host plant resistance and how rapidly such changes can occur.

Large variation in pathogenicity has been reported among the isolates of *S. graminicola*. Isolates from Africa and India vary in pathogenicity and have been shown to be sexually compatible (Ball, 1983; Idris and Ball, 1984; Ball *et al.*, 1986; King, 1992; Werder and Ball 1992; Thakur, 1995; Thakur and Rao, 1997; Thakur *et al.*, 1998; 1999; 2002; Sivaramakrishnan *et al.*, 2003). In India, increased virulence in *S. graminicola* has been attributed to large-scale cultivation of a number of genetically homogeneous single-cross F_1 hybrids (Thakur, 1999; Thakur *et al.*, 1999). The recent virulence analysis of 15 *S. graminicola* isolates from various parts of India, inoculated on putative host differentials, classified them into five major pathotype groups distributed in hybrid-intesive states of Rajasthan, Maharashtra, Gujarat, Madhya Pradesh, Karnataka, and Andhra Pradesh (Sivaramakrishnan *et al.*, 2003). Increased virulence diversity in the pathogen population has shortened the useful life of several commercial hybrids and farmers have suffered heavy economic losses due to reduced grain and fodder yields.

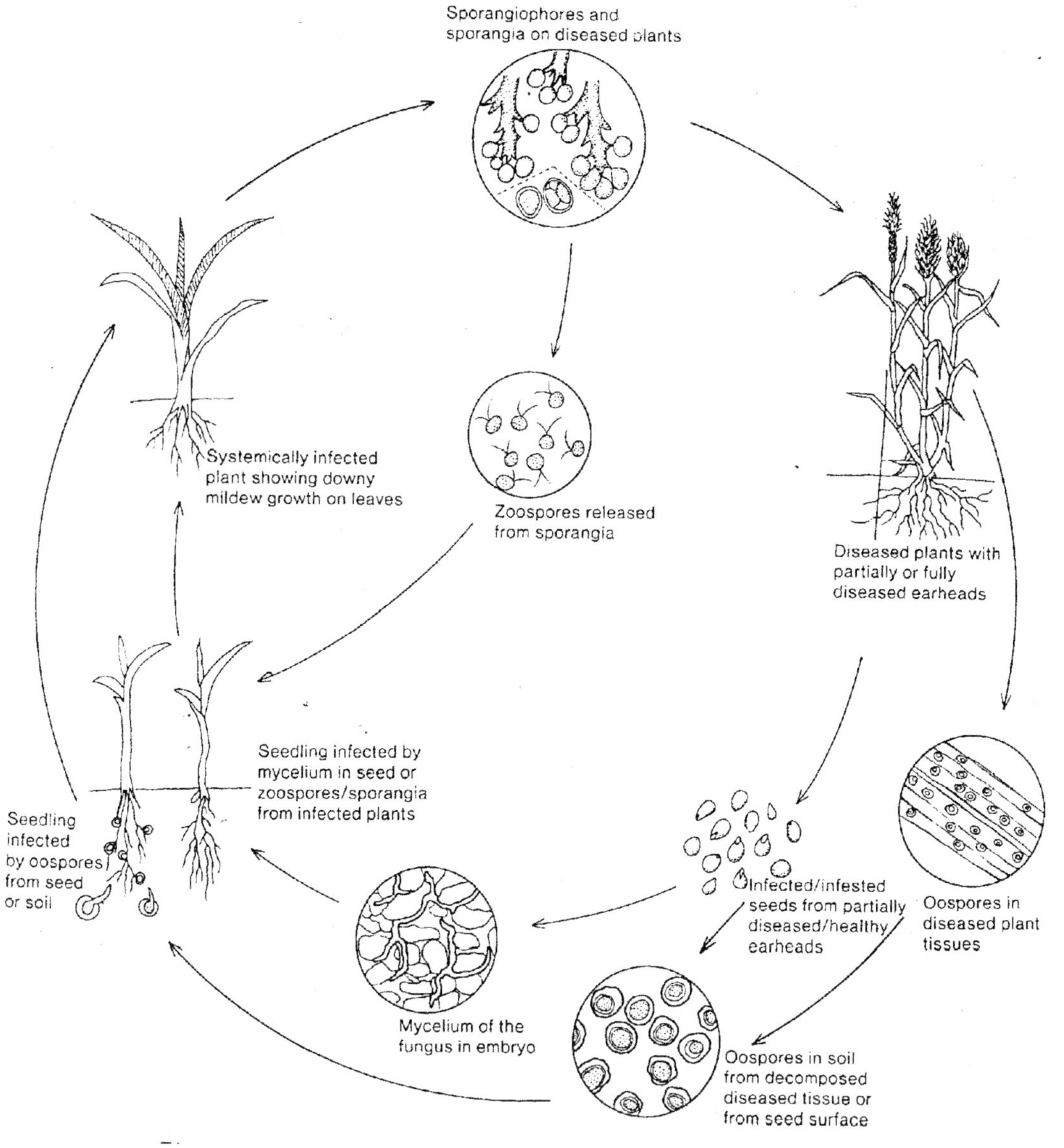

Fig. 22.2: The Infection Process and Disease Cycle in Pearl Millet Downy Mildew (*Source*: Chahal *et al.*, 1994)

Monitoring virulence changes in the pathogen population, identifying resistance to specific and multiple pathotypes, and directing breeding programs towards strategic utilization and deployment of resistance genes form the basis of long-term downy mildew disease management in pearl millet.

DNA markers have been used to characterize pathogen isolates and to determine the genetic structure of pathogen populations and its relationship to virulence diversity. The molecular methods have the advantages of precise scoring, economy when conducted on a large scale, and the simultaneous measurement of variation at several loci including those with co-dominant alleles. At ICRISAT, DNA fingerprinting with Simple Sequence Repeats (SSRs) has been used to detect genetic diversity among pathotypes of *S. graminicola* (Sastry *et al.*, 1995). Of the SSRs tested, $(GATA)_4$ identified the most polymorphism amongst the host genotype-specific pathogen isolates of *S. graminicola*. The same SSRs also were used to confirm the emergence of a new virulent pathotype specifically adapted to recently introduced commercial hybrid cultivar MLBH 104 (Thakur *et al.*, 1999). In a study on genetic basis of host-pathogen specificity, Sastry *et al.* (2001) demonstrated a change in virulence of a *S. graminicola* pathotype on a pearl millet line 700651 with a corresponding change in RAPDs and DNA fingerprinting profiles in two extreme asexual generations of the pathotype. More recently, Amplified Fragment Length Polymorphism (AFLP) markers also have been used to detect genetic variation among *S. graminicola* isolates from India (Sivaramakrishnan *et al.*, 2003). We often can score 50-70 bands with each primer-pair combination and the proportion of polymorphic bands per primer pair ranges from 35-60 per cent. Currently, five major genetic groups of pathotypes of *S. graminicola* are known to exist in India, which are not exactly the same as reported for the virulence groups above (Sivaramakrishnan *et al.*, 2003).

Management Practices

Various disease management practices, such as chemical fungicide, cultural methods, biocontrol and host plant resistance have been used to control downy mildew in pearl millet. Of these, chemical seed treatment and host plant resistance have been most effective and common, and thus only these are discussed here.

Chemical Management

The systemic fungicide metalaxyl [methyl DL-(2, 6-dimethyl phenyl)-N (2, methoxyacetyl)-N-alaninate] (trade name Apron® SD 35 from Ciba-Geigy Corporation) released in the 1970s has been widely used to control downy mildew in pearl millet. Seed treatment with metalaxyl at 2 g a.i. kg^{-1} seed controls the disease effectively for up to 35-40 days after crop emergence (Williams and Singh, 1981; Dang *et al.*, 1983). Later, depending on the weather conditions and the susceptibility of the cultivar(s), downy mildew symptoms may appear on late tillers and the panicles may be malformed. Seed treatment with Apron® SD 35 followed by a foliar spray with Ridomil® MZ 75 WP (metalaxyl 8 per cent + mancozeb 64 per cent) has been recommended for complete control of the disease (Singh and Shetty, 1990). However, this treatment is neither economical nor feasible for most farmers. In the 1980s Ciba Geigy released Apron Plus® (metalaxyl + carboxin + furathiocarb), which was highly effective as a seed treatment to control downy mildew and increased pearl millet yields by 30 per cent in Mali (Scheuring *et al.*, 2002). More recently, Syngenta Crop Protection commercially released Apron Star®, which is composed of mefonoxam (fungicide-effective against downy mildews), difenoconazole (fungicide- effective against seedling *Fusarium* spp. and seed-borne smuts) and thiamethoxam (insecticide- effective against seedling soil and sucking insects) has been found to be highly effective in enhancing crop establishment, controlling seedling diseases and insects, controlling downy mildew and thereby increasing grain yield in Mali (Scheuring *et al.*, 2002). However, this product is not yet commercialized in India. Seed treatment with Ridomil® MZ 72 WP has been a common practice for private seed companies in India. This treatment extends the economic life in farmers' fields of a susceptible/tolerant pearl millet hybrid by 2-3 years.

Host Plant Resistance

Considerable progress has been made over the past 30 years in managing downy mildew through host plant resistance in India. Research at ICRISAT and in partnership with the AICPMIP- and UK-based scientists to improve screening systems and to identify and use host-plant resistance has been very successful (Hash *et al.*, 1997; 1999; Hash and Witcombe, 2002). Effective disease management can be obtained with cultivars that have durable resistance. In order to breed cultivars with durable resistance we need to have: availability of a large number of genetically diverse germplasm accessions, an effective screening technique to clearly identify resistance, knowledge of inheritance and genetics of resistance, and effective breeding methods to utilize the resistance. Major gene resistance is usually most efficiently exploited either to prevent the pathogen population from easily adapting to or to minimize the selection pressure exerted on the pathogen population. Various components involved in the pearl millet downy mildew resistance breeding program at ICRISAT-Patancheru are briefly discussed below.

Screening Techniques

Over the past two decades, ICRISAT has developed highly effective field and greenhouse screening techniques for pearl millet downy mildew (Singh *et al.*, 1993a). These continue to be refined (Jones *et al.*, 2001) and are being used routinely and extensively worldwide.

Field Screening Technique

The field screening technique has major components of infector rows, test rows, indicator rows, and a simple disease scoring system (Williams *et al.*, 1981). A highly susceptible line or a mixture of two or more susceptible lines sown before the test material on every ninth row is used as infector or inoculum donor. At the 1–2 leaf stage, infector rows are spray-inoculated with a fresh sporangial suspension (1×10^5 spores mL^{-1}) of *S. graminicola* during early evening (1800–1900 h). Furrow irrigation is provided 8–10 hr before inoculation to ensure the high humidity necessary for successful infection. High humidity is maintained by mist or furrow irrigation until one week after inoculation. When the infector rows develop more than 50 per cent downy mildew incidence, the test rows are sown in the intervening eight rows. Test materials are generally sown 3–4 weeks after the infector rows. Control rows of known disease reaction (susceptible to highly resistant) are sown with the test materials at regular intervals, generally after every 10 to 15 rows. Highly susceptible indicators are sown alternately with the controls. Downy mildew incidence levels on these lines indicate the general disease pressure and its distribution, and thus the effectiveness of the nursery. Disease incidence on susceptible indicator lines should be about 70 per cent for the field screening to be reliable.

For a systemic disease like downy mildew, a plant showing symptoms even on a single leaf is scored as susceptible, and percentage disease incidence is the most appropriate measurement. Disease scoring is done twice, first at 30 days after sowing the test material, and second after 60 days. All diseased plants during the first count should be removed or marked by plastic pegs to avoid under-counting due to infected plants that die or disappear prior to the second count. At the second count, downy mildew may appear also as green ears on plants that had no foliar symptoms (latent infection), or on nodal tillers.

Greenhouse Screening Technique

The technique involves inoculation of pot-grown seedlings at the one-to-two-leaf-stage of development. Inoculation is done either by placing a drop of sporangial suspension (1×10^6 sporangia mL^{-1}) at the tip of each seedling or by spray inoculation with a hand sprayer. Inoculation at 20°C with freshly prepared, chilled (2–4°C) inoculum, is followed by overnight incubation in a chamber at 20°C with >95 per cent RH. Pots are then kept on greenhouse benches at 25±2°C. Seedlings are evaluated for downy mildew incidence two weeks after inoculation. This permits easy screening of large numbers of plants of many different test materials against different pathotypes in a small space and in a short period of time. Inclusion of appropriate indicator and control entries with each test is essential to ensure effectiveness of screening and data interpretation. Greenhouse screening, besides being rapid, precise, and cost effective, considerably reduces the size of the field disease nurseries required for routine screening of early generation breeding materials. However, field confirmation of resistance of advanced breeding products must still be done through multilocation testing to confirm resistance durability.

Resistance Sources

ICRISAT has the world's largest collection of pearl millet germplasm (more than 21500 accessions) in its gene bank at Patancheru. During the past two decades about 3500 germplasm accessions were screened for downy mildew resistance and many accessions with high degrees of resistance were identified (Singh, 1990; 1995; Singh *et al.*, 1997). Identified accessions were tested at downy mildew hot-spots in Asia and Africa through the collaborative International Pearl Millet Downy Mildew Nursery (IPMDMN), and a number of accessions with stable resistance were identified. Pearl millet lines that exhibited high levels of resistance in multiyear multilocational field tests in IPMDMN were identified as stable resistant lines. These include P7 (ICML 12), SDN 503 (ICML 13), 700251 (ICML 14), 700516 (ICML 15), and 700651 (ICML 16) (Singh *et al.*, 1990a). In addition, breeding lines, including IP 16438, IP 16762, P310-17, and P1449-3 also exhibited stable resistance.

Several elite inbred lines developed ICRISAT–Patancheru have been found to have downy mildew resistance. These include, dwarf inbred pollinator ICMP 85410 (Hash and Witcombe, 1994; Talukdar *et al.*, 1996); tall pollinator ICMP 423 (Rai *et al.*, 1994) that has been converted into a male-sterile line (ICMA 90111) based on the EGP 261 cytoplasm (Rai and Hash, 1993); and a topcross pollinator ICMR 312 (Witcombe *et al.*, 1996), which is genetically heterogeneous for genes controlling its downy mildew resistance (Talukdar and Singh, 1993).

Resistant lines have also been developed through pure line selection of variability within susceptible parents. Selection within S1-S2 progenies of 7042 (IP 2696), a landrace from Chad, provided a downy mildew resistant line ICML 22 (7042 DMR) (Singh *et al.*, 1992). This line has desirable traits for earliness, photoperiod insensitivity, and fertility restoration (Singh *et al.*, 1994) and it is being used in breeding programs in India. However, this resistance is not effective to the downy mildew pathogen populations in western Africa.

A unique type of resistance called "recovery resistance"–a phenomenon in which systemically infected plants outgrow the disease to produce healthy panicles-was discovered by Singh and King (1988). Pearl millet lines with this trait do not prevent spore germination and penetration, disease symptom development, or sporulation, thus allowing the pathogen to complete its life cycle. Plants having this type of resistance quickly recover from the disease, and subsequently behave like conventional resistant genotypes. Recovery resistance has been discovered in many pearl millet accessions and breeding lines. Through pedigree breeding the level of recovery resistance was increased to more than 95 per cent in a selection from ICMB 1–maintainer of ICMA 1 (81A), which is widely used in commercial hybrid seed production in India (Singh and Talukdar, 1996). Other sources of this type of resistance include SDN 503, P1449, and ICMB 841.

Inheritance of Resistance

Most studies on the inheritance of pearl millet downy mildew resistance have not been conclusive due to the fact that both the pathogen and host are allogamous and highly variable (Thakur *et al.*, 1992). Segregation for host plant resistance has generally shown continuous variation (Singh *et al.*, 1980; Basavaraju *et al.*, 1981; Dass *et al.*, 1984). There is clear evidence that the A_1 cytoplasm is not associated with susceptibility or resistance to pearl millet downy mildew (Anand Kumar *et al.*, 1983; Yadav *et al.*, 1993; Yadav, 1994; Yadav, 1996). However, there is ample evidence for genes in the nucleus controlling host plant reaction to this disease. Except in one case where resistance was reported to be recessive (Singh *et al.*, 1978), resistance is generally inherited as a dominant trait and variation in segregating populations is continuous (Singh *et al.*, 1993a). In few cases where clear Mendelian segregations have been observed, the numbers of dominant genes governing resistance have been one, two or even three (Deswal and Govila, 1994; Hash *et al.*, 2003). Singh and Talukdar (1998) showed a major gene resistance in a pearl millet germplasm IP 18292 against Patancheru isolate of *S. graminicola*. However, most studies on the inheritance of resistance to downy mildew suggest the presence of minor genes along with any purported major genes. In general, nonadditive gene action has been reported for much of the heritable variability present (Deswal and Govila, 1994; Kataria *et al.*, 1994). Such nonadditive gene action can contribute substantially to general combining ability (GCA), since parents having dominant resistance can be expected to have high GCA for this trait when compared with more susceptible parents. Recent studies on molecular marker based genetic linkage maps for pearl millet are yielding interesting results (Jones, 1994; Liu *et al.*, 1994; Hash *et al.*, 1995; Jones *et al.*, 1995; Breese *et al.*, 2002; Jones *et al.*, 2002; Hash and Witcombe, 2002) that will facilitate genetic manipulation of disease resistance.

Breeding for Downy Mildew Resistance

Both conventional and molecular breeding approaches have been followed to breed for downy mildew resistance in pearl millet.

Conventional Breeding

Conventional breeding, including pure-line selection (selection within partially inbred lines), pedigree selection, backcrossing, induced mutation, and recurrent selection procedures has all been used with varying degrees of success (Hash *et al.*, 1997, 1999; Hash and Witcombe, 2002). Pedigree and recurrent selection (Andrews *et al.*, 1985; Rattunde and Witcombe, 1993; Weltzien and King, 1995) are the most widely and successfully used methods for improving downy mildew resistance in pearl millet.

Pure line selection from within-line variability and pedigree selection within segregating populations: In pure line selection a large population of a susceptible line is grown in the disease nursery or inoculated in the greenhouse. Disease-free plants are selfed and the progenies of these are screened and resistant plants identified. The process of selection and selfing is repeated for several generations until some progenies with desired levels of resistance combined with uniform morphological traits of the original line have been identified. A notable success with this method was the development of ICMB 841 by selection within a seedlot of 5141B (Singh *et al.*, 1990c). Although it now appears that ICMB 841 is in fact the product of an outcross of the original 5141B with an unidentified disease resistance donor

(Hash *et al.*, 1997, 1999), the superior combining ability of 5141B was retained and combined with downy mildew resistance in ICMB 841. ICMA 841, cytoplasmic male-sterile line counterpart of ICMB 841, has been exploited commercially in India as a female parent of several F_1 hybrids, including ICMH 423, Pusa 23, and Pusa 322. Of these, Pusa 23 is presently the most widely grown public-bred hybrid in India (Bhatnagar, 2003). Recent experience at ICRISAT suggests that with a combination of strict greenhouse seedling disease screens and pure line selection, it may sometimes be possible to effect selection for improved resistance of elite inbreds that have been bred or maintained by pedigree-bulk procedures. For example, selections of ICMB 89111 and ICMP 451 with markedly higher levels of downy mildew resistance than the original lines have been identified (Hash and Thakur, unpublished). However, it is expected that several generations of pedigree selection among the segregating progeny of crosses (including backcrosses) involving well-characterized resistance sources and agronomically elite hybrid parental lines would normally be more likely to prove effective in generating agronomically elite pearl millet hybrid parental lines with high levels of downy mildew resistance (Hash *et al.*, 1997; 1999).

Recurrent selection: With refinement in the greenhouse screening method, recurrent selection has been highly effective for improving downy mildew resistance. At ICRISAT-Patancheru, about 600 S_1 progenies per population are screened in the greenhouse each selection cycle and about 50 per cent of these progenies having higher downy mildew susceptibility are rejected. The remaining 50 per cent are evaluated for agronomic traits in single-replication augmented design trials conducted at several sites. About 50 agronomically superior downy mildew-free S_1 progenies selected from these field trials are then recombined the following season-either by manual full-sib crossing or by random mating in isolation-to produce the cycle bulk population for the next round of recurrent selection. If full-sibs are made, these can be screened in the greenhouse, and remnant seed of the most resistant ones are sown in the target environment to identify agronomically superior progenies (Hash *et al.*, 1997; 1999). These procedures have been highly effective in improving or maintaining levels of downy mildew resistance in ICRISAT's pearl millet composite populations (Rattunde and Witcombe, 1993; Weltzien and King, 1995).

Molecular Breeding

Molecular approaches can be employed either in the pyramiding of two, or more resistance genes in a host line through marker-assisted selection (MAS), or through genetic modification of existing resistance genes. MAS uses as selection criteria 'tags' having high heritability that are genetically linked to portions of the genome controlling characters of interest. These markers can be morphological traits, proteins including isozymes, or DNA markers such as restriction fragment length polymorphisms (RFLPs), random amplified polymorphic DNAs (RAPDs), amplified fragment length polymorphisms (AFLPs), sequence tagged microsatellites (STMSs and SSRs), and others. MAS provides an excellent alternative to time consuming and laborious testcrosses required to pyramid genes exhibiting epistasis (Gale and Witcombe, 1992; Paterson *et al.*, 1991). It also provides a system for rapidly incorporating a number of genetically different resistances into a common elite inbred parent background. The resulting near-isogenic lines (including those with pyramided resistances) can then, in theory, be used to produce a wide array of hybrid cultivar types that are genetically heterogeneous for their disease resistance complements, but are relatively uniform (compared to open-pollinated pearl millet cultivars) for other agronomic performance traits (Witcombe and Hash, 2000; Hash and Witcombe, 2002).

DNA markers for quantitative trait loci controlling resistance of ICMP 85410, P7-3 and ICMP 451-P6 against downy mildew isolates from India and western Africa have been identified (Jones, 1994;

Jones *et al.*, 1995, 2002; Breese *et al.*, 2002). In order to use MAS in improvement of pearl millet downy mildew resistance, it is necessary to have a donor parent with identified linked markers for some of its resistance genes; an elite parent with inadequate resistance, which has different marker alleles than the donor parent and lacks these resistance genes (at least the specific alleles linked to markers in the donor genome), and a system for efficiently determining genotype (or at least phenotype in case of dominantly inherited markers) at the marker loci in individual segregating plants derived from crosses between the resistance source and elite parent.

To date, DNA markers have been identified for about 60 different putative downy mildew resistance QTLs in pearl millet (Breese *et al.*, 2002; Hash and Witcombe, 2002). Using RFLP-based marker-assisted backcrossing (MABC), several mapped QTLs have been transferred to backgrounds of elite inbred parental lines of a popular single-cross hybrid HHB 67 (843A × H 77/833-2). Hybrid performances of these 'improved parents' were compared with those of their recurrent parents, 843B and H 77/833-2, in greenhouse and field downy mildew screens and have proven superior to the recurrent parent lines (Hash *et al.*, 2003). Similarly, ICMR 01004 and ICMR 01007, BC_4 derivatives of elite pollinator H 77/833-2 into which two DM resistance QTLs from resistance donor ICMP 451-P6 were independently transferred, performed as well or better than the recurrent parent H 77/833-2 for downy mildew resistance and also for grain and stover yield combining ability. Marker-assisted introgression of the downy mildew resistance QTLs from elite donor parent ICMP 451-P6 was accompanied by favorable linkage drag for several agronomic traits, including 1000-grain mass, panicle length, plant height, and improved resistance to rust. Hybrids based on crosses involving ICMR 01004 and ICMR 01007 have exhibited marginal grain yield superiority and substantially better downy mildew resistance than the control hybrid HHB 67 (Hash *et al.*, 2003).

Conclusions

During the past three decades considerable knowledge has been generated on biology, epidemiology and management of pearl millet downy mildew in India. Well-refined field screening and greenhouse screening techniques have enhanced the efficiency of resistance breeding programs. A number of well-defined resistance sources are available and several of these have successfully been utilized in breeding downy mildew resistant hybrid parental lines at ICRISAT and in the Indian national program. Although downy mildew has been apparently contained and no major epidemic has occurred during the past 15 years, we need to continue monitoring the performance of major commercial hybrids in farmer's fields and the associated virulence shifts in *S. graminicola* populations. Use of DNA markers will be helpful in identifying specific resistance genes in host differentials and genetic changes in pathogen populations leading to evolution of new virulences.

Both conventional and molecular breeding approaches have been successful in breeding resistant hybrids that have contributed significantly in improving productivity and production of pearl millet in India. Because of the narrow genetic base of the highest yielding single-cross hybrids and highly variable nature of the pathogen, evolution of new virulences matching resistances in the host is bound to occur. A number of genetic markers for downy mildew resistance QTLs have been identified and a demonstration marker-assisted backcross program has succeeded in developing more downy mildew and rust resistant versions of highly popular hybrid HHB 67. The new hybrids have improved resistance to downy mildew (and in some cases rust), superior agronomic traits, and in some cases enhanced drought tolerance (when one of the parents has been improved for this by marker-assisted backcrossing) compared to the original HHB 67. Thus it appears that molecular breeding approaches offer tremendous opportunities to improve elite hybrids for important characters and combinations of characters that are otherwise difficult and expensive to assess. Monitoring virulence shifts in the pathogen population, identification of resistance, and introgression of resistance into agronomic elite hybrid parental lines using appropriate breeding methods combined with judicious use of seed dressing chemical fungicide should form the strategy of downy mildew management in pearl millet for the foreseeable future.

REFERENCES

Anand Kumar, K., Jain, R.P. and Singh, S.D. (1983). Downy mildew reactions in pearl millet lines with and without cytoplasmic male sterility. *Plant Dis.*, **67**: 663-665.

Andrews, D.J. (1987). Breeding pearl millet grain hybrids. In: *Hybrid seed production of selected cereal oil and vegetable crops*, (Eds., Feistritzer, W.P. and Kelly, V.), pp. 83-109, FAO Plant Production and Protection Paper 82. Food and Agriculture Organization of the United Nations: Rome, Italy.

Andrews, D.J., King, S.B., Witcombe, J.R., Singh, S.D., Rai, K.N., Thakur, R.P., Talukdar, B.S., Chavan, S.B. and Singh, P. (1985). Breeding for disease resistance and yield in pearl millet. *Field Crops Research*, **11**: 241-258.

Ball, S.L. (1983). Pathogenic variability of downy mildew (*Sclerospora graminicola*) on pearl millet host cultivar reactions to infection by different pathogen isolates. *Ann. Appl. Biol.*, **102**: 257-264.

Ball, S.L., Pike, D. J. and Burridge, C. V. (1986). Characterization of populations of *Sclerospora graminicola. Ann. Appl. Biol.*, **108**: 519-526.

Basavaraju, R., Safeeulla, K.M. and Murty, B.R. (1981). Genetic variance and heritability for resistance to downy mildew in pearl millet. *Indian Journal of Genetics and Plant Breed.*, **41**: 137-143.

Bhatnagar, S.K. (2003). Coordinator's review of All India Coordinated Pearl Millet Improvement Project 2002-2003, pp. 7, Indian Council of Agricultural Research, Mandor, Jodhpur, India.

Breese, W.A., Hash, C.T., Devos, K.M. and Howarth, C.J. (2002). Pearl millet genomics – an overview with respect to breeding for resistance to downy mildew. In: *Sorghum and Millets Pathology 2000*, (Ed., Leslie, J.F.), pp. 243-246, Iowa State Press, Ames, Iowa, USA.

Chahal, S.S., Thakur, R.P. and Mathur, S.B. (1994). *Seed-borne disease and seed health Testing of pearl millet*, Pp. 72, Danish Government Institute of Seed Pathology for Developing Countries, Copenhagen, Denmark.

Crute, I.R. 1981. The host specificity of Peronosporaceous fungi and the genetic relationship between host and parasite. In: *The Downy Mildews*, (Ed., Spencer, D.M.), pp. 237-253, Academic Press, New York City, New York.

Dang, J.K., Thakur, D.P. and Grover, R.K. (1983). Control of pearl millet downy mildew caused by *Sclerospora graminicola* with systemic fungicides in an artificially contaminated plot. *Annals App. Biol.*, **102**: 99-106.

Dass, S., Kapoor, R.L., Paroda, R.S. and Jatasra, D.S. (1984). Gene effects for downy mildew (*Sclerospora graminicola*) resistance in pearl millet. *Indian J. Genet. Plant Breed.*, **44**: 280-285.

Deswal, D.P. and Govila, O.P. (1994). Genetics of resistance to downy mildew (*Sclerospora graminicola*) in pearl millet (*Pennisetum glaucum*). *Indian J. Agric. Sci.*, **64**: 661-663.

Gale, M.D. and Witcombe, J.R. (1992). DNA markers and marker-assisted applications in plant breeding, with particular reference to pearl millet breeding. In: *Biotechnology and crop improvement in Asia*, (Ed., Moss, J.P.), pp. 323-332. International Crops Research Institute for the Semi-Arid Tropics: Patancheru, A.P. 502 324, India.

Hash, C.T. and Witcombe, J.R. (1994). Pearl millet mapping populations at ICRISAT. In: *Use of Molecular Markers in Sorghum and Pearl Millet Breeding for Developing Countries*, (Eds., Witcombe, J.R. and Duncan, R.R.), pp. 69-75, Overseas Development Administration, London, U.K.

Hash, C.T. and Witcombe, J.R. (2002). Gene management and breeding for downy mildew resistance. In: *Sorghum and Millets Diseases* (Ed., Leslie, J.F.), pp. 27-36, Iowa State Press, USA.

Hash, C.T., Cavan, G.P., Bidinger, F.R., Howarth, C.J. and Singh, S.D. (1995). Downy mildew resistance QTLs from a seedling heat tolerance mapping population. *Intl. Sorg. Millets Newsl.*, **36**: 66-67.

Hash C.T., Sharma, A., Kolesnikova-Allen, M.A., Serraj, R., Thakur, R.P., Bidinger, F.R., Bhasker Raj, A.G., Rizvi, S.M.H., Beniwal, C.R., Yadav, H.P., Yadav, Y.P., Srikant, Bhatnagar, S.K., Yadav, R.S., Howarth, C.J., Breese, W.A. and Witcombe, J.R. (2003). Marker-assisted breeding to improve pearl millet hybrid HHB 67: Lab to field. In: *Abstracts and Final Program, In the Wake of the Double Helix: From the Green Revolution to the Gene Revolution*, pp 82, Bologna, Italy, May 27–31, 2003.

Hash, C.T., Singh, S.D., Thakur, R.P. and Talukdar, B.S. (1999). Breeding for disease resistance. In: *Pearl Millet Breeding* (Eds., Khairwal, I.S., Rai, K.N., Andrews, D.J. and Harinarayana, G.), pp. 337-379, Oxford & IBH, New Delhi, India.

Hash, C.T., Witcombe, J.R., Thakur, R.P., Bhatnagar, S.K., Singh, S.D., and Wilson, J.P. (1997). Breeding for pearl millet disease resistance, In: *Proceedings of the International Conference on Genetic Improvement of Sorghum and Pearl Millet, 22-27 Sep. 1996*, pp. 337-372, INTSORMIL Publication 97-5. Lincoln, NE, USA.

Idris, M.O., and Ball, S.L. (1984). Inter- and intra-continental sexual compatibility in *Sclerospora graminicola*. *Plant Pathol.*, **33**: 219-223.

Jones, E. (1994). Mapping quantitative trait loci for resistance to downy mildew in pearl millet. Ph.D. thesis, School of Biological Sciences, pp. 19, University of Wales: Bangor, U.K.

Jones, E.S., Breese, W.A. and Shaw, D.S. (2001). Inoculation of pearl millet with the downy mildew pathogen, *Sclerospora graminicola*: chilling inoculum to delay zoospore release and avoid spray damage to zoospores. *Plant Pathol.*, **50**: 310-316.

Jones, E.S., Breese, W.A., Liu, C.J., Singh, S.D., Shaw, D.S. and Witcombe, J.R. (2002). Mapping quantitative trait loci for resistance to downy mildew in pearl millet: field and glasshouse screens detect the same QTL. *Crop Sci.*, **42**: 1316-1323.

Jones, E.S., Liu, C.J., Gale, M.D., Hash, C.T. and Witcombe, J.R. (1995). Mapping quantitative trait loci for downy mildew resistance in pearl millet. *Theor. Appl. Genet.*, **91**: 448-456.

Kataria, R.P., Yadav, H.P., Beniwal, C.R. and Narwal, M.S. (1994). Genetics of incidence of downy mildew (*Sclerospora graminicola*) in pearl millet (*Pennisetum glaucum*). *Indian J. Agric. Sci.*, **64**: 664-666.

King, S.B. (1992). World review of pearl millet diseases: knowledge and future research needs. In: *Sorghum and Millets Diseases: A Second World Review*, (Eds., de Milliano, W. A. J., Frederiksen, R.A. and Bengston, G.D.), pp. 95-108, ICRISAT, Patancheru (A.P.), India.

Liu, C.J., Witcombe, J.R., Pittaway, T.S., Nash, M., Hash, C.T., Busso, C.S. and Gale, M.D. (1994). An RFLP-based genetic map of pearl millet (*Pennisetum glaucum*). *Theor. Appl. Genet.*, **89**: 481-487.

Michelmore, R.W., Pawar, M.N. and Williams, R.J. (1982). Heterothallism in *Sclerospora graminicola*. *Phytopathology*, **72**: 1368-1372.

Paterson, A.H., Tanksley, S.D. and Sorrells, M.E. (1991). DNA markers in plant improvement. *Advances in Agronomy*, **46**: 39-90.

Rai, K.N. and Hash, C.T. (1993). A new source of cytoplasmic-nuclear male sterility in pearl millet. In: *Annual Report 1992* (Semi-formal publication), pp. 67-68, Cereals Program, International Crops Research Institute for the Semi-Arid Tropics: Patancheru, 502 324, A.P., India.

Rai, K.N., Talukdar, B.S., Singh, S.D., Rao, A.S., Rao, A.M. and Andrews, D.J. (1994). Registration of ICMP 423 parental line of pearl millet. *Crop Sci.*, **34**: 1430.

Rattunde, H.F.W. and Witcombe, J.R. (1993). Recurrent selection for increased grain yield and resistance to downy mildew in pearl millet. *Plant Breed.*, **110**: 63-72.

Sastry, J.G., Ramakrishna, W., Sivaramakrishnan, S., Thakur, R. P., Gupta, V.S. and Ranjekar, P.K. (1995). DNA fingerprinting detects genetic variability in the pearl millet downy mildew pathogen (*Sclerospora graminicola*). *Theor. Appl. Genet.*, **91**: 856-861.

Sastry, J.G., Sivaramakrishnan, S., Rao, V.P., Thakur, R. P., Singru, R.S., Gupta, V.S. and Ranjekar, P.K. (2001). Genetic basis of host-specificity in *Sclerospora graminicola*, the pearl millet downy mildew pathogen. *Indian Phytopath.*, **54**: 323-328.

Scheuring, J.F., Katile, S.O. and Kollo, I.A. (2002). Boosting pearl millet yields with Apron Plus and Apron Star seed treatments. In: *Sorghum and Millets Diseases*, (Ed., Leslie, J.F.), pp. 47-49, Iowa State Press, USA.

Shetty, H. S. and R. Ahmad. (1981). Physiologic specialization in *Sclerospora graminicola*. *Indian Phytopath.*, **34**: 307-309.

Singh, F., Singh, R.K., Singh, R.M. and Singh, R.B. (1978). Genetic analysis of downy mildew (*Sclerospora graminicola*) resistance in pearl millet [*Pennisetum typhoides* (Burm.) S. & H.]. *Zeitschrift für Pflanzenzüchtung*, **81**: 54-59.

Singh, F., Singh, R.M., Singh, R.B. and Singh, R.K. (1980). Genetic studies of downy mildew resistance in pearl millet. In: *Trends in Genetical Research of Pennisetums*, (Eds., Gupta, V.P. and Minocha, J.L.), pp. 171-172, Wesley Press: Ludhiana, India.

Singh, S.D. (1990). Sources of resistance to downy mildew and rust in pearl millet. *Plant Dis.* **74**: 871-874.

Singh, S.D. (1995). Downy mildew of pearl millet. *Plant Dis.*, **79**: 545-550.

Singh, S.D. and King, S.B. (1988). Recovery resistance to downy mildew in pearl millet. *Plant Dis.*, **72**: 425-428.

Singh, S.D., and Shetty, H.S. (1990). Efficacy of systemic fungicide metalaxyl for the control of downy mildew (*Sclerospora graminicola*) of pearl millet (*Pennisetum glaucum*). *Indian J. Agric. Sci.*, **60**: 575-581.

Singh, S.D. and Talukdar, B.S. (1996). Recovery resistance to downy mildew in pearl millet parental lines ICMA 1 and ICMB 1. *Crop Sci.*, **36**: 201-203.

Singh, S.D. and Talukdar, B.S. (1998). Inheritance of complete resistance to pearl millet downy mildew. *Plant Dis.*, 82: 791-793.

Singh, S.D. and Williams, R.J. (1980). The role of sporangia in the epidemiology of pearl millet downy mildew. *Phytopathology*, **70**: 1187-1190.

Singh, S.D., Alagarswamy, G., Talukdar, B.S. and Hash, C.T. (1994). Registration of ICML 22 photoperiod insensitive, downy mildew resistant pearl millet germplasm. *Crop Sci.*, **34**: 1421.

Singh, S.D., Ball, S.L. and Thakur, D.P. (1987b). Problems and strategies in the control of downy mildew. In: *Proceedings of the International Pearl Millet Workshop, 7-11 April 1986*, ICRISAT Center, India, (Eds., Witcombe, J.R. and Beckerman, S.R.), pp. 161-172, International Crops Research Institute for the Semi-Arid Tropics: Patancheru, 502 324, A.P., India.

Singh, S.D., King, S.B. and Malla Reddy, P. (1990a). Registration of five pearl millet germplasm sources with stable resistance to downy mildew. *Crop Sci.*, **30**: 1164.

Singh, S.D., Singh, P., Rai, K.N. and Andrews, D.J. (1990c). Registration of ICMA 841 and ICMB 841 pearl millet parental lines with A1 cytoplasmic-genic male sterility system. *Crop Sci.*, **30**: 1378.

Singh, S.D., King, S.B. and Werder, J. (1993a). Downy Mildew Disease of Pearl Millet. Information Bulletin no. 37. ICRISAT, Patancheru, 502 324, A.P., India.

Singh, S.D., Lal, S. and Pande, S. (1993b). The changing scenario of maize, sorghum, and pearl millet diseases. In: *Pests and Pest Management in India – The Changing Scenario*, (Eds., Sharma, H.C. and Veerabhadra, M.), pp. 130-139, Plant Protection Association of India: (NPRTI), Rajendranagar, Hyderabad, 500 030, A.P., India.

Singh, S.D., Singh, P., Andrews, D.J., Talukdar, B.S. and King, S.B. (1992). Reselection of a pearl millet cultivar utilizing residual variability for downy mildew reaction. *Plant Breed.*, **109**: 54-59.

Singh, S.D., Wilson, J.P., Navi, S.S., Talukdar, B.S., Hess, D.E. and Reddy, K.N. (1997). Screening Techniques and Sources of Resistance to Downy Mildew and Rust in Pearl Millet. Information Bulletin no. 48. ICRISAT, Patancheru, 502 324, A.P., India.

Sivaramakrishnan, S., Thakur, R.P., Kannan, S. and Rao, V.P. (2003). Pathogenic and genetic diversity among Indian isolates of *Sclerospora graminicola* from pearl millet. *Indian Phytopath.*, **56**: 392-397.

Talukdar, B.S. and Singh, S.D. (1993). Stability and heritability for downy mildew resistance in two pearl millet pollinators. In: *Annual Report 1992* (Semi-formal publication), pp. 70-71, Cereals Program, International Crops Research Institute for the Semi-Arid Tropics: Patancheru, 502 324, A.P., India.

Talukdar, B.S., Prakash Babu, P.P., Rao, A.M., Ramakrishna, C., Witcombe, J.R., King, S.B. and Hash, C.T. (1996). Registration of ICMP 85410 dwarf, downy mildew resistant, restorer germplasm line of pearl millet. *Crop Sci.*, **36**: In press.

Thakur, R.P. (1995). Status of international sorghum anthracnose and pearl millet downy mildew virulence nurseries. In: *Disease Analysis through Genetics and Biotechnology: Interdisciplinary Bridges to Improved Sorghum and Millet Crops*, (Eds., Leslie, J.F. and Frederiksen, R.A.), pp. 75-92, Iowa State University Press, Ames, Iowa.

Thakur, R.P. (1998). Diseases of pearl millet and their management. In: *Diseases of Field Crops and Their Management* (Ed., Thind, T.S.), pp. 53-77, National Agricultural Technology Centre, Ludhiana – 141 004, India

Thakur, R.P. (1999). Pathogen diversity and plant disease management. *Indian Phytopath.*, **52**: 1-9.

Thakur, R.P. and Rao, V.P. (1997). Variation in virulence and aggressiveness among pathotypes of *Sclerospora graminicola* on pearl millet. *Indian Phytopath.*, **50**: 41-47.

Thakur, R.P., Rao, V.P., and Hash, C.T. (1998). Emergence of a highly virulent pathotype of *Sclerospora graminicola* at Jodhpur, Rajasthan, India. *Intl. Sorg. Millet Newsl.*, **39**: 140-142.

Thakur, R.P., Rao, V.P., Sastry, J.G., Sivaramakrishnan, S., Amruthesh, K.N. and Barbind, L.D. (1999). Evidence for a new virulent pathotypes of *Sclerospora graminicola* on pearl millet. *J. Mycol. Plant Pathol.*, **29**: 61-69.

Thakur, R.P., Magill, C.W., Sivaramakrishnan, S., Hash, C.T., Shetty, H.S. and Hess, D.E. (2002). Variability in *Sclerospora graminicola*, the pearl millet downy mildew pathogen. In: *Sorghum and Millets Diseases*, (Ed., Leslie, J.F.), pp. 51-56, Iowa State press, Ames, Iowa.

Thakur, R.P., Shetty, K.G. and King, S.B. (1992). Selection for host-specific virulence in asexual populations of *Sclerospora graminicola*. *Plant Pathol.*, **41**: 626-632.

Werder, J. and Ball, S.L. (1992). Variability of pearl millet downy mildew (*Sclerospora graminicola*) and its implications for breeding for stable and durable resistance. *Trop. Pest Manage.*, **38**: 43-47.

Weltzien R.E. and King, S.B. (1995). Recurrent selection for downy mildew resistance in pearl millet. *Plant Breed.*, **114**: 308-312.

Williams, R.J. and Singh, S.D. (1981). Control of pearl millet downy mildew by seed treatment with metalaxyl. *Annals Appl. Biol.*, **97**: 263-268.

Williams, R.J., Singh, S.D. and Pawar, M.N. (1981). An improved field screening technique for downy mildew resistance in pearl millet. *Plant Dis.*, **65**: 239-241.

Witcombe, J.R. and Hash, C.T. (2000). Resistance gene deployment strategies in cereal hybrids using marker-assisted selection: Gene pyramiding, three-way hybrids, and synthetic parent populations. *Euphytica*, **112**: 175-186.

Witcombe, J.R., Rao, M.N.V.R., Talukdar, B.S., Singh, S.D. and Rao, A.M. (1996). Registration of ICMR 312 pearl millet topcross pollinator germplasm. *Crop Sci.*, **36**: 471.

Yadav, O.P. (1994). Influence of A1 cytoplasm in pearl millet. *Plant Breed. Abstr.*, **64**: 1375-1379.

Yadav, O.P. (1996). Downy mildew incidence of pearl millet hybrids with different male-sterility inducing cytoplasms. *Theor. Appl. Genet.*, **92**: 278-280.

Yadav, O.P., Manga, M.V. and Gupta, G.K. (1993). Influence of A1 cytoplasmic substitution on downy mildew incidence in pearl millet. *Theor. Appl. Genet.*, **87**: 558-560.

23

Indian Contributions to Aerobiology of Fungal Plant Pathogens: An Overview

B.P.R. VITTAL

Centre for Advanced Studies in Botany, University of Madras, Chennai – 600 025, India

Aerobiology is concerned with the liberation, dispersal and deposition of spore of plant pathogenic fungi and so can make valuable contributions to understanding the epidemiology of plant diseases. Infection agents may be detected and quantified in aerosols and their persistence in air and relationship with disease spread can be determined. In India many workers studied the air mycoflora of crop fields with varied objectives. While many workers concentrated on saprophytic moulds, a few have, however, studied the aerobiology of pathogens causing crop disease. Some of these studies are briefly reviewed below.

Early Studies

In India, after Cunningham's (1873) pioneering work, systematic studies on aeromycology were initiated by plant pathologists. One of the earliest studies was that of Mehta (1952) on cereal rusts. He trapped uredospores of the three rusts of wheat and barley using aeroscope; with vertically placed glass slides at 62 stations located in different parts of the country. With balloons, kite and apparatus specially designed for catching spores; from the upper air and aeroplane, he collected data on the incidence of these spores at different heights up to about 4800 ft and reported the presence of teleutospore; of *P. graminis avenae* and *P. purpurea* in the catche. Other notable contributions during this period were those of Padmanabhan *et al.* (1953) who investigated the incidence of the conidia of *Helminthosporium oryzae*, pathogen causing leaf spot of rice, in the air over paddy fields at Cuttack from February to August 1950 and then from April 1951 to January 1955, using aeroscope. They reported that the occurrence of *H. oryzae* spores in air was the lowest in the period from May to August while October to December was the period with maximum catches. Using similar aeroscopes the prevalence of airborne conidia of *H. oryzae* at Chinsurah, throughout the year for 5 years, was studied by Sen Gupta and Chattopadhyay (1963) with a view to knowing the existing relationship between the number of conidia trapped and the extent infection in the field in a particular season. Subsequently aerobiological studies on a variety of plant pathogenic fungi have been carried out which were reviewed by Sreeramulu (1967) and Tilak (1998). The interactions between the pathogen, host-plant, environment, time and space have been critically discussed by Nagarajan and Muralidharan (1995) in their book on 'Dynamics of Plant Disease'.

Rust and Smut Diseases

Extensive studies on the epidemiology of stem rust of wheat caused by *Puccinia graminis tritici* were made by Nagarajan (1973). The black or stem rust is known to appear in epidemic form in southern, central and eastern parts of the country where relatively higher temperatures prevail during the crop season. Joshi *et al.* (1974) and Joshi (1976) have provided details on the spread of stem rust pathogen. Disease surveys carried out since 1967–68 have demonstrated that in the case of black rust the main focus of infection is South India and that the northern hills play a very insignificant part, if any, in the epidemics of black rust (Joshi *et al.*, 1971, 1974). The survey data collected over a period of 14 years have established that black rust appears first in parts of South or Central India or in the foot hills of the Nilgiri hills. As season advances there is gradual rise in temperature from January onwards and black rust moves northwards. The first build up of brown rust (*Puccinia recindita*) inoculum seem to take place in the plains of Karnataka in South which then moves northwards towards Maharashtra and Madhya Pradesh. At the same time the inoculum of foot hills of Bihar and Uttar Pradesh moves to Northern plains. Quite often, the spores of black and brown rusts are introduced directly in one jump from the South Indian foci, to Central India. Epidemiological studies have shown that from South and Central India the spores are wind-borne and rain-deposited. They are transported by upper winds from the Nilgiris and Palni hills and washed down over Central India by rain. The movement of the rust inoculum from Nilgiri hills towards north has been designated as the "*Puccinia Path*" (Joshi, 1982). As far as northern India is concerned, central Nepal can be considered the most important focus for brown rust (Joshi *et al.*, 1974, 1977). Joshi *et al.* (1977) and Nagarajan (1977) have concluded that the uredospores can be dispersed particularly in the eastern regions of North India up to a distance of 270 km from Central Nepal.

Considerable work has been done in India on the aerobiology of *Puccinia arachidis*, the groundnut rust pathogen (Mallaiah and Rao, 1982; Murdhankar and Pande, 1991; Benagi and Kulkarni, 1995). Mallaiah and Rao (1982) observed that in the rainy season, the pattern of rust incidence in the field and air borne urediniospore concentrations varied greatly depending on the rain fall and weather conditions. In winter, urediniospore concentrations increased steadily as the crop reached maturity and in summer, they appeared mainly towards the end of the crop period, with the onset of monsoon rains. Airborne urediniospores showed a clear diurnal periodicity pattern with peak occurrence around noon (Fig. 23.1). Incidence in the air was maximum when temperatures were 29–31°C and relative humidity 75–85 per cent. Numbers decreased with increasing height above the ground level with a sharp reduction above 2 m. The pattern of urediniospore incidence in the air closely followed

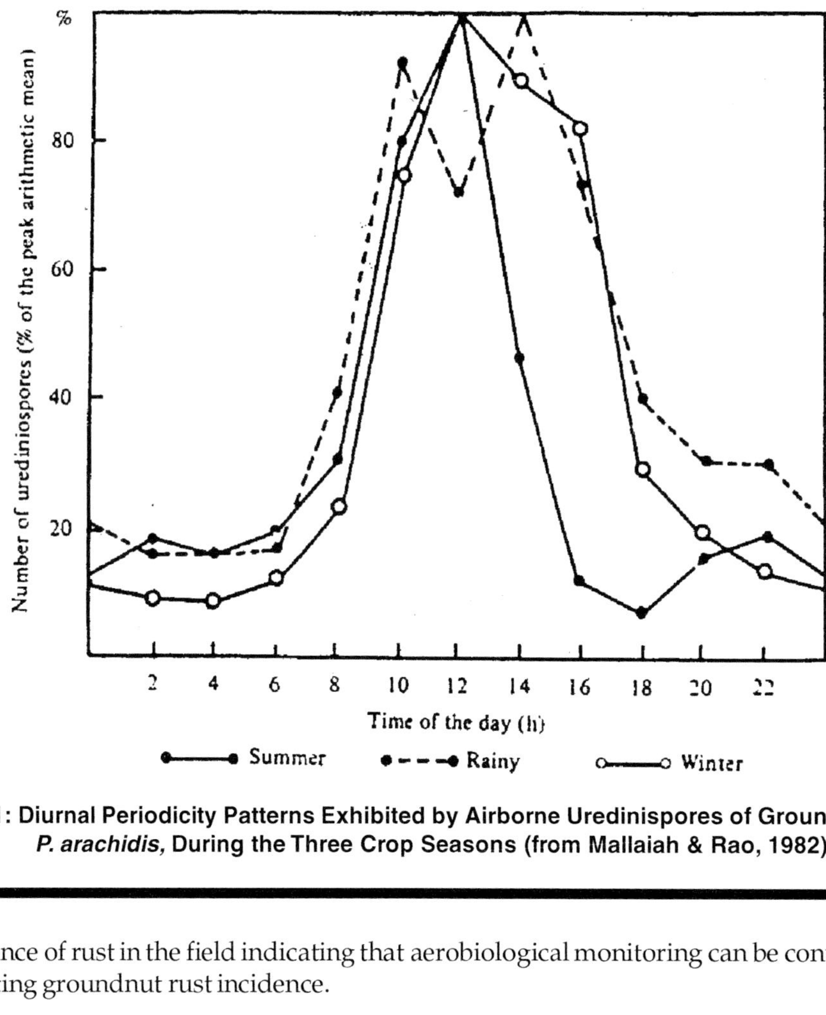

Fig. 23.1: Diurnal Periodicity Patterns Exhibited by Airborne Uredinispores of Ground Nut Rust, *P. arachidis*, During the Three Crop Seasons (from Mallaiah & Rao, 1982)

the prevalence of rust in the field indicating that aerobiological monitoring can be conveniently used for forecasting groundnut rust incidence.

The occurrence of the uredospores of sugarcane rust fungi (*Puccinia kuehnii* and *P. erianthi*) within and above sugarcane crop was investigated by Sreeramulu and Vittal (1971). They reported that the uredospores occur in the air in November, December and January, and further observed that the prevalence of dry conditions lead to higher concentrations in the air.

Among the smut diseases, the aerobiology of *Ustilago scitaminea* Syd., causal organism of whip smut of sugarcane, was studied in detail by Sreeramulu and Vittal (1972). They sampled the air within and above sugarcane plots planted with smut-moculated setts. The first seasonal peak for smut spores was observed in June–August corresponding with the emergence of whips resulting from primary infection. A second and larger peak was recorded in October coinciding with the second flush in whip production due to secondary infection. Spore concentrations recorded above the crop were 5–8 times lower than those inside and showed only one seasonal peak occurring after October. Both the numbers of dispersion units (spore clumps) and of total spores showed a day-time dispersal maximum, occurring before noon inside the crop, and at noon above the crop (Fig. 23.2). Rapid shedding of spores from the

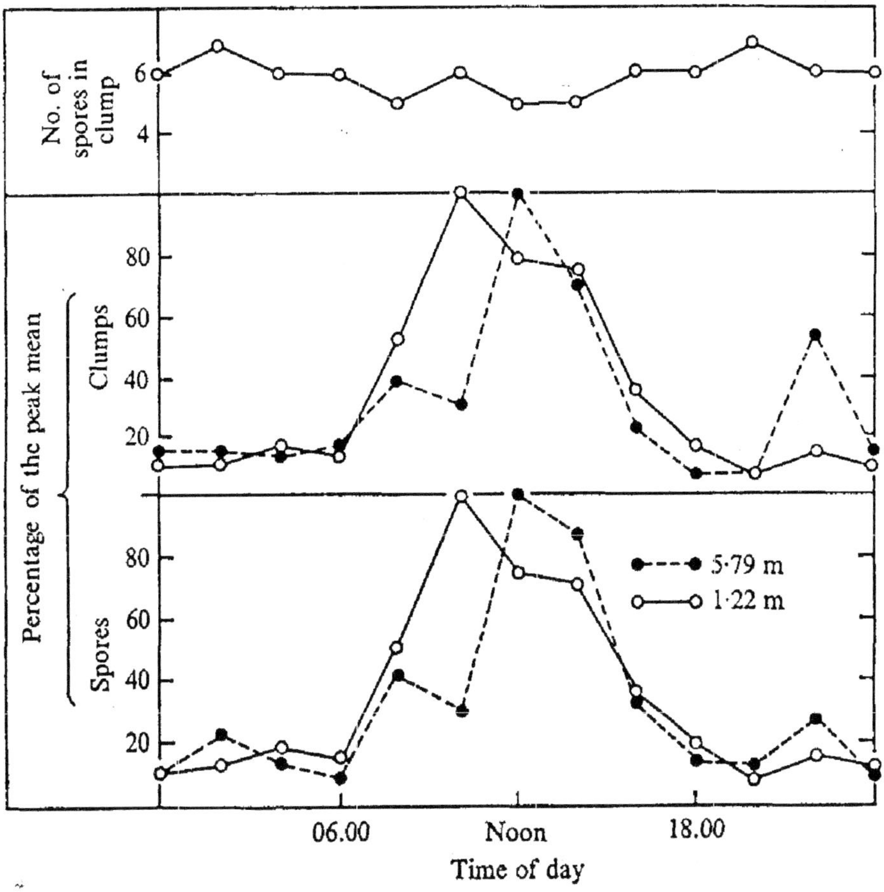

Fig. 23.2: Diurnal Variations the Clumping of Spores and in Concentration of Spores and Clumps of the Sugarcane *Ustilago scitaminera* in the Air (from Sreeramulu & Vittal, 1972)

whips was generally encouraged by dry and reduced by wet weather, and was affected by the intensity and duration of rainfall. Maximum dispersal occurred at 22–24° C and 50–60 per cent r.h. It was also observed that dispersal normally occurs up to a distance of 20 to 40 m from a known source.

Downy Mildews

A systematic study on the aerobiology of *Peronosclerospora sorghi*, causal fungus of sorghum downy mildew, was carried out by Shenoi and Ramalingam (1979) at Mysore. They reported the airborne dispersal of both conidia and oospores against the earlier belief that oospores were not airborne. According to them conidia which are specialised for active liberation are produced in great numbers and dispersed locally at night under conditions which are also favourable for their immediate germination and foliar infection; the oospores which are better adapted to survive dry conditions are shed into the air when turbulence is more and are carried farther contaminating the soil. While the conidial population in air was highest during the vegetative phase of the crop, the oospores were caught on the traces mainly during the post-flowering and ripening stages. The role of air and water in the dispersal of conidia and oospores of *Peronosclerospora sorghi* was investigated by Rajasab *et al.* (1979). The dispersal of conidia took place in the post midnight hours (Fig. 23.3). Oospores were caught on the traces in small numbers. They exhibited a diurnal pattern with the catches spread over all the hours during daytime but with high a numbers between 10 and 16 hr. Oospores were carried by normal winds to short distances and by whirlwinds over longer distances. Rain water carried the buyont oospores with its flow and increased inoculum levels in low lying areas.

Tikka Disease of Groundnut

The aerobiology of *Cercospora* species, *viz. C. personata* (*Cercosporidium personatum*) and *C. arachidicola* causing tikka disease of groundnut was studied by several workers. Sreeramulu (1970) reported that the conidia of *C. arachidicola* appeared first in the air about a month after the emergence of seedlings and continued to occur in low numbers throughout. But conidia of *C. personata,* which appeared a week later, became more dominant and assumed greater importance as the season advanced. The curve depicting the course of development of the epidemic, as represented by the increase in the quantity of inoculum of *C. personata* available in the air over the field, is sigmoid with a prolonged lag-phase leading to a short exponential phase. Conidia of both the species showed typical forenoon pattern of diurnal periodicity with the daily peaks recurring at 10 a.m. and higher numbers between 8 a.m. and 2 p.m. More or less similar observations were made by Mallaiah and Rao (1980) and Tilak

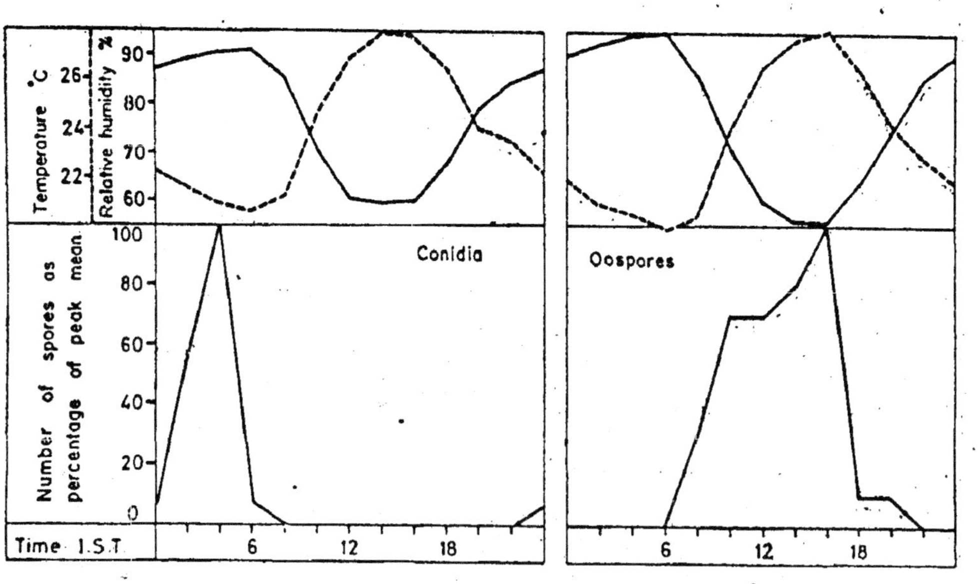

Fig. 23.3: Mean Circulation Periodicity of *Perenosclerospora sorghi* Conidia and Oospores in Air, Related to Weather (from Shenoi & Ramalingam, 1979)

(1988). Mallaiah (1989) while stressing the importance of aerial dispersal under field conditions reported that the conidia of *Cercospora* spp. occur in air at a wide range of temperatures and relative humidity. The detailed studies on the aerobiology of diseases of groundnut carried out by Tilak and co-workers have clearly brought out the relationship between the spore load of pathogens in air, growth stages of the crop, meteorological factors and disease incidence and its subsequent spread High humidity and low temperature were found to favour spread of *Cercospora personata* and *C. arachidicola* ut (Tilak, 1988). Spores of *Cercospora* spp. causing tikka disease of groundnut were found to account for only 2.05 per cent of the total airspora with high percentage in khariff than in rabi (Kalkar and Patil, 1997). The seasonal peak for *Cercospora koepkei*, causal agent of yellow spot disease of sugarcane was noticed in early November after 3–4 weeks of disease outbreak (Sreerarnulu *et al.*, 1971).

Blast Disease of Rice

The aerobiology of the rice blast fungus, *pyricularia oryzae* has been worked out by a few workers in India (Muralidharan and Venkatarao, 1980; Sreeramulu and Ramalingam, 1966). Stray conidia which survive on grasses act as a source of primary inoculum. From each blast lesion number of conidia are liberated which become airborne. Foliar infections occur rapidly during seedling and tillering. Rice blast pathogen is highly sensitive to fluctuation of relative humidity, temperature and sunlight and the disadvantages are overcome by diurnal periodicity in sporulation. For the appearance of spore load in air congenial meteorological factors are prerequisite and the spore load prevails in the atmosphere 35 days in advance, before the first appearance of disease on host plant.

Miscellaneous Studies

In addition to detailed and specific studies described above aerobiological studies of some more pathogens of crop plants were investigated. The Periodicity in the airborne spores of the rice false smut fungus, *Ustilaginoidea virens*, was studied by Sreeramulu and Vittal (1966). They reported that the maximum incidence of spores in the air occurs over a period of about four weeks which coincides with the 'heading' phase of the rice plants, indicating that both early and late infections are possible. On normal dry days the airborne spores of the fungus exhibited a regular and sharply defined diurnal periodicity with the daily maximum recurring at 22.00 hr. (Fig. 23.4). On the other hand, the stackburn fungus, *Trichoconis padwickii* was maximum during the period when the crop alters into ripening phase (Sreeramulu and Vittal, 1966). The importance of rain or dew droplets and drips from the leaf canopy on the spread of midrib lesions under field condition has been emphasized by Sreeramulu and Vittal (1970) in their studies on incidence and spread of red rot lesions caused by *Colletotrichum falcatum* on midribs of sugarcane leaves. They further reported that the prevalence of wet weather conditions is more important for midrib lesion initiation and spread than the age of the leaf or even the differences in the relative susceptibility of the cane variety. Splash dispersal is reported to be an important mechanism for certain airborne plant pathogens (Tilak, 1987). Rain causing mechanical disturbance to the infected leaves are found to release more number of uredospores from rust pustules on coffee leaves (Rajasab and Rajendran, 1983).

Tilak and Babu (1984) made an aerobiological survey over bajra fields with special reference to certain airborne diseases of bajra, especially, blast disease caused by *pyricularia penniseti*. The effects of weather on the perennation and prediction of apple scab, caused by *Venturia inaequalis* was studied by Gupta (1988).

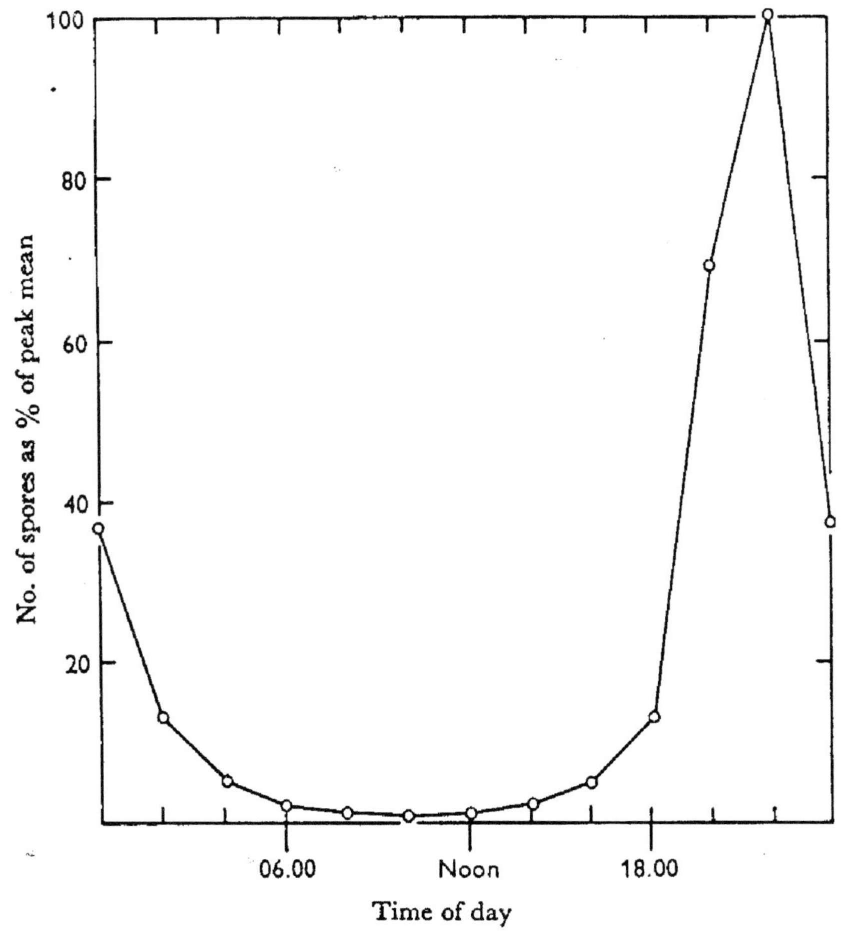

Fig. 23.4: Diurnal Periodicity Curve Exhibited by the Spores of Rice False Smut Fungus, *Ustilaginoidea virens* (from Sreeramulu & Vittal, 1966)

Conidia of *Alternaria porri*, causal agent of purple blotch of onion, were numerous in the air when the crop was in 7–8 leaf stage (Chawda and Rajasab, 1994). The peak conidial concentration was favoured by moderate temperature and relative humidity. Increasing incidence of conidia of *Alternaria dauci*, causing leaf blight of carrot, were reported with progress of disease in field (Channabasavaraj Gudige and Narayana Reddy, 1997). The aerobiology of rust disease of mulberry caused by *Peridiospora mori* was studied by Prasad *et al.* (1994). The peak release of mulberry rust spores was noticed during day time with low light intensity. The urediniospores were found to be more in air above the plant canopy what the atmosphere was dry. The spores were noticed in the air over mulberry gardens from August and the spore load increased up to December declining thereafter coinciding with disease severity. Maximum number of conidia of *Alternaria alternata*, causal agent of leaf and stem-spots of sunflower, were trapped when the crop was either in flowering or just crossed that stage (Ramchander Rao, 1993a,b). While studying the epidemiology of early blight of tomato caused by *Alternaria solani*, Naik and Sinha (1997) reported that the conidial dispersal takes place throughout the growing season from the primary spots and high relative humidity favours conidial dispersal. While studying the significance of airborne *Drechlsera setariae* conidia to disease development in pearl millet crop, Raveesha and Shetty (1997) reported that the conidia were abundant in the air from the early stages of the crop. *D. setariae* seem to colonise both in the sporulating and non-sporulating phases of the downy mildew infected leaves. If *D. setariae* colonises in the sporulating phase, it may help in reducing the build up of downy mildew inoculum preventing downy mildew epidemics.

Air mycoflora over fields of cereals, pulses, vegetables, oil seed and cash crops at Imphal were studied by Singh and co-workers. The effect of intercultivation practices such as weeding, rogueing, hoeing, foliage disturbance and green pod picking in the release of uredospores of *Uromyces fabae* has been emphasized by Singh and Singh (1992). They suggested that mechanical disturbances in the crop canopy during interculture releases the spore clouds creating a possibility of intense disease development. The occurrence of the spores of *Helminthosporium maydis, Cercospora sorghi* and *Puccinia sorghi* causing leaf blight, leaf spot and rust disease over a maize field suggested that the growth phase and favourable meteorological parameters play a significant role in causing diseases than the concentration pathogenic spores in the air (Singh and Dorycanta, 1992). Rainfall and high relative humidity are reported to initiate infection of sunflower by the rust fungus, *Puccinia helianthi* (Tilak and Ramachander Rao, 1989).

Vertical Profiles

When a spore cloud is generated from source, the spores get diffused both horizontally and vertically. Spore concentration generally decreases with height, and the decrease is normally logarithmic, although the picture varies in different conditions of air stability. A few studies have been devoted to monitor the spore levels within and above crop fields. Concentrations of spores of smut, rust, yellow spot was reported to be much higher within sugarcane field when compared with above crop (Sreeramulu and Vittal, 1969). Vertical profiles of spore concentrations within and above canopy in sorghum fields were investigated by Subba Reddi and Ramakrishna (1978). These studies revealed maximum concentrations within the crop when compared with higher levels. Further, spore concentration declined with increase in height from the ground level. In their studies on vertical distribution of conidia of *Peronosclerospora sorghi* over sorghum fields, Rajasab *et al.* (1979) could not detect any gradient in their concentration.

REFERENCES

Benagi, V.I. and Kulkarni, S. (1995). Epidemiology and aerobiology of rust of groundnut caused by *Puccinia arachidis* Speg. *Indian J. Aerobiology*, **8**: 1-6.

Chawda, H.T. and Rajasab, AH. (1997). Day to day variations in the incidence of *Alternaria porri* conidia over a purple blotch infected onion field. In: *Aerobiology: Proc. of 5th International Conference*, Bangalore, 1994, 221-227.

Channabasavaraj Gudige, C.A and Narayana Reddy, C. (1997). Aeromycoflora over carrot field with special reference to air-borne inoculum of *Alternaria dauci*. In: *Aerobiology: Proc. 5th International Conference*, Bangalore, 1994, 243-248.

Cunningham, D.D. (1973). *Microscopic examination of air*, Government Printer, Calcutta.

Gupta, G.K. (1988). Effects of weather on the perennation and prediction of apple scab, *Venturia inqequalis*. In: *Perspectives in Mycology and Plant Pathology*, M.P.H. Delhi, pp. 353-367.

Kalkar, S.A and Patil, G.V. (1997). Fungal airspora of a groundnut field at Nagpur. In: *Aerobiology*–Proc. 5th International Conference, Bangalore, 1994, pp. 249-254.

Joshi, L.M. (1976). Recent contributions towards epidemiology of wheat rusts in India. *Indian Phytopath.*, **29**: 1-16.

Joshi, L.M. (1982). Wheat rusts management: Present knowledge and future prospects. *Kavaka*, **10**: 1-12.

Joshi, L.M., Nagarajan, S. and Srivastava, K.D. (1977). Epidemiology of brown and yellow rust of wheat in North India I. Time and place of appearance. *Phytopath. Z.*, **90**: 116-122.

Joshi, L.M. Saari, E.E., Gera, S.D. and Nagarajan, S. (1974). Survey and epidemiology of wheat rusts in India In: *Current trends in Plant Pathology*, (Ed. S.P. Raychaudhuri and J.P. Verma), Bot. Dept., Lucknow Univ., India, pp. 150-159.

Mallaiah, K.V. (1989). Aerobiology of *Cercospora* specific pathogenic to crop plants. In. *Environment Ecology and Aerobiology*, (Ed. S.T. Tilak), Today and Tommorrow Printers and Publishers, New Delhi, pp. 67-78.

Mallaiah, K.V. and Rao, A.S. (1980). Aerobiology of two Specific of *Cercospora* pathogenic to groundnut. *Proc. Int. Acad. Sci.*, **46B**: 215-222.

Mallaiah, K.V. and Rao, A.S. (1982). Aerial dissemination of urediniospores of groundnut rust. *Trans. Br. Mycol. Soc.*, **78**: 21-28.

Mehta, K.S. (1952). Further studies on cereal rusts in India. *ICAR Sci. Monogr.*,**18(2)**.

Murdhankar, S.V. and Pandey, B.N. (1991). Aerobiological and epidemiological approach to groundnut rust. *Indian J. Aerobiology*, 19-22.

Muralidharan, K. and Valkatarao, G. (1980). A simple method of forecasting outbreak of rice blast. *Indian Phytopath.*, **33**: 560-564.

Nagarajan, S. (1973). Studies on uredospore transport of Puccinia graminis tritici and the epidemiology of stem rust of wheat in India. Ph.D. Thesis. Univ., Delhi, pp. 125.

Nagarajan, S. (1977). Meteorology and forecasting of epidemics of diseases. *Sym. Basic Sci. Agric.*, INSA, New Delhi, **1**: 136-143.

Nagarajan, S. and Muralidharan, K. (1995). *Dynamics of Plant Diseases*. Allied Publishers, p. 245.

Naik, M.K. and Sinha, P. (1997). Epidemiology of early blight of tomato. *Aerobiology*–Proc. 5th International Conference, Bangalore, 1994.

Padmanabhan, S.Y., Ganguly, D. and Balakrishnan, M.S. (1953). *Helminthosporium* disease of rice II. Source and development of seedling infection. *Indian Phytopath.* **6**: 96-105.

Prasad, K.V., Dayakar Yadav, B.R. and Sullia, S.B. (1997). Aerobiology of *Peridiopsora mori* causing mulberry rust. In: *Aerobiology*–Proc. 5th International Conference, 1994, Bangalore, pp. 261-272.

Rajasab, A.H. and Rajrodran, C. (1983). Dispersal of *Hemeleia vastatrix* uredospores and spread of coffee leaf rust. *J. Coffee Res.*, **13**: 120-124.

Rajasab, A.H., Shenoi, M.M. and Ramalingam, A. (1979). Epidemiology of sorghum downy mildew III. Dispersal and deposition of inoculum. *Kavaka*, **7**: 63-68.

Ramachander Rao, K.S. (1993a). Alternaria leaf spot of sunflower I. Epidemiology and airborne conidia. *Indian J. Aerobiology*, **6**: 45-48.

Ramachander Rao, K.S. (1993b). Alternaria leaf spot of sunflower II. Meteorological factors associated with disease incidence and atmospsheric spore load. *Indian J. Aerobiology*, **6**: 49-52.

Raveesha, K.A. and Shetty, H.S. (1997). Significance of air-borne *Drechslera setariae* to disease development in pearl millet crops. *Aerobiology*–Proc. 5th International Conference, Bangalore, 1994, pp. 255-260.

Sen Gupta, C. and Chattopadhyay, S.B. (1963). Proc. 50th Indian Sci. Cong., Delhi. *Agric. Sci. Abst.*, **78.**

Shenoi, M.M. and Ramalingam, A. (1979). Epidemiology of sorghum downy mildew. II. Circadian and seasonal periodicities in conidia and oospores. *Proc. Indian Acad. Sci.*, **88B**: 95-102.

Singh, N.I. and Dorycanta, H. (1992). Aerobiology of crop diseases of Manipur VIII. Fungal airspora over maize field in Senapati district. *Indian J. Aerobiol.*, Sp. Vol.: 141-144.

Singh, S.R. and Singh, N.I. (1992). Aerobiology and epidemiology of rust of broad bean (*Vicia faba* L.) II. Role of interculture practices in the release of uredospores of *Uromyces fabae. Indian J. Aerobiol.* Spl. Volume: 133-136.

Sreeramulu, T. (1967). Aerobiology in India. *J. Scientific & Industrial Research*, **26**: 474-480.

Sreeramulu, T. (1970). Conidial dispersal in two species of *Cercospora* causing tikka leaf- spots on groundnut (*Arachis hypogaea* L.). *Indian J. Agric Sci.*, **40**: 173-178.

Sreeramulu, T. and Ramalingam, A. (1966). A two-year study of the air spora of a paddy field near Visakhapatnam. *Indian J. Agric. Sci.*, **36**: 111-132.

Sreeramulu, T. and Vittal, B.P.R. (1966). Periodicity in the air-borne spores of the rice false smut fungus, *Ustilaginoidea virens. Trans. Br. Mycol Sac.*, **49**: 443-449.

Sreeramulu, T. and Vittal, B.P.R. (1966). Some aerobiological observations on the rice stackbum fungus, *Trichoconis padwickii. Indian Phytopathology*, **19**: 215-221.

Sreeramulu, T. and Vittal, B.P.R. (1970). Incidence and spread of red rot lesions on midribs of sugarcane leaves. *Plant Disease Reporter*, **54**: 226-231.

Sreeramulu, T. and Vittal, B.P.R. (1971). Periodicity in the uredospore content of air within and above a sugarcane field. *J. Indian Bot. Soc.*, **50**: 39-44.

Sreeramulu, T. and Vittal, B.P.R. (1972). Spore dispersal of the sugarcane smut (*Ustilago scitaminea*). *Trans. Br. Mycol. Soc.*, **58**: 301-312.

Sreeramulu, T. Vittal, B.P.R. and Ramakrishna, V. (1971). Aerobiology of *Cercospora koepkei Kruger causing the yellowspot disease of sugarcane. Indian J. Agric. Sci.*, **41**: 655-662.

Subba Reddi, C. and Ramakrishna, V. (1978). Vertical profiles of spores over a sorghum field. *Phytopath. Z.*

Tilak, S.T. (1984). Aerobiology and cereal crop diseases. *Rev. Trop. Pl. Path.*, **1**: 329-354.

Tilak, S.T. (1986). Aerial transport of plant pathogens. In: *Vistas in Plant Pathology*, (Eds. Verma, A. and Verma, J.P.), Malhotra Publishing House, New Delhi, pp. 237-248.

Tilak, S.T. (1988). Aerobiology and epidemiology of groundnut disease. In: *Perspectives in Mycology and Plant Pathology.* Prof. B.S. Mehrotra Comm. Vol., MPH, New Delhi, pp. 336-352.

Tilak, S.T. (1998). *Aerobiology.* Satyajeet Prakashan, Pune, 469 pp.

Tilak, S.T. and Babu, M. (1984). Aerobiology and epidemiology of certain diseases of bajra. *Ind. J. Bot.,* **7**: 77-83.

Tilak, S.T. and Mane, D.A. (1981). Aerobiological approach to rust disease of bajra. In: *Proc. Nat. Conf Env. Biopoll.,* (Ed. S.T. Tilak), Vaijayanti Prakashan, Aurangabad, pp. 247-250.

Tilak, S.T. and Pande, B.N. (1989). Aerobiology and epidemiology of bajra diseases. In: *Recent researches in Ecology and Environment Pollution,* (Ed. S.T. Tilak), Today and Tomorrows Printers and Publishers, New Delhi, pp. 79-96.

Tilak, S.T. and Ramachander Rao, K.S. (1987). Significance of air monitoring of ergot at two places in Marthwada. *Indian J. Bot.,* **10**: 105-109.

Tilak, S.T. and Ramachander Rao, K.S. (1989). Aerobiology of sunflower disease. In: *Recent Researches in Ecology and Environmental Pollution,* (ed. S.T. Tilak), Today and Tomorrow Publishers, New Delhi, pp. 41-66.

24

Trees of Religious Importance from Amarakosha

BRAHMANAND DESHPANDE, *Mahamahopadyaya Vidyabhushana*
'Aitihasik' 14, Anupam Colony, Shreyanagar, Aurangabad – 431 005, India

Amukha

The rule of the Imperial Guptas (C. 300-500 AD) is aptly appreciated as the Golden Period or the Classical age of the ancient Indian History. It witnessed and unprecedented progress in different spheres of culture and was really a trend-setter. *Amarakosha,* the lexicograph turned a legend, was composed in this period around 300-350 AD (Weber). The real title of this unique lexicograph is NÈmalixgÈnu„Ènakosha but it is popularly known as *Amarakosha* after its composer AmarasiŠha. Very little is known about this author except for the two possibilities:

1. He was a *kshatriya* as the suffix *siŠha* in his name indicates.
2. He was *Bauddha* because while listing the deities he has placed Lord Buddha in the first position.

Amarakosha is meticulously composed. Divided in three sections (*kÈ‡Ças*) and sub- sections (*Vargas*), it enlists names of deities, demons, planets, mountains, rivers along with flora, fauna, royal officers, weapons, utensils, castes, professions, architectural units, fabrics, mercandiles and what not? The fifth *varga* of second *kÈ‡Ça* deals with flora. Lack of place compells us to allude to religious trees only.

Pippala

Amarasimha alludes to Pippala (*Ficus religiosa* Linn, Moraceae) Bodhidruma, Chaladala, Pippala, Ku¤jarÈsana and A„vattha. It was under this tree that Bhagavana Buddha attained *sambodhÏ* or the proper enlightenment at Bodhagaya on the bank of the river Nira¤jarÈ. Mahendra and Sa×ghamitrÈ, the son and daughter of Emperor Asoka took the sapling of this tree to AnurÈdhÈpura in Sri Lanka and planted it there (Jain). Pippala or BodhÏ-*v*, *iksha* is thus sacred for the Buddhists. It became the symbol of *sambodhÏ* and later on of Buddha himself. In Hinayan a sect of Buddhism image worship was not practised. Buddha was represented by symbols like the Wheel of Religion (*Dharmachakra*), the turban of Buddha (*ush‡i„a*), his foot prints, throne etc. But Bodhi-tree became the most popular symbol for the Master.

Pippala adores nearly every Hindu temple. Ladies circumambulating (*pradakshi‡È*). This holy tree is a common sight in the temple complex. The world–famous KailÈsa cave (Cave No. 16 at Ellora, District Aurangabad, Maharashtra) contains a beautiful image of ƒiva in meditating pose (YogÏ„wara ƒiva). He is Seated cross-legged under a Pippala-tree with serene look on his face.

Udumbara

Amarasimha has listed Jantuphala, Yaj¤È×ga, Hemadugdhaka, as the synonyms for Udumbara or Umbara (*Ficus recemosa*. Linn. Moraceae). This tree is very sacred for the Hindus. Stools and chairs made from its wood are used by *gurus, ÈchÈryas* and holy persons. Its dry branches, and twins are burnt in the Sacrificial pyre, (*yaj¤a*) and spoons and ladels used in sacrifice are made from its wood and hence the synonym *yajnanga*; DattÈtreya, symbolising the holy trinity of the Hindus *i.e.* BrahmÈ, Vish‡u and ƒiva is largely worshipped in Maharashtra, Karnataka and Gujarat. Umbara tree is associated with DattÈtreya. Its fruits contain germs and hence the appellation Jantuphala. Because of these living organisms, the fruit of the Umbara is said to be the miniature replica of the globe. Hindus worship this holy tree and abhorr its destruction in any way.

Nimba

Nimba (*Azadirchta indica* A. Juss Meliaceae) the ubiquitous tree of India, is alluded to by the synonyms PÈribhadra, Nimbataru, MandÈram Parijataka, Arish–a, Sarvatobhadra, Hi×guniryÈsa, Malaka and Pichumanda. The last name arrests our attention at once. Pichu means leucoderma (albinism) and leprosy. The regular consumption of the leaves, flowers and bark of this tree controls this disease and hence it is called as Pichumanda.

Nimba is associated with SÊrya (the sun god). The Jainad (District Adilabad, Andhra Pradesh) inscription of Jagaddeva ParamÈra (EI) opens with a *sloka*, purporting to register that the Sun God should be worshiped on Sunday with the Nimba flowers. The inscription purports to the register the construction of the temple of NimbÈditya by PadmÈvatÏ the spouse of LolÈrka, the minister of Jagaddeva ParamÈra. This proves the close association of the Nimba tree with SÊrya (¶ditya). Chakradhara, the founder of MahÈnubhava Sect. (C. 1230-1275) toured about 14 Districts of Maharashtra and visited more than 250 places. During his sojourn at Pratishhana (Modern Paithan, District Aurangabad, Maharashtra) he visited the temple of NimbÈditya. A ruined temple of NimbÈditya can be located at Nandur (Taluka Shevagaon, District Ahmednagar, Maharashtra). NimbÈrka, the famous expounder of DvaitÈdvaita school of Vedanta is said to have born at Mungi (Taluka Shevagaon, District Ahemadanagar, Maharashtra), 15 kms South West of Paithan and surely this village has a rich growth of more than thousand Nimba trees.

Vata

For Va–a, VaÇa or BaÇa (*Ficus bengalensis*, Linn. Moraceae) *Amarakosha* has only two synonyms *viz.* Nyagrodha and BahupÈt. The latter, probably, suggests the prop-roots of this majestic tree. Va–a is the most sacred tree for the Hidus. In Uttar Pradesh, Bihar, and Bengal the village folk plant this tree with a pompous religious ceremony. The large Vata tree on the sacred confluence of the rivers Ga×gĚ and Yamuna at PrayĚga is called Akshayava–a or Cha–uka-Va–a. Several lithic and copper-plate charters purport to record the donations to temple-establishments or *BrĚhma ×as* by royal personages after a sacred bath in the confluence and a worship of Akshaya or Cha–uka Va–a.

The Nagardhan copper-plate charter of Svamiraja and NannarĚja of the RĚsh–raku–a dynasty of Vidarbha (CII). (K.E.322 573 A.D.) purports to record the donations to some Brahmanas when the royal brothers were camping on the banks of JĚhnavĪ (Ga‡gĚ) near Cha–uka-va–a.

The vast expanse and numerous prop-roots give the Va–a trees a majestic look and the mysterious and sinister darkness at the bottom. Hence the belief that Va–a is a favourite abode of evil and vicios spirits, like the Yakshas or VetĚls. Undressed stones, smeared with vermillion and popularly known as *tĚnda Äas* are kept on the platforms (pĚra) built round trunks of these trees and were worshipped as images (sic) of these evil spirits.

Va–a is especially associated with Yakshas and Yakshi‡Īs. Vamana *purĚnĚ* extols that Ma‡ibhadra, the prominent among the Yakshas is specially associated with va–a tree and always resides near it. At times even the Va–a tree is said to be a manifestation of Yaksha.

The Vaghali stone inscription (Taluka Chalisgaon, District Jalgaon, Maharashtra) of the YĚdava ruler SeÊ‡anchandra II (EI), furnishes interesting details of the association of Yakshas and Yakshinis with the Va–a tree. Here the field donated to the *BrĚhma ‡as* is called as Va–a-yakshi‡i-va–a. The record is dated in ƒaka 991, ¶shĚdha, solar eclipse (July 21, 1069 A.D.)

GĚthĚsaptĚsati is' the anthology of poems composed by diverse poets and compiled by the SĚtavĚhana ruler HĚla around 1st century A.D. It is an ornament of the PrĚkrit language. GĚthĚ No. 372 of this monumental works speaks of the stone-slab kept under the Va–a tree and worshipped as Va–a-yaksha (Jogalekar).

Govindaprabhu (C. 13th A.D.) is reverred by the followers of the MahĚnubhĚva Sect. His biography, *Govindaprabhucharitra*, composed by MhĚmbha–a (Deshpande) speaks of Va–ajambĚ (bhĚ) at BhĚnakheÇe. (Taluka Chandur, District Amaravati, Maharshtra) JambhĚ and JambhĚ-JambhalĚ, the Yaksha couple was very popular. Their robust images are found in the AjĚnta, Ellora and Aurangabad caves.

Va–a is believed to be the most primitive tree. In the primordial waters, the Supreme Being is represented as floating on a Va–a leaf.

Kadamba

Kadamba (*Neolamarckia cadamba*, Roxb. Bosser, Rubiaceae) is alluded to by AmarasiŠha as NĪpa, PriyĚka, Kadamba and Halipriya. The last appellation underlines its association with Hali (= one who sports *hala* or plough as his weapon) or BalarĚma, the elder brother of K‚ishna. The water of the river YamunĚ were rendered poisonous by the great serpent KĚliyĚ. Lord K‚ishna climbed a Kadamba trees and jumped into the river to chastise this venomous evil.

The flowers of Kadamba look exactly like a drumstick, with a long stem and globular head, with thousands of miniscule flowers of exquisite fragrance on it. *GĚthĚ* No. 177 of *GĚthĚsapta„atĪ*, composed by the poetess Asuladdhi, (Satvahana) compares them with the small stone balls in the catapult of God Madana (God of Love, Indian Cupid).

Exquisitely delicious wine is distilled from these flowers and called as KÈdambarÏ. BÈ‡abha—a, the great prose-writer, named the heroine of his vastely popular and intricate long story as Kadambari and hence all the episodical works are called as KÈdambarÏ in Marathi. The Kadambas of Vanavasi (C. 300-600 A.D.) worshipped this tree as the totem of their family.

Chi¤cha

Chi¤chÈ (*Tamarindus indica* Linn. Caesalpiniaceae) is alluded to as Tintidi, *Chi¤chÈ* and AmlikÈ in *Amarakosha.* Its religious importance is not widely known. Tamarind trees were planted around the temples of Siva. This cluster was known as Ti‡Çivana or Di‡Çiravana. Pandharpur (District Solapur, Maharashtra) is famous for its Vi—hala temple (Local Information). There is a Di‡Çiravana at Pandharpur also. ChÏchÈÏ or ChÏchÈe, a lesser goddess at Paithan figures in *LÏlachÈritra* also (Kolte).

ƒÈla

ƒÈla, Sala or SÈlai (*Shorea robusta,* Gaertn Dipterocarpaceae) is also called as Sarja, KÈr„ya, A„vakar‡aka and Sasyasambara by AmarasiŠha.

This tree is closely associated with Lord Buddha. It is a popular belief with the Buddhist that the Master was born at Lumbini under a Sala tree. MÈyÈvatÏ his mother was in an advanced stage of pregnancy and was processing from Kapilavastu, the capital of Sakyas to DevadahÈ, her parental town. In the garden at LumbinÏ the labour occurred under a ƒÈla tree. A place called Rummindei, on the border of Uttar Pradesh and Nepal answers to Lumbini. A sculptured panel found here depicts the scene of the Nativity of Lord Buddha.

Lord Buddha travelled far and wide to preach his gospel. He became an octagenarian. While travelling from PÈvÈpurÏ, the capital of Malla republic to Ku„inagara, the capital of the other tribe of the Mallas, he took ill. On the outskirts of the city of Ku„inagara the Master breathed his last between two ƒala trees. These two happennings in the life of the Master attracted the attentions of sculptors and painters. At Ajanta (Cave No. 2) we have a beautiful painting of the Nativity event where the ƒala tree is clearly discernable while in cave no. 26 a beautifully sculpted panel on the left wall shows his MahÈparinirvÈ‡a on a grandiose scale. Here also the sculptor has done full justice to the two ƒÏla trees showing their trunks, branches and flowers with minute details.

BhÊrjav,iksha

BhÊrjav,iksha (*Betula utilis* D. Don. Betulaceae) has two apt synonyms *viz.* CharmÏ and M,idutvak as its bark was widely used for writing. Even after the introduction of paper, the use of BhÊrjapatra was not abandoned. To the contrary sacred texts, prayers, psalms, ritualistic formulae, charms, sacred and magic figures (*yantras*), hymns and even the chants and figures of black magic were written on the BhÊrjapatras only. These leaves were stored in sacred places like temples or hermitages (*ma—has*).

ShamÏ

Shami (*Prosopis spicigera* Linn. Mimosaceae) is also called as Saktuphala and ƒiva. This tree commands unique religious importance. It is associated with warfare and worshipped on the VijayÈ Da„amÏ day for the success of military campaigns. The Pa‡davas have kept their weapons on this tree when they were of exile for twelve long years. The tree sy£bolises the war deity and is held as sacred.

Bilva

Bilva (*Aegle marmelos* Linn. Rutaceae) or popularly known as Bela is provided with ƒÈnÇilya, ƒailusha, MÈlura and ƒrÏphala as synonyms. No need to dilate upon its close association with ƒiva,

PÈrvatÏ or UmÈ in her MÈhesvarÏ form is shown sporting in a Bilva fruit in her hand. Bilva is called ƒriphala. In Maharashtra Coconut is erroneously, called as ƒrÏphala.

REFERENCES

Weber. *History of Sanskrit Literature*, p. 106.

Jain, Kailash Chandra. *Malawa Through Ages*, p. 104

EI., XXII, pp. 654 E.I./pp

CII., IV, Part II, pp. 654 f.f.

Deshpande, Brahmanand, *Saptapar‡Ï*, p. 7.

EI., II, pp. 221-228.

Jogalekar, S.A. (Ed). Hala Satvahana, *GÈthÈsapta„atÏ*, *Gatha* No. 372.

Deshpande, Brahmanand (Ed.). MhÈimbha–a, *Givindaprabhucharitra*, p. 118.

HÈla SÈtvÈhana, *op. cit*, *GÈthÈ* No. 177.

Local Information.

Kolte, V.B. (Ed.). MhÈimbhata, *LÏlacharitra*, 3rd Edition, *LilÈ* No. 3, 395, 464.

Abbreviations

CII: Corpus Inscriptionum Indicarum

EI: Epigraphia Indica

KE: Kalachuri Era.

25

Viruses Infecting Chilli/ Capsicum in India

SATYA PRAKASH & S.P.S. TOMER

India Agricultural Research Institute, Regional Station, Agricultural College Estate, P. O. Shivajinagar, Pune – 411 005, India

ABSTRACT

Chilli/capsicum (*Capsicum annuum* L.) is an important solanaceous spice/vegetable cash crop of India. It is a host of a variety of pathogen *i.e.* fungi, bacteria, viruses and phytoplasma. Some 19 viruses have been reported. Most of them are characterized and grouped and are distributed throughout the growing areas of the country. Most of the viruses are aphid transmitted and produces a variety of symptoms in chilli/capsicum. Cent percent loss is reported where infection occurred at early ontogenetic stage. Detection of these viruses can be done on the basis of symptomatic reaction on indicator hosts, serology and molecular study. Various methods of management, desired primarily at the vectors of the viruses are described herein. This chapter reviews the viruses infecting chilli/capsicum, methods of detection, their management and future research needs.

Key words: *Capsicum annuum, virus, phytoplasma, indicator host, serology, molecular techniques, insect borne, non-insect borne, management*

Chilli/Capsicum (*Capsicum annuum* Linn.) is one of the most important spice/vegetable crops of India. India ranks first in it's cultivation and export in the world. At present, it is grown all over the country of which, 57 per cent of the area is occupied in Andhra Pradesh, Karnataka and Maharashtra States. The total area under chilli/capsicum in India is about 90,38,00 ha. and the production is 4800 metric tonnes with an average yield of 9057 kg/ha. (Anonymous, 1997). This crop has been found to be susceptible to a variety of diseases caused by fungi, bacteria, viruses and phytoplasma (Anonymous, 1992; Singh, 1993; Singh and Singh, 2000; Satya Prakash, 2001). Diseases of viral origin have been recognized as a serious threat to successful cultivation of this crop. They cause enormous losses in respect of marketable fruit quality and quantity. So for, about 72 viruses have been reported to infect naturally as well as artificially to this crop world over (Table 25.1). In India, 19 different viruses have been reported to occur as natural infection (Satya Prakash , 2001; Satya Prakash and Singh, 2004). Cent percent losses have been reported in case where infection occurred at early ontogenetic stage (Prasada Rao, 1976; Deol and Rataul, 1978 a; Satya Prakash, 2001). These viruses can be detected by symptomatic reaction on indicator hosts, serology and molecular techniques. For the prevention or reduction of virus infection, particularly for the viruses transmitted by aphids in a non-persistent manner the following management practices have been reported with varying degrees of success in farmers' field: granular insecticide, systemic insecticide and mineral oil sprays, mulching, insect traps, white nylon net, sticky yellow polythene sheets, cultivation of non-susceptible barrier crops and growing resistant varieties (Green and Kim, 1991, Satya Prakash *et al.*, 2002b).

Table 25.1: Worldwide Distribution of Chilli/Capsicum Viruses

Taxonomic Group	*Virus*	*Geographical Distribution*	*References*
FILAMENTOUS VIRUSES			
Potyviruses	Potato virus 'Y' (PVY)	Worldwide	Joshi and Bhargava (1962), Bidari and Reddy (1990), Green and Kim (1991), Green (1992), Brioso (1996), Satya Prakash (2001), Satya Prakash and Singh (2004)
	Tobacco etch virus (TEV)	Worldwide	Bidari and Reddy (1986), Green and Kim (1991), Green (1992), Satya Prakash (2001), Satya Prakash and Singh (2004)
	Pepper veinal mottle virus (PVMV)	West Africa (Ghana, Nigeria, Ivory Coast), India, USA, France	Green and Kim (1991), Nova-Womdim and Atibalentja (1993), Satya Prakash (2001), Satya Prakash and Singh (2004)
	Chilli veinal mottle virus (CVMV)	Asia (Indonesia, Malaysia, Philippines, Thailand, Sri Lanka, Taiwan, Korea, India)	Duriat (1996), Green and Kim (1991), Green (1992), Satya Prakash (2001), Satya Prakash *et al.* (2002a), Satya Prakash and Singh (2004)
	Pepper vein banding virus (PVBV)	India	Bidari and Reddy (1990), Green and Kim (1991), Ravi (1991)
	Pepper mild mosaic virus (PMMV)	Venezuela	Green and Kim (1991)
	Pepper mottle virus (PeMtV)	North America (Florida, New Mexico, Texas, Arizona and California), Central America (El Salvador), USA, Thailand, India	Sandhu and Chohan (1979), Green and Kim (1991), Sutic *et al.* (1999)

Contd...

Table 25.1–Contd...

Taxonomic Group	*Virus*	*Geographical Distribution*	*References*
	Pepper severe mosaic virus (PeSMV)	Argentina	Green and Kim (1991)
	Peru tomato virus (PTV)	Peru	Green and Kim (1991)
	Chilli vein banding mottle virus (CVbMV)	Thailand	Siriwong *et al.* (1995)
	Ethiopian pepper mottle virus (EPMV)	Ethiopia	Hiskias *et al.* (1999)
Carla viruses	Potato virus M (PVM)	USSR, India	Green and Kim (1991)
	Potato virus S (PVS)	USSR, India	Misra *et al.* (1979), Green and Kim (1991)
Potex viruses	Potato aucuba mosaic virus (PAMV)	Worldwide	Green and Kim (1991), Todorovic *et al.* (1997)
	Potato virus X (PVX)	Worldwide	Nagaich *et al.* (1968), Rao *et al.* (1970), Green and Kim (1991), Satya Prakash and Singh (2004)
ROD-SHAPED VIRUSES			
Tobamo viruses	Tobacco mosaic virus (TMV)	Worldwide	Chamberlain (1947), Prasada Rao and Reddy (1971), Bidari and Reddy (1990), Green and Kim (1991), Brioso (1996), Garborjanyi *et al.* (1998), Sutic *et al.* (1999), Satya Prakash (2001), Satya Prakash and Singh (2004)
	Pepper mild mottle virus (PeMdMtV)	North America, Australia, Japan, Denmark, Iceland, England, France, Greece, Italy, Netherlands, Spain, Taiwan, Zambia, Canada, China, Argentina	Green and Kim (1991), Xiang *et al.* (1994), Ndunguru and Kapooria (1996), Beczner *et al.* (1997)
	Tomato mosaic virus (ToMV)	Worldwide	Green and Kim (1991), Satya Prakash and Singh (2004)
	Bell pepper mottle virus (BPeMtV)	Argentina	Green and Kim (1991)
	Tobacco mild green mosaic virus (TMGMV)	Worldwide	Green and Kim (1991)
	Dulcamara yellow fleck virus (DYFV)	Hungary	Green and Kim (1991)
Tobra viruses	Tobacco rattle virus (TRV)	USA, Italy, China, Brazil, Japan, Europe	Green and Kim (1991), Tian and Zhang (1995), Satya Prakash and Singh (2004)
	Pepper ringspot virus (PRSV)	Brazil	Robinson and Harrison (1987), Brioso (1996)
ISOMETRIC VIRUSES			
Luteo viruses	Beet western yellows virus (BWYV)	Europe, USA, Japan	Green and Kim (1991)
	Pepper vein yellows virus (PVYV)	Japan	Yonaha *et al.* (1995)
	Potato leaf roll virus (PLRV)	Brazil	Brioso (1996)

Contd...

Table 25.1–Contd...

Taxonomic Group	*Virus*	*Geographical Distribution*	*References*
Tobacco necrosis and Satellite virus	Tobacco necrosis virus (TNV)	Worldwide	Green and Kim (1991)
Tombus viruses	Tomato bushy stunt virus (TBSV)	USA, Europe, North Africa	Green and Kim (1991)
	Moroccan pepper virus	Europe, North Africa	Green and Kim (1991)
Faba virus	Broad bean wilt virus (BBWV)	Argentina, Egypt, Japan, Morocco, Italy, Bulgaria, China	Green and Kim (1991), Hristova and Tordanova (1995), Satya Prakash and Singh (2004)
Nepo viruses	Tobacco ring spot virus (TRSV)	North America, India	Bidari and Reddy (1990), Green and Kim (1991)
	Tomato ring spot virus (TomRV)	USA	Green and Kim (1991)
	Tomato black ring virus (TBRV)	Europe	Green and Kim (1991)
Tymo virus	Belladona mottle virus (BMV)	Europe, USA	Green and Kim (1991)
Alfamo virus	Alfalfa mosaic virus (AMV)	Worldwide	Hosszu *et al.* (1975), Pop (1988), Green and Kim (1991)
Cucumo viruses	Cucumber mosaic virus (CMV)	Worldwide	Anjaneyulu and Apparao (1967), Bidari and Reddy (1990), Green and Kim (1991), Hristova and Kim (1995), Brioso (1996), Atencio *et al.* (1997), Satya Prakash (2001), Satya Prakash and Singh (2004)
	Tomato aspermy virus (TAV)	USA, Europe, New Zealand, Japan	Green and Kim (1991), Natsuaki *et al.* (1994)
Ilar virus	Tobacco streak virus (TSV)	USA, New Zealand, Argentina, Europe, Japan	Green and Kim (1991)
Cryptic virus	Red pepper cryptic virus (RPCV)	Japan	Green and Kim (1991)
Gemini viruses	Curly top virus (CTV)	North America, Turkey	Green and Kim (1991)
	Tobacco leaf curl virus (TLCV)	Japan, Europe, USA, India, Sri Lanka	Mishra *et al.* (1963), Green and Kim (1991), Green and Kalloo (1994), Satya Prakash and Singh (2004)
	Tomato yellow leaf curl virus (TYLCV)	Egypt, Thailand, West Indies and Spain	Shalaby *et al.* (1997), Roye *et al.* (1999), Samretwanich *et al.* (2000), Satya Prakash and Singh (2004)
	Pepper mild tigre virus (PMTV)	Mexico	Green and Kim (1990), Green and Kalloo (1994)
	Chino del tomate virus (CdTV)	Mexico	Green and Kim (1991)
	Serrano golden mosaic virus (SGMV)	Mexico, USA	Green and Kim (1991), Green and Kalloo (1994)
	Tomato dwarf leaf curl virus (ToDLCV)	West Indies	Roye *et al.* (1999)

Contd...

Table 25.1–Contd...

Taxonomic Group	*Virus*	*Geographical Distribution*	*References*
Hybrigemini viruses	Tomato pseudo curly top virus	Brazil	Brioso (1996)
	Beet curly top virus (BCTV)	Mexico, USA	Leyendecker (1950), Menzies and Giddings (1953)
	Pepper yellow vein mosaic virus (PeYVMV)	England, Holland, Hungary, Indonesia	Green and Kim (1991), Green (1992), Duriat (1996)
ENVELOPED VIRUS			
Tospo viruses	Tomato spotted wilt virus (TSWV)	Worldwide	Prasada Rao *et al.* (1987), Green and Kim (1990), Green (1992), Vicchi and Talame (1994), Kimura *et al.* (1997), Satya Prakash (2001), Satya Prakash and Singh (2004)
	Impatiens necrotic spot virus (INSV)	Italy	Vicchi *et al.* (1999)
	Tomato chlorotic spot virus (TCSV)	Brazil	Brioso (1996)
	Groundnut ringspot virus (GRSV)	Brazil	Brioso (1996)
Rhabdo virus	Egg plant mottle dwarf virus	Italy	Roggero *et al.* (1995)
Other Not Well Characterized viruses	Tomato stolbur	Romania, Bulgaria, Turkey and USSR	Satya Prakash and Singh (2004)
	Pepper vein chlorosis	Korean Republic	Kim *et al.* (1990)
	Bell pepper dwarf mosaic	India	Chowfla *et al.* (1985)
	Arizona pepper	Arizona (Tucson)	Nelson and Wheeler (1972)
	Chilli mosaic	India, Netherlands	Jha and Raychaudhuri (1956), Rast and Maat (1986)
	Puerto Rican pepper mosaic	Puerto Rico	Perez *et al.* (1974)
	Aster ringspot	Florida	Anderson (1954)
	Lucerne mosaic	Romania	Pop (1988)
	Chilli puckering	Indonesia	Duriat (1996)
	Brinjal mosaic	India	Naqvi and Mahmood (1975)
	Chilli leaf curl	Sri Lanka, India	Green and Kim (1991)
	Bean distortion dwarf (BDDV)	China	Xi *et al.* (1982)
	Sweet potato mosaic (SPMV)	—	Elmer (1957)
	Plantago mosaic (PMV)	Bulgaria	Kovachevsky (1963)
	Potato spindle tuber	—	O'Brien and Raymer (1964)
	Lettuce big vein (LeBVV)	Netherlands	Rast (1992)
	Green vein banding	Cuba	Lopez Cardet and Blanco (1972)
	Launaea mosaic	India	Naqvi *et al.* (1976)
	Marigold mottle (MMV)	India	Naqvi *et al.* (1978)
	Pepper yellow vein (PYVeV)	U.K., Netherlands	Fletcher *et al.* (1987), Rast (1992).

CHILLI/CAPSICUM VIRUSES

Genus–Potyvirus (Family–Potyviridae)

Potato Virus Y (PVY)

Occurrence

Potato virus Y (PVY) is one of the most common potyviruses infecting Chilli/Capsicum throughout the country (Bidari and Reddy, 1990; Satya Parkash, 2001; Raj Verma *et al.*, 2002).

Symptomatology

Infected plants exhibit mosaic, mottle, dark green vein-banding, and vein-clearing of the leaves (Figs. 25.1 &25. 2). Young leaves show prominent more severe symptoms than older ones (Jeyarajan and Ramakrishnan, 1969; Bidari and Reddy, 1990; Satya Prakash, 2001). Other symptoms, such as leaf crinckling, leaf distortion, and growth stunting are also common. Fruits from diseased plant show mottling, deformity, and are small in size (Satya Prakash, 2001). Yield loss up to 100 per cent has been reported (Prasada Rao, 1977; George *et al.*, 1995).

Host Range

On sap inoculation, PVY infects several plant species belonging to various taxonomic families (*i.e.*, Amaranthaceae, Chenopodiacea and Solanaceae). Some indicator/diagnostic hosts for the virus are: *Chenopodium* spp., *Nicotiana glutinosa* and *Solanum demissum* × *S. tuberosum* 'A6' hybrid. *Datura stramonium* is immune to PVY (Bidari and Reddy, 1990; Satya Prakash, 2001).

Transmission

Potato virus Y is transmitted by sap inoculation and by several aphid species *viz. A. craccivora*, *A. gossypii*, *A. fabae*, *H. setariae* and *M. persicae* (Bidari and Reddy, 1990; Satya Prakash, 2001, Satya Prakash and Singh, 2004).

Physical Properties

Thermal inactivation point, dilution end point and longevity *in-vitro* of the virus lies between 60-65°C, 1:1000-1:5000 and 2-4 days, respectively (Bidari and Reddy, 1990; Satya Prakash, 2001).

Electron Microscopy

Virions are flexuous rods measuring 690-800 × 12-13 nm in size. Cytoplasmic inclusions consists "pin wheel" and "scrolls" (Fig. 25.3) typical of potyvirus.

Pepper Veinal Mottle Virus (PVMV)

Occurrence

In India, for the first time PVMV was reported by Prasada Rao (1976). Geographically, it is distributed throughout the chilli/capsicum growing areas of Karnataka, Uttar Pradesh and Maharashtra (Prasada Rao, 1976; Bidari and Reddy; 1990; Satya Prakash, 2001; Raj Verma *et al.*, 2002).

Symptomatology

Pepper veinal mottle virus infected plants exhibit vein-clearing and mosaic mottling (Fig. 25.4) on just emerging leaves as an initial symptom. The subsequently produced leaves elicit distortion and extremely reduced leaf lamina leading to the formation of fern-leaf structure (Prasada Rao, 1976; Bidari and Reddy, 1990; Satya Prakash, 2001). It also causes *zig*-zag appearance and green longitudinal streaks on stem (Fig. 25.5). Other symptoms include shortening of internodes and stunted growth of plant (Prasada Rao, 1976; Singh, 1993).

Transmission

The virus is readily transmitted by mechanical sap inoculation and by various species of aphids (*viz. Myzus persicae*, *Aphis gossypii* and *A. craccivora*) (Bidari and Reddy, 1990; Satya Prakash, 2001).

Host Range

Experimentally, pepper venal mottle virus infects very few plant species belonging to families Chenopodiaceae and Solanaceae (Satya Prakash, 2001).

Physical Properties

Dilution end point, thermal inactivation point, and longevity *in vitro* of the virus are between 1: 1000-1:10000, 55-65°C, and 24-48 hrs (at room temperature), respectively (Bidari and Reddy, 1990; Satya Prakash, 2001).

Electron Microscopy

Virus particles of PVMV are filamentous flexuous rods measuring 648-776 × 12 –19 nm in size (Prasada Rao, 1976; Bidari and Reddy, 1990; Satya Prakash, 2001). It induces cytoplasmic cylindrical inclusions as "pin wheels".

Tobacco Etch Virus (TEV)

Occurrence

Tobacco etch virus (TEV) was first reported in India by Bidari and Reddy (1983) from Karnataka. Now the virus occurs naturally on Chilli/Capsicum in Karnataka, Uttar Pradesh and Maharashtra (Bidari and Reddy, 1983; Satya Prakash, 2001; Raj Verma *et al.*, 2002).

Symptomatology

TEV infected plants exhibit following symptoms: dark green mottling, mild chlorosis, faint yellow flecking, vein-clearing, severe stunting and wilting of plants. Infected fruits are small in size and exhibit shrivelled areas and discolouration (Bidari and Reddy, 1986; 1990; Satya Prakash, 2001).

Transmission

The virus is transmitted by sap inoculation and aphids, such as *A. craccivora*, *A. gossypii* and *M. persicae* in a non-persistent manner (Satya Prakash, 2001).

Host Range

TEV infects some members of Chenopodiaceae and Solanaceae (Bidari and Reddy, 1990; Satya Prakash, 2001). *Datura metel* and *D. stramonium* are diagnostic hosts for the virus.

Physical Properties

Thermal inactivation point, dilution end point and longevity *in–vitro* of the virus lies between 60-70°C, 1: 1000-1:5000 and 24-48 hrs, respectively (Bidari and Reddy, 1990; Satya Prakash, 2001).

Electron Microscopy

Particle morphology of the virus is flexuous filamentous rods measuring 623 × 12-13 nm in size. It induces cytoplasmic inclusions consisting of "pin-wheels" and "laminated aggregates" (Satya Prakash, 2001).

Pepper Vein Banding Virus (PVBV)

Occurrence

Pepper vein banding virus (PVBV) was first reported in India by Prasada Rao (1976) from Karnataka. The virus is epidemically distributed throughout the growing areas of Karnataka (Ravi, 1991).

Symptomatology

PVBV infected plants exhibit faint vein–clearing on young leaves followed by dark green continuous vein-banding (Prasada Rao,1976; Bidari, 1982), which later show various types of mosaic-mottling and in some cases they become thread-like. Internodal length is shortened giving a more stunted appearance. Infected plants blossom poorly (Bidari, 1982; Ravi, 1991).

Transmission

The virus is transmitted through mechanical sap inoculation and also by several aphids namely, *A. craccivora*, *A. gossypii* and *M. persicae* (Prasada Rao, 1976; Bidari and Reddy, 1990, Ravi, 1991).

Host Range

Experimentally, pepper vein banding virus infects systemically to several plant species belonging to family Solanaceae. It elicits necrotic local lesions along with veinal necrosis in *N. unguiculata* L. (Bidari and Reddy, 1990; Ravi, 1991).

Physical Properties

Thermal inactivation point, dilution end point and longevity *in-vitro* of the virus are between 55-60°C, 1:5000-1:10000 and 8-24 hrs. (21-26°C), respectively (Prasada Rao, 1976; Bidari and Reddy, 1990).

Electron Microscopy

Virus particles are flexuous rods measuring 648-759 × 13 nm (Prasada Rao, 1976; Bidari and Reddy, 1990; Ravi, 1991 and Ravi *et al.*, 1997), and the coat protein has a molecular weight of 30,000 Daltons (Ravi *et al.*, 1997).

Chilli Veinal Mottle Virus (CVMV)

Occurrence

Chilli veinal mottle virus was first observed during survey in 1997 and 1998 in Eastern Uttar Pradesh by Satya Prakash *et al.* (2001).

Symptomatology

CVMV infected plants show dark green mottling and distortion of leaves (Fig. 25.6). Fruits formed on infected plants are malformed (Satya Prakash, 2001; Satya Prakash *et al.*, 2001, 2002a).

Transmission

The virus is readily sap transmitted and also by several aphid species *i.e. A. craccivora*, *A. gossypii*, *A. spiricola* and *Myzus persicae*, in a non-persistent manner (Satya Prakash, 2001; Satya Prakash *et al.*, 2001, 2002a).

Host Range

Artificially, it infects to a very few plant species of family Solanaceae. The virus induce systemic local lesions on *Nicotiana sylvestris* , which is a diagnostic host.

Physical Properties

Dilution end point, thermal inactivation point and longevity *in vitro* of the virus are between 1:5000–1:10000, 65–70°C and 2–4 days, respectively (Satya Prakash *et al.*, 2002a).

Electron Microscopy

CVMV induces cytoplasmic inclusion as "pinwheel". The virus particles are flexuous rods measuring 750 × 12 nm in size (Satya Prakash, 2001, Satya Prakash *et al.*, 2001, 2002a).

Pepper Mottle Virus (PeMtV)

Symptomatology

In India it was reported by Sandhu and Chohan (1979). The virus was restricted to few areas of the Punjab state. Infected plants elicit mottle, is the key symptom.

Transmission

It is transmitted by aphid, but not by seeds of infected plants.

Host Range

Host range of the virus is restricted to *C. annuum*, vegetable marrow, *D. metel, P. floridana, N. glutinosa*, tobacco, *S. nigrum* and egg plant. It also produces local lesions on inoculated leaves of *C. amaranticolor*.

Physical Properties

Thermal inactivation point lies between 50-55°C, dilution end point between 1: 500–1: 1000 and longevity *in vitro* of the virus is up to 3 days at room temperature.

Electron Microscopy

The virus particles are of filamentous rods measuring 737 nm in length (Sharma and Sharma, 1999).

Genus–Cucumovirus (Family– Bromoviridae)

Cucumber Mosaic Virus (CMV)

Occurrence

In India, this virus was first reported by Jha and Raychaudhury (1956) and later by Anjaneyulu and Apparao (1967). Cucumber mosaic virus occurs wherever Chilli/Capsicum are grown at large scale, *viz.* Andhra Pradesh, Tamil Nadu, Karnataka, Maharashtra, Uttar Pradesh, Haryana and Bihar (Satya Prakash, 2001). It is widespread and possibly the most damaging of all viruses infecting Chilli/ Capsicum. It reduces quality significantly, and fruit yield by 60 to 100%, especially in the early plant infections (Deol and Rataul, 1978a; George *et al.*, 1995).

Symptomatology

Common symptoms of virus infection are mosaic mottling, yellow discolouration, vein-clearing, curling, distortion and filiformity (Figs. 25.8 & 25.9), inward rolling, puckering and oak-leaf pattern on leaves, and reduction in internodal length (Anjaneyulu and Apparao, 1967; Prasada Rao, 1976; Bidari and Reddy, 1990; Kiranmai *et al.*, 1997; Satya Prakash, 2001). Fruits exhibit malformation, warty swelling (Fig. 25.10) and chlorotic spots (Prasada Rao, 1976; Satya Prakash, 2001).

Transmission

The virus is readily transmitted by mechanical sap inoculation and by various aphids namely, *A. craccivora*, *A. gossypii*, *A. citricola*, *Macrosiphum euphorbiae* and *Myzus persicae* (Jha and Raychaudhury, 1956; Bidari and Reddy, 1990; Kiranmai *et al.*, 1997; Satya Prakash, 2001).

Host Range

Host range of this virus is very wide. It infects several plant species of family Chenopodiaceae, Cucurbitaceae and Leguminosae. CMV induces characteristic systemic mosaic in *Cucumis sativus* (Prasada Rao, 1976; Bidari and Reddy, 1990; Satya Prakash, 2001). Kiranmai *et al.* (1997) reported an unusual strain of CMV, which did not infect *C. sativus*. It produces systemic symptoms on *Vinca rosea*, *Tagetus indica*, *Luffa acutangula*, *Momordica charantia*, *C. annuum*, *Nicotiana glutinosa* (Fig. 25.11), *N. rustica*, *N. tabacum* vars., *Petunia hybrida*, *Physalis floridana* and *P. minima*. It also induces local symptoms as chlorotic local lesions on *Cajanus cajan*, *Phaseolus aureus*, *P. mungo* and *Vigna unguiculata*.

Physical Properties

The virus has thermal inactivation point between 55-70°C, dilution end point between 1:200-1:10,000 and longevity *in-vitro* between 1-15 days (Jha and Raychaudhury, 1956; Bidari and Reddy, 1990; Kiranmai *et al.*, 1997; Satya Prakash, 2001).

Electron Microscopy

Virus particles are spherical measuring between 27-30 nm in diameter (Prasada Rao, 1976; Bidari and Reddy, 1990; Kiranmai *et al.*, 1997; Satya Prakash, 2001). Molecular weight of coat protein and ds RNA are between 24.5-25.7 × 10^3d and 2.48-2.52 × 10^6d, respectively (Kiranmai *et al.*, 1997).

Genus–Potexvirus

Potato Virus X (PVX)

Occurrence

In India, potato virus X was first reported by Nagaich *et al.* (1968) and subsequently by Rao *et al.* (1970). The virus was found to occur naturally in a few chilli growing areas of the country.

Symptomatology

PVX induces systemic mosaic and necrosis of leaves, and bushyness and growth stunting of plants (Nagaich *et al.*, 1968; Rao *et al.*, 1970). Characteristic symptom, necrotic local lesions on inoculated leaves of *Capsicum pendulum* L., *Capsicum frutescens* L. (var. green and bull nose) and *C. microcarpum* has also been reported (Rao *et al.*, 1970).

Transmission

Potato virus X is transmitted by sap, soil and beetles (*i.e. Epilachna* and *Coccinella*) (Nagaich *et al.*, 1972).

Host Range

The virus infect systemically to *Capsicum annuum* L., *Solanum nigrum* L., *S. melongena* L., *Physalis floridana* Rydb., *Lycopersicon esculentum* Mill. and *Petunia hybrida* Vilm. It produces local necrotic lesions followed by systemic symptoms in *Nicotiana tabacum* (vars. White Burley, Harrison's special and N.N. Strain). PVX also causes necrotic local lesions on inoculated leaves of *Datura stramonium* and *Gomphrena globosa* (Rao *et al.*, 1970; Rishi and Dhawan, 1989).

Physical Properties

The virus has a dilution end point between 1:10000-1:100000, thermal inactivation point between 65-70°C and longevity *in-vitro* between 18-20 days (Rishi and Dhawan, 1989).

Electron Microscopy

The virus particles are flexuous rods measuring 512 × 12 nm in size (Sharma and Sharma, 1999).

Genus–Tobamovirus

Tobacco Mosaic Virus (TMV)

Occurrence

In India, TMV occurs throughout the chilli/capsicum-cultivated field of the country, because of its easiest mode of transmission (Satya Prakash, 2001).

Symptomatology

The virus causes leaf mottling (greenish-yellow) and curling, and irregular expansion of leaves. Tobacco mosaic virus often causes a brown streaking of some of the branches, resulting yellowing and dropping of the leaves. At times, plants affected by this virus may show a general yellowing of many of the leaves with little streaking of the stem (Singh, 1993; Satya Prakash, 2001). The common (green mottle) form of tobacco mosaic often causes little, if any, mottled symptoms on the fruit. However, fruits on streaked branches are wrinkled and withered (Fig. 25.12) (Singh, 1993).

TMV–Pant Nagar produces typical mosaic mottling (Fig. 25.10), puckering, blistering and smalling of leaves (Kamra and Dubey, 1975 b). A TMV isolate reported from Karnataka produces necrotic lesions, leaf dropping of inoculated leaves and death of some of the plants. Survived plants show yellowish irregular patches on leaves, stunting and bear few flower and fruits. Fruits are thin-skinned, smaller and mottled (Bidari and Reddy, 1990)

An unusual strain of TMV causing brilliant yellow mottle of the leaves has been reported from Maharashtra (Capoor and Sawant, 1987). It causes yellow spotting of some fruits (Fig. 25.13). The spots are irregular in shape and not noticeable, sunken or wrinkled.

Transmission

Tobacco mosaic virus is transmitted by sap, seed (up to 21%) and also by contact (Singh, 1993; Satya Prakash, 2001). Some strains of TMV infecting chilli are reported to be transmissible through soil and grafting (Singh, 1993).

Host Range

It infects several plant species belonging to Amaranthaceae, Chenopodiaceae, Cucurbitaceae, Leguminosae and Solanaceae families (Bidari and Reddy, 1990; Satya Prakash, 2001). TMV induces characteristic necrotic local lesions on inoculated leaves of *Gomphrena globosa* and *Nicotiana glutinosa* (Fig. 25.14) (Kamra and Dubey, 1975 b; Prasada Rao, 1976; Nagai *et al.*, 1981; Rishi and Dhawan, 1989; Bidari and Reddy, 1990; Singh, 1993; Capoor and Sawant, 1987; Satya Prakash, 2001).

Physical Properties

Thermal inactivation point, dilution end point and longevity *in-vitro* of the virus lies between 90-95°C, 1:1,00,000–1:10000000000 and 1-36 months, respectively (Prasada Rao, 1976; Sharma and Dubey, 1977; Kapoor and Sawant, 1987; Rishi and Dhawan, 1989; Bidari and Reddy, 1990; Satya Prakash, 2001).

Electron Microscopy

TMV particles isolated from Chilli/Capsicum have been reported as rigid rods measuring on an average 250-300 × 15-18nm in size (Kamra and Dubey, 1975a, b; Prasada Rao, 1976; Capoor and Sawant, 1987; Bidari and Reddy, 1990; Satya Prakash, 2001).

Genus-Tospovirus

Tomato Spotted Wilt Virus (TSWV)

Occurrence

TSWV was first reported in India by Bidari and Reddy (1984). It is distributed throughout the Chilli/Capsicum growing areas of Karnataka, Assam and Eastern Uttar Pradesh (Bidari and Reddy, 1990; Upadhyaya, 2000; Satya Prakash, 2001).

Symptomatology

TSWV incite a variety of symptoms on chilli/Capsicum *viz.* concentric zonation on leaves, mosaic mottling on terminal leaves with dense coalescence of small rings, spots and concentric rings, veinal necrosis (Fig. 25.15), sudden yellowing and browning, necrotic spots, and necrosis of terminal shoots (Nagai, 1984; Bidari and Reddy, 1984, 1990; Singh, 1993; Satya Prakash, 2001). Die-back symptom in chilli has also been reported (Upadhyaya, 2000). Fruits exhibit chlorotic spots, green or red areas surrounded by yellow halos (Fig. 25.16), large necrotic blotches and at times concentric rings, necrosis and abortion of flowers are common (Nagai, 1984; Bidari and Reddy, 1984; Singh, 1993; Satya Prakash, 2001).

Transmission

The virus is transmitted by sap and nymphs of thrips species (*i.e. Thrips tabaci, Frankliniella schultzei* and *Scirtothrips dorsalis* (Bidari and Reddy, 1990; Rathaiah *et al.*, 1997; Upadhyaya, 2000; Satya Prakash, 2001).

Host Range

TSWV infects several plant species belonging to Amaranthaceae, Boraginaceae, Chenopodiaceae, Compositae, Cucurbitaceae, Leguminosae and Solanaceae families (Bidari and Reddy, 1990; Satya Prakash, 2001).

Physical Properties

TSWV is unstable in extracted plant sap (1-4 hrs. at room temperature). It has very low thermal inactivation point (45-50°C) and dilution end point (1:500–1:1000) (Prasada Rao, 1976; Bidari and Reddy, 1990; Satya Prakash, 2001).

Electron Microscopy

Virions are spherical and enveloped measuring 70-90 nm in diameter (Fig. 25.17) (Bidari and Reddy, 1990; Satya Prakash, 2001).

Genus–Nepovirus (Family–Comoviridae)

Tobacco Ring Spot Virus (TRSV)

Occurrence

The virus was first reported by Prasada Rao (1976) from Karnataka.

Symptomatology

TRSV produces vein clearing followed by netting and chlorotic rings (Fig. 25.18) on the leaves of Chilli/Capsicum. Late infected leaves show mosaic-mottling and narrowing (Prasada Rao, 1976; Bidari and Reddy, 1990).

Transmission

It is transmitted by sap and by nematode.

Host Range

The host range of the virus is very wide belonging to families Chenopodiaceae, Compositae, Cucurbitaceae, Leguminosae and Solanaceae.

Physical Properties

Dilution end point, thermal inactivation point and longevity *in-vitro* of the virus lies between 1:5000–1:10000, 60–65°C and up to 48 hrs at room temperature, respectively.

Electron Microscopy

Virions are isometric measuring 27-28 nm in size (Bidari and Reddy, 1990).

Genus–Geminivirus (Family–Geminiviridae)

Tobacco Leaf Curl Virus (TLCV)

Occurrence

Tobacco leaf curl virus (TLCV) on chilli was first reported in India by Mishra *et al.* (1963). Now it has become one of the most economically important viral disease of chilli/capsicum, possible occur throughout the country.

Symptomatology

The characteristic symptoms of the virus infection are: vein clearing, curling, blistering, puckering and heavy crinkling of the leaves with chlorosis and yellowing , and swelling of the veins (enations). Infected plants are stunted and often bushy in appearance (Mishra *et al.*, 1963; Seth and Dhanraj , 1972; Venkatesh *et al.*,1997). Incidence up to 100 per cent in farmers field are not uncommon. Losses up to 80 per cent have been reported in many parts of northern India (Singh *et al.*, 1979).

Transmission

The virus is transmitted by whitefly (*Bemisia tabaci*).

Host Range

The virus infects experimentally to *C. annuum* and 34 other plant species including *L. esculentum, Beta vulgaris , Carica papaya , Cyamopsis tetragonoloba, Sesamum indicum , P. vulgaris* and *Petunia hybrida* (Pal and Tandan, 1937; Valand and Muniyappa, 1992).

Electron Microscopy

The virus particles are geminate or paired measuring 15-20 × 25-30 nm (Sharma and Sharma, 1999).

Methods for Detection

Biological (Infectivity Vioassay)

Chilli viruses can be detected by using indicator or diagnostic or differential hosts *viz., Capsicum annuum* var. California Wonder, *Chenopodium amaranticolor, Cucumis sativus, D. metel, D. stramonium, Gomphrena globosa, Lycopercicon esculentum, Nicandra physaloides, Nicotiana glutinosa, N. sylvestris, N. tabacum* var. White Burley, *Phaseolus vulgaris, Solanum demissum* × *S. tuberosum* 'A6' (Prasada Rao, 1976; Bidari, 1982; Satya Prakash, 2001).

Serological

Following serological methods are used for detection of viruses infecting Chilli/Capsicum:

1. Agarose Gel Diffusion Test
2. Ring Precipitin Test
3. Immuno Electrophoresis
4. Immuno Fluorescence
5. Enzyme Linked Immunosorbent Assay (ELISA)–Direct and indirect:
   (a) Direct Antigen Coating (DAC).
   (b) Double Antibody Sandwich (DAS).

Molecular

Molecular techniques which can be used for the detection and characterization of Chilli/Capsicum viruses are as follows:

1. PCR (Polymerase Chain Reaction).
2. RT-PCR (Reverse Transcription-Polymerase Chain Reaction).
3. SDS-PAGE (Sodium Dodecyl Sulphate-Polyacrylamide Gel Electrophorosis).
4. Agarose Gel Electrophoresis for Mr. determination of nucleic acid.

Management

Unlike fungi and bacteria, viruses, so far, cannot be controlled by using chemicals. Indirect control measures and plant and vector resistant thus remain the only practical methods for controlling viruses. To effectively control virus attacks, the virus has to be correctly identified and its epidemiology and ecology should be understood. Some of the methods most commonly used to control viruses are presented below.

Insect-Borne Viruses

1. Nursery should be grown under nylon net and the nursery bed should be covered with net at the time of sowing.
2. Spray the nursery bed with either monocrotophose (0.05%) or metasystox (0.025%) (Hanada *et al.*, 1997) or oxydemetonmethyl (0.02%) at 10 day intervals to control insect vectors.
3. Soil application of carbofuron or disyston or thymet–10g @ 1.5 Kg a.i./hector at the time of sowing seeds in the nursery bed.

4. Application of above insecticide in the field at the time of transplanting seedlings followed by spraying of either monocrotophos or oxydemetonmethyl at 10 day intervals.
5. Virus incidence can be reduced by sowing barrier crops like maize, sunflower, jowar, bajra or sorghum around Chilli/Capsicum plot at least 60 days in advance (Deol and Rataul, 1978b; Avilla *et al.*, 1996; Hanada *et al.*, 1997; Fereres, 2000; Fajinmi and Oladiran, 2000; Satya Prakash *et al.*, 2002b).
6. Spray of mineral oils @ 1.5–2.8% have been found effective in reducing the disease incidence (Nitzany, 1966; Sharma and Varma, 1982; Loebenstein *et al.*, 1970; Zitter and Osaki, 1978).
7. The virus incidence and spread can be reduced by application of reflective mulches (silver plastic mulch) at the time of transplanting of seedlings (Satya Prakash *et al.*, 2002b).
8. Application of yellow sticky polythene sheets erected on the windward side of chilli/ capsicum field helps to reduce virus incidence by trapping the vectors.
9. Adjust the sowing time of the crop so that it does not coincide with high vector population, particularly at the seedling stage, when the plants are particularly susceptible to virus infection.
10. Removal of volunteer plants emerging at the edges of a planting or within a field since they may be potential virus carriers.
11. Use of resistant varieties.
12. Biological control of virus vectors through predators.
13. Rouging of infected plants from the field as soon as they are noticed.

Non-Insect-Borne Virus Diseases

1. Selection of seeds from healthy looking plants.
2. Removal of infected plants from the field.
3. Florida-VR-2 resistant to common strain of TMV, Pant C1 and C2 are tolerant to TMV (Konai and Nariani, 1983), Punjab Lal is resistant to TMV. Varieties Tambel-1, Tambel-2, Tam mild clute-1 and 2, Tam mild Japeno-1 and Hidalgo are resistant to chilli mosaic caused by TMV. PSP-11, a new cultivar of *C. frutescence* has been found resistant to TMV (Tewari and Viswanath, 1988).
4. Disinfection of pepper seeds in Trisodium phosphate (TSP) for 2 hrs. @ 100 gm./liter eliminates the externally borne TMV from the seed (Rast and Stijger, 1987).

Future Research Needs

1. Extensive survey of the Chilli/Capsicum growing areas throughout the country for the presence of viruses.
2. Establishment of well equipped virus laboratories throughout the country to carryout molecular studies for proper diagnosis of viruses.
3. Development of proper and effective management techniques feasible to the farmers.
4. Breeding multiple resistant varieties using conventional breeding techniques.
5. To provide basic information regarding identification and management of virus diseases of Chilli/Capsicum and availability of resistant varieties to the farmers.

REFERENCES

Anderson, C.W. (1954). The aster ringspot virosis of Central Florida. *Phytopathology*, **44**: 87-92.

Anjaneyulu, A. and Apparao, A. (1967). Natural occurrence of cucumber mosaic virus on chilli in India. *Indian Phytopath.*, **20**: 380-381.

Anonymous (1992). The Wealth of India. Vol. 3: Ca-Ci. pp. 680, New Delhi.

Anonymous (1997). FAO Year book Vol. 52, Rome, FAO of United Nations, 1998, pp. 132-135.

Atencio, F.A., Gracia, O., Mendoza, E.E.A., Zandomeni, R. and Grau, O. (1997). Detection of both sub-groups I and II of cucumber mosaic cucumovirus and their satellite RNAs on pepper in Argentina. *Plant Disease*, **81**: 695.

Avilla, C., Collar, J.L., Dugue, M., Hernaiz, P., Martin, B. and Fereses, A. (1996). Barrier Crops as a method of control of non-persistent viruses in pepper. *Baletin de Sanidad Vegetal, Plagas*, **22**: 301-307.

Beczner, L., Rochon, D.M. and Hamilton, R.I. (1997). Characterization of an isolate of pepper mild mottle tobamovirus occurring in Canada. *Canadian Journal of Plant Pathology*, **19**: 83-88.

Bidari, V.B. (1982). Distribution and epidemiology of chilli viruses in the Karnataka. Ph.D. Thesis, Deptt. of Plant Pathology, UAS, Bangalore, Karnataka, India.

Bidari, V.B. and Reddy, H.R. (1983). Prevalence of chilli viruses in Dharwad district. *Plant Pathology Newsletter*, **1**: 11-12.

Bidari, V.B. and Reddy, H.R. (1984). Tomato spotted wilt virus-A new record on chilli in India. *Current Science*, **53**: 1045-1046.

Bidari, V.B. and Reddy, H.R. (1986). Natural occurrence of tobacco etch virus (TEV) on chilli in India. *Current Science* , **55**: 254-256.

Bidari, V.B. and Reddy, H.R. (1990). Identification of naturally occurring viruses on commercial cultivars of chilli. *Mysore Journal of Agricultural Sciences*, **24**: 45–51.

Brioso, P.S.T. (1996). Diseases caused by viruses in *Capsicum*. *Informe Agropeculario* (Belo Horizonte), **18**: 74-80.

Capoor, S. P. and Sawant, D. M. (1987). Studies on bell pepper yellow mosaic virus. *Indian Phytopath.*, **40**: 443-450.

Chamberlain, E.E. (1947). Tomato streak. *Bull. N.Z. Dep. Sci. industr. Res.*, **281**, 11pp.

Chowfla, S.C., Pal, J., Bhardwaj, S.V. and Baruah, B.P. (1985). Bell pepper dwarf mosaic–a new disease in Himachal Pradesh. *Indian Journal of Plant Pathology*, **3**: 252-254.

Deol, G.S. and Rataul, H.S. (1978a). Influence of plant age at inoculation by cucumber mosaic virus on disease incidence and yield of chilli (*Capsicum annuum* Linn.). *Indian Journal of Plant Protection*, **6**: 82-83.

Deol, G.S. and Rataul, H.S. (1978b). Role of various crops in reducing the incidence of cucumber mosaic virus in chilli (*Capsicum annuum* L.). *Indian J. Entomol.*, **40**: 261-264.

Duriat, A.S. (1996). Management of pepper viruses in Indonesia: problems and progress. *IARD Journal*, **18**: 45-50.

Elmer (1957): Sweet potato mosaic (Abstr.). *Phytopathology*, **47**: 9.

Fajinmi, A. A. and Oladiran, A. O. (2000). Efficacy of maize intercrop in the control of viral disease (s) of pepper. *Capsicum and Eggplant Newsletter*, **19**: 93-96.

Fereres, A. (2000). Barrier crops as a cultural control measure of non-persistently transmitted aphid-borne viruses. *Virus Research*, **71**: 221-231.

Fletcher, J.T., Wallis, W. A. and Davenport, F. (1987). Pepper yellow vein, a new disease of sweet pepper. *Plant Pathol.*, **36**: 180-184.

Gaborjanyi, R., Horvath, J., Kovacs, J. and Kazinczi, G. (1998). Role of virus-and phytopalasma infections in pepper decline in Hungary: an overview. *Acta Phytopathologica et Entomologica Hungarica*, **33**: 261-268.

George, T.E., Anand, N. and Singh, S.J. (1995). Influence of potato virus 'Y' (PVY) and cucumber mosaic virus (CMV) on growth and yield of bell pepper (*Capsicum annuum* L.). *Indian Cocoa, Arecanut and Spices Journal*, **19**: 1-4.

Green, S.K. (1992). Viruses in Asia/Pacific region. *Proc. Conference on Chilli Pepper Production in the Tropics*, Kuala Lumpur, 13-14 October 1992. *MARDI-AVRDC-MAPPS*, pp. 98-129.

Green, S.K. and Kalloo, G. (1994). Leaf curl and yellowing viruses of pepper and tomato: an overview. *Asian Vegetable Research and Development Centre. Technical Bulletin*, **21**: 51.

Green, S.K. and Kim, J.S. (1991). Characteristics and control of viruses infecting peppers: a literature review. *AVRDC Technical Bulletin*, **18**: 60.

Handa, A., Chowfla, S.C. and Thakur, P.D. (1997). Barrier crops and Insecticidal sprays for managing a mosaic disease complex in chilli. In: *IPS–Golden Jubilee International Conference*, Nov.10-15, 1997, New Delhi , India. p. 295.

Hiskias, Y., Lesemann, D.E. and Vetten, H.J. (1999). Occurrence, distribution and relative importance of viruses infecting hot pepper and tomato in the major growing areas of Ethiopia. *Journal of Phytopathology*, **147**: 5-11.

Hristova, D. and Kim, E.S. (1995). Strains of cucumber mosaic virus spread on pepper in Bulgaria. *Rasteniev "dni Nauki"*, **32**: 84 89.

Hristova, D. and Tordanova, A. (1995). Comparative study of Bulgarian and German isolates of broad bean wilt virus. *Rasteniew dni Nauki*, **32**: 96-98.

Jeyarajan, R. and Ramakrishnan, K. (1969). Potato virus 'Y' on chilli (*Capsicum annuum* L.) in Tamil Nadu, India. *Madras Agric. J.*, **56**: 761-766.

Jha, A. and Raychaudhuri, S.P. (1956). Mosaic disease of chilli (*Capsicum frutescens* L.). *Indian J. Agric. Sci.*, **26**: 217-222.

Joshi, R.D. and Bhargava, K.S. (1962). A vein banding mosaic disease of chillies (*Capsicum frutescens* L.). *Indian J. Microbiology*, **2**: 29-34.

Kamra, B.S. and Dubey, G.S. (1975a). Particle morphology of a TMV strain from chilli. *Indian Phytopathology* (1975 publ. 1976), **28**: 96.

Kamra, B.S. and Dubey, G.S. (1975b). Studies on a mosaic disease of chilli. Indian *Phytopathology*, **28**: 212-217.

Kim, J.S., Kim, S.K., Lee, S.H. and Lee, M.W. (1990). A pepper vein chlorosis virus causing stem necrosis and vein chlorosis on red pepper in Korea. *Korean Journal of Plant Pathology*, **6**: 376-381.

Kimura, H., Okitsu, M., Yamanaka, K., Ikka, T., Takahashi, O., Miyazaki, Y. and Matsui, T. (1997). Detection of TSWV on plants grown in the surrounding area of green pepper cultivation in Ibaraki Prefecture–Tomato spotted wilt tospovirus on *Capsicum annuum*. *Proceedings of the Kanto–Tosan Plant Protection Society*, **44**: 41-42.

Kiranmai, G., Sreenivasulu, P. and Nayudu, M.V. (1997). Characterization of cucumber mosaic cucumovirus isolates naturally infecting three Solanaceous vegetable crops in Andhra Pradesh. *Indian Phytopathology*, **50**: 416-425.

Konai, M. and Nariani, T. K. (1983). Screening of chilli varieties for virus resistance. Vegetable Science **10**: 134-136.

Kovachevsky, I.C. (1963). Untersuchungen iiber dasWegerichmosaik in Bulgarian. *Phytopath. Z.*, **49**: 126-146.

Leyendecker, P.J. (1950). Plant disease survey for new Mexico. *Plant Dis. Reptr.*, **34**: 39-40.

Loebenstein, G., Alper, M. and Levy, S. (1970). Field tests with oil sprays for the prevention of aphid spread viruses in peppers. *Phytopathology*, **60**: 212-215.

Lopez-cardet, Y. and Blanco, N. (1972). Green vein banding of pepper. *Revista de Agricultura, Cuba*, **5**: 30-36.

Mishra, M.D., Raychaudhuri, S.P. and Jha, A. (1963). Virus causing leaf curl of chilli. *Indian J. Microbiol.*, **3**: 73-76.

Misra, A., Chaudhuri, D.P., Mishra, P.K. and Jha, A. (1979). Electron microscopical and serological investigation on 'Indian chilli mosaic virus'. *J. Plant Dis.* and *Protection*, **86**: 39-42.

Nagai, H. (1984). Melhoramento genetico do pimentao e pimenta, visando a resistencia a viroses. *Informe Agropecuario*, **10**: 55-58.

Nagai, Y., Takeuchi, T. and Tochihara, H. (1981). A new mosaic disease of sweet pepper caused by peper strain of tobacco mosaic virus. *Annals of the Phytopathological Society of Japan*, **47**: 541-546.

Nagaich, B.B., Upadhya, M.D., Prakash, O. and Singh, S.J. (1968). Cytoplasmically determined expression of symptoms of potato virus X crosses between species of *Capsicum*. *Nature, Lond.*, **220**: 1341-1342.

Naqvi, Q.A., Haseeb, A., Saxena, S.K. and Mahmood, K. (1978). Influence of marigold mottle virus (MMV), the reniform nematode, *Rotylenchulus reniform's* and a stunt nematode on the growth of chilli and on nematode multiplication. *Acta Bot.*, **6**: 174-175.

Naqvi, Q.A. and Mahmood, K. (1975). Reaction of different varieties of brinjal, tomato and pepper to the new strain of brinjal mosaic virus. *Geobias*, **2**: 189-190.

Naqvi, Q.A., Mahmood, K. and Saxena, S.K. (1976). Varietal response of vegetables against Launaea mosaic virus. *Geobios*, **3**: 203-205.

Natsuaki, T., Wang, W.Q. and Okuda, S. (1994). Characterization of chrysanthemum mild mottle virus isolates from sweet pepper (*Capsicum annuum*) and Petunia (*Petunia hybrida*). *Bulletin of College of Agriculture, Utsunomiya University*, **15**: 11-19.

Ndunguru, J. and Kapooria, R.G. (1996). Pepper mild mottle tobamovirus found infecting cultivars of *Capsicum annuum* in Zambia. *Bulletin OEPP*, **26**: 725-728.

Nelson, M.R. and Wheeler, R.E. (1972). A new virus disease of pepper in Arizona. *Plant Disease Reporter*, **56**: 731-735.

Nitzany, F. E. (1966). Tests for the control of field spread of pepper viruses by oil sprays. *Plant Dis. Reptr.*, **50**: 150-160.

Nova-Womdim, R. and Ativalentja, N. (1993). Identification and characterization of a pepper veinal mottle virus strain in Cameroon. *Capsicum and Eggplant Newsletter*, **12**: 69-72.

O'Brien, Muriel, J. and Raymer, W.B. (1964). Symptomless hosts of the potato spindle tuber virus. *Phytopathology*, **54**: 1045-1047.

Pal, B.P. and Tandon, R.K. (1937). Types of tobacco leaf curl in Northern India. *Indian J. Agric. Sci.*, **7**: 363-393.

Perez, J.E., Irizarry, H. and Cortes-Monlorr, A. (1974). Present status of virus infections of peppers in Puerto Rico. *Journal of Agriculture of the University of Puerto Rico*, **58**: 137-139.

Pop, I.V. (1988). Yellow fleck of *Capsicum* recently observed in Romania, caused by lucerne mosaic virus. *Analele Institutului de Cercetari pentru protectia Plantelor*, **21**: 11-14.

Prasada Rao, R.D.V.J. (1976). Characterization and identification of some chilli mosaic viruses. Ph.D. thesis University of Agricultural Sciences, Hebbal, Bangalore, Karnataka, India.

Prasada Rao, R.D.V.J. (1977). Effect of potato virus Y on growth and yield of chilli. *Indian Journal of Plant Protection*, **5**: 199-200.

Prasada Rao, R.D.V.J., Chakrabarty, S.K., Rajeswari, R. and Reddy, A.S. (1987). Occurrence of tomato spotted wilt virus on cowpea (*Vigna unguiculata* (L.) Walp.), chilli (*Capsicum annuum* L.) and brinjal (*Solanum melongena* L.). *Indian Journal of Plant Protection*, **15**: 117-119.

Raj Verma, Singh, S.J., Singh, R.K. and Satya Prakash (2002). A severe outbreak of viruses in Polyhouse grown Capsicum (*Capsicum annuum* L.) in Pune. In: *XIV Annual Conference and National Seminar on "Management of Viral Diseases with emphasis on Global Trade and WTO Regime"*, Jan.18-20, 2002, Hebbal, Bangalore, p. 26. (Abst.)

Rao, K.N., Apparao, A. and Reddy, D.V.R. (1970). A ringspot strain of potato virus X on chilli (*Capsicum annuum*). *Indian Phytopath.*, **23**: 69-73.

Rao, M.H.P. and Reddy, H.R. (1971 Pub. 1972). A new strain of tobacco mosaic virus on tomato (Lycopersicon esculentum). *Indian Phytopath.* (1971 published 1972), **24**: 672-678.

Rast, A. Th.B. (1992). Host range comparison of the causal agents of pepper yellow vein and lettuce big vein. *Netherlands Journal of Plant Pathology*, **98**: 325-328.

Rast, A.Th.B. and Maat, D.Z. (1986). Yellow mosaic strain of capsicum mosaic virus. *Mededelingen Vande Faculteit Landbouwwetenschappen Rijksuniversiteit Gent*, **51**: 791-797.

Rast, A. Th. B. and Stijger, C.M. (1987). Disinfection of pepper seed infected with different strains of capsicum mosaic virus by trisodium phosphate and dry heat treatment. *Plant Pathol.*, **36**: 583-588.

Rathaiah, Y., Saikia, A.K. and Borua, S. (1997). Yield and reaction to leaf and stem necrosis of some chilli varieties. (Abst.) *National Symposium on Major Diseases of Crop Plants in Eastern India and Their Management*, Assam Agril. Univ., Jorhat, p.11.

Ravi, K.S. (1991). Studies on pepper vein banding virus and other components of Murda Syndrom in chilli (*Capsicum annuum* L.). Ph.D. thesis University of Agricultural Sciences, Hebbal, Bangalore,

Karnataka, India.

Ravi, K. S., Joseph, J., Nagaraju, N., Prasad, S. K., Reddy, H. R. and Savithri, H. S. (1997). Characterization of a pepper vein-banding virus from chilli pepper in India. *Plant Disease*, **81**: 673-676.

Rishi, N. and Dhawan, P. (1989). Characterization and identification of viruses causing mosaic diseases of *Capsicum* in nature. *Indian Phytopath.*, **42**: 221-226.

Robinson, D.J. and Harrison, B.D. (1987). CMI/AAB Descriptions of Plant Viruses No. 347.

Roggero, P., Milne, R.G., Masenga, V., Ogliara, P. and Stravato, V.M. (1995). First reports of eggplant mottle dwarf rhabdovirus in cucumber and in pepper. *Plant Disease*, **79**: 321.

Roye, M.E., Wernecke, M.E., McLaughlin, W.A., Nakhla, M.K. and Maxwell, D.P. (1999). Tomato dwarf leaf curl virus, a new bipartite geminivirus associated with tomatoes and peppers in Jamaica and mixed infection with tomato yellow leaf curl virus. *Plant Pathologyl*, **48**: 370-378.

Samretwanich, K., Chiemsombat, P., Kittipakorn, K. and Ikegami, M. (2000). A new Geminivirus Associated with a Yellow Leaf Curl Disease of Pepper in Thailand. *Plant* Dis., **84**: 1047.

Sandhu, K.S. and Chohan, J.S. (1979 Pub. 1980). Studies on the Characterisation of mottle disease of chilli (*C. annuum*) in the Punjab. *Indian J. Mycology and Plant Pathology*, **9**: 177-182.

Satya Prakash (2001). Studies on virus diseases of chilli (*Capsicum annuum* Linn.). Ph. D. Thesis submitted to the Deptt. of Botany, D.D.U. Gorakhpur Univ. Gorakhpur, Uttar Pradesh, India, p. 283.

Satya Prakash and Singh, S.J. (2004). Viral diseases of chilli/capsicum in India, their management and future research needs. *Vegetable Science*. (In Press).

Satya Prakash, Singh, S.J. and Upadhyaya, P.P. (2001). Identification and characterization of chilli veinal mottle potyvirus occurring naturally on chilli (*Capsicum annuum* L.) in India. In: *National Symposium on "Eco-friendly Approaches for Plant Disease Management"*, Jan. 22-24, 2001, Chennai, p. 19 (Abst.).

Satya Prakash, Singh, S.J., Singh, R.K. and Upadhyaya, P.P. (2002a). Distribution, incidence and detection of a potyvirus on chilli from Eastern Uttar Pradesh. *Indian Phytopath.*, **55**: 294 –298.

Satya Prakash, Singh, S.J. and Singh, R.K. (2002b). Non-chemical approach to combat insect-transmitted virus diseases in chilli. In: *"Symposium on Major Plant Diseases and their Management (Western chapter IPS)"* Central Institute for Cotton Research Nagpur, October 29-30, 2002, pp. 35-36.

Seth, M.L. and Dhanraj, K.S. (1972). A new strain of tobacco leaf curl causing enation in chilli (*Capsicum annuum* L.). *Phytopath. Z.*, **73**: 365-370.

Shalaby, A.A., Nakhala, M.K., Shafia, M.S., Mazyad, H.M. and Maxwell, D.P. (1997). Molecular characterization of tomato yellow leaf curl geminivirus (TYLCV) isolated from pepper collected in Egypt. *Annals of Agricultural Science, Moshtohai*, **35**: 819-831.

Sharma, I. and Dubey, G.S. (1977). Studies on a mosaic disease of bell pepper *Capsicum annuum* L. *Biovigyanam*, **3**: 1-6.

Sharma, O.P. and Sharma P.N. (1999). Virus diseases of chillies and okra. In: Diseases of Horticultural Crops–Vegetables, Ornamentals and Mushrooms (Eds. Varma, L.R. & Sharma, R.C., 1999), Indus Publishing Co., New Delhi, pp. 410-424.

Sharma, S. R. and Varma, A. (1982). Control of virus diseases by oil sprays. *Zbl. Microbiol.*, **137**: 329-347.

Singh, D. and Singh, S. J. (2000). Chilli little leaf–a new phytoplasma disease in India. *Indian Phytopath.*, **53**: 281-282.

Singh, S.J. (1993). Viral and Mycoplasmal Diseases of Vegetable Crops. Technical Bulletin No.9 (published by R.M. Pandey), p.142. IIHR, Bangalore, India.

Singh, S.J., Sastry, K.S. and Sastry, K.S.M. (1979). Efficacy of different insecticides and oil in the control of leaf curl virus disease of chillies. *J. Plant Dis. and Prot.*, **86**: 253-256.

Siriwong, P., Kittipakorn, K. and Ikegami, M. (1995). Characterization of chilli-vein-banding mottle virus isolated from pepper in Thailand. *Plant Pathology*, **44**: 718-727.

Sutic, D. Dragoljub, Ford, E. Richard and Tosic, T. Malisa (1999). *Hand Book of Plant Viruses*. CRC Press, Boca Raton London, New York, Washington D.C., pp. 126-134 (chilli viruses).

Tewari, V. P. and Viswanath, S. M. (1988). Note on breeding for resistance in chillies–today and tomorrow. *New Botanist*, **15**: 185-186.

Tian Wenhui and Zhang Shuping (1995). Identification of tobacco rattle virus on sweet pepper. *Journal of Hebei Agricultural University*, **18**: 32-34.

Todorovic, B., Tosic, M. and Neskovic, S. (1997). Aucuba mosaic virus on potatoes in Yugoslavia. *Acta Horticulture*, **462**: 353-356.

Upadhyay, S. (2000). Management of chilli leaf and stem necrosis caused by tomato spotted wilt virus. *J. Mycol. Pl. Pathol.*, **30**: 159-162.

Valand, G.B. and Muniyappa, V. (1992). Epidemiology of tobacco leaf curl virus in India. *Ann. Appl. Biol.*, **120**: 257-267.

Venkatesh, H.M., Muniyappa, V., Ravi, K.S., Viraktamath, C.A. and Krishnaprasad, P.R. (1997). Detection, Transmission and Management of Chilli Leaf Curl Disease. *Ist Natl. Sym. On Pest Mgmnt. Hort. Crops*, 15-17 Oct., 1997, Bangalore, India.

Vicchi, V., Fine, P. and Cardont, M. (1999). Presenza del virus della maculatura necrotica dell impatieno (INSV-Tospovirus) su colture Orticole in Emilia–Romagna. *Informatore Fitopatologia*, **49**: 53-55.

Vicchi, V. and Talame, M. (1994). Severe infections of TSWV on Capsicum crops in greenhouse in Marche. *Informatore Agrario*, **50**: 71-73.

Xi, Z., Xu, S. and Mang, K. (1982). A small ssDNA containing virus–bean distortion dwarf virus. *Acta Microbiologica Sinica*, **22**: 293-299.

Xiang Benchun, Xie Hao, Cui Xing Ming, Li Chun, Liu Suping, Xi Dehui and Yi Yuqi (1994). Isolation and identification of pepper mild mottle tobamovirus in Xinjiang. *Chinese Journal of Virology*, **10**: 240-245.

Yonaha, T., Toyosato, T., Kawano, S. and Osaki, T. (1995). Pepper vein yellows virus, a novel luteovirus from bell pepper plants in Japan. *Ann. of the Phytopathological Soc. of Japan*, **61**: 178-184.

Zitter, T. A. and Osaki, H. Y. (1978). Aphid-borne vegetable viruses controlled with oil sprays. *Proc. Florida State Hort. Soc.*, **91**: 287-289.

26

Epidemiology and Integrated Management of Fruit-Rot Diseases of *Trichosanthes dioica*

A.K. ROY & ANJAN KRISHNA

University Department of Botany, T.M.B. University, Bhagalpur – 812 007

India amongst developing countries occupies second position for the production of horticulture produces particularly top fruits & cucurbits. However, their production was tripled down as after independence emphasis was laid on self sufficiency in production and ultimately green revolution paved the way. Moreover, in order to increase food base as per our dietary requirement and minimize malnutrition due emphasis is being given on production of fruits and vegetables since during the last more than a decade.

Fruits and vegetables are rich sources of vitamins and minerals but are short lived and highly perishable. In general, a period of 10 days after post harvest is found optimum to deteriorate the quality which subsequently is hasten up when these are victimized by microbes. Damage like blemishes or brushes during harvest, handling or transport make these perishables more prone to microbial infection. Growers and Merchants would wish to discard decaying produce but the infections that initiate decay are not usually visible or predictable at the time of harvest, and overt symptoms only after ripening, a stage usually equated with edibility in fruit crops. Smith *et al.* (1964) recorded more than 250 parasitic diseases of fruits and vegetables in U.S.A. that causes decay and blemishes after harvest and before they reach to the consumers. The losses due to diseases greatly depend upon nature of fruit crops, methods of harvesting and storage conditions.

Pointed gourd (*Trichosanthes dioica.* Roxb.) a common cucurbitaceous vegetable is cultivated all over the warmer regions of India, particularly in U.P., Bihar, West Bengal and Assam. Its cultivation requires hot humid climate with planting time in early Feb-Apr and late May-July. However, a well drained sandy-loam soil is well suited for its cultivation in Bihar. Alluvial strips of the river Ganga are invariably used for cultivation of these vegetable on commercial scale. Among twenty cultivated types of pointed gourd (parwal) fruits, the two important forms *i.e.* Dandalia and Hilly green are found to cultivate commonly in this area. These green perishables are also prone to microbial infection which causes both pre- and post harvest losses. Some earlier contributions on fruit rot diseases of Parwal have been made by Pant (1951), Srivastava *et al.* (1964), Choudhary (1968), Prakash *et al.* (1973) and Waraitch *et al.* (1975).

A bulk of harvested fruits (Parwal) in this area is spoiled every year but no concrete suggestions have been pointed out as yet for preventing and minimizing the losses resulted in the field, during storage and marketing on minimum cost. The Parwal mainly the product of Diaraland, covers almost all markets of North-Bihar especially in summer season which attracts the attention of cultivars for the better production in the field and protection in the storage.

The main objectives of the present write-up are to focus on losses of *T. dioica* fruits in field, storage and marketing due to fungal infection. Epidemiological studies of fruit-rots (soft) have also been taken into consideration. An integrated approach for the disease management has been reviewed to filter out sustainable perfect measures.

Fruit-Rots and Fungi

Extensive survey of different fields and markets of Bihar state was conducted for the collection of diseased fruits of Parwal. There are twenty cultivated varieties of Parwal *viz.*, Safed dhari, Dariapur local, Dandalia, Hilly safeda, Hilly green, Sabouria, Mirdangia, Guthlia, Rahimpuri, Safeda, Chitkabra, Tirkolwa, Sabur local, Gopalganji, Dholi safeda, Sawpulia safeda, Punjabi safed, Sonpuri green and Sonpuri safed. Pointed gourd variety is named after the name of locality where they are growing and on the basis of their morphological features. However, the common types are Dandalia (TDD), plae green without marking; Hilly green (TDH), longitudinal white marking; Hilly safeda (TDS) light green with white bands and Sonpuri (TDSON), dark green with distinct white longitudinal bands.

The centres where pointed gourd fruits are being stored for 1-3 days have also been surveyed. The method employed for transportation of fruits from field to market is traditional as a result making them more prone to fungal invasion. A general survey of a market and field reveals that about 15-20 percent fruits were ravaged by various fungi every year. During plucking, packing in gunny bags, transportation and marketing of fruits various kinds of injuries are inevitable which make them more prone to fungal invasion. Moreover, insects sometimes also injure the fruits and open scope for saprophytic fungi to attack. Generally, infection occurs during post harvest handling or storage. Sometimes single infected fruit becomes source of infection during storage and transit. However, Sahu Kritagyan and Singh (1980) reported 15 per cent loss due to *Fusarium* sps. and 5 per cent loss due to *Alternaria* rot under field condition. Fruits lying on the ground in the field are in direct contact with soil pathogens, which get entrance through artificial injuries. Agarwal and Prasad (1978) isolated forty three fungi from rhizosphere, rhizoplane and phyllosphere association with *T. dioica* fruit rot.

In all, nine fungi were recorded causing soft rot diseases in *T. dioica* fruits. Out of these fungi, *F. equiseti, F. semitectum, C. lunata, P. aphanidermatum* were of common occurrence and often noticed throughout the rainy season whereas *F. semitectum* was observed in early season. *P. aphanidermatum* appeared in virulent form just after rains whereas *A. alternata and C. gloesporioides* were found to be moderately virulent. *A. niger and A. flavus* were associated as secondary invaders with the soft rot fruits.

Of all, *F. equiseti* rot was most devastating and was found to ravage a bulk of fruits every year. In 1985, *F. equiseti* soft rot appeared in epidemic form during rainy season followed with warm and humid weather. The appearance of the disease was recorded at every stage of the fruit which resulted in softening and discoloration of the tissue. It was recorded as infectious disease and even a single diseased fruit was sufficient to cause the rot of bulk of fruits under storage and market. It was locally called by the farmers as *cholera* of parwal, due to mass mortality and also as *"Galwa"* disease due to softening of fruits.

Epidemiology

Epidemiological is concerned with the determinants of diseases and with factors that influence their frequency and distribution with population. A large number of literatures available confirm that the frequency of disease development is dependent upon environmental factors *viz.*, temperature humidity, light, wind and moisture. One of these or more than that are considered to be responsible for perpetuation of the pathogen form one crop season to other, to build up the early inoculums, to the formation of spores, dissemination of inoculums and primary infection, to the production of secondary and of over wintering stage. Sindhan and Bose (1980), Bambawale and Bedi (1982) and Kulkarni *et al.* (1982) studied the epidemiology of angular spot of french bean, early blight of potato and brown leaf spot of rice in Punjab and Karnataka respectively. However, reports on the disease of cucurbitaceous vegetables in general and parwal in particular in relation to environmental factors are meagre. Recorded data reveal that the incidence of different pathogens is related to the change in metrological factors.

Altogether nine fungi were found to be associated with both Dandalia and Hilly green variety of pointed ground fruits. *F. equisti* and *F. moniliforme* were noticed as dominant fungi from June to September whereas *F. semitectum* and *C. lunata* were present in early period *i.e.* from May to July. The percentage occurrence of *A. alternata* was scanty. *C. gloeosporioides* was found to be present upto August and it was isolated only from Dandalia variety in July, Fungal incidence was maximum when temperature was 27.9°C., RH 84.5 per cent and rainfall 438.0 mm. *A flavus* and *A. niger* were invariably isolated from diseased fruits during Aug-Oct months, however, *P. aphanidermatum* was isolated only from diseased fruits of Hilly green variety.

Different weather conditions have been reported favourable for the development of early blight of potato and tomato (Wager, 1945). Harrison *et al.* (1965) observed that the severity of the disease on potato was as good in cool and dry weather as it was under warm and moist conditions .Saharan and Kaushik (1981) recorded the magnitude of temperature, humidity and rainfall for outbreak of powdery mildew epidemic in Haryana. Sindhan and Bose (1981) found that anthracnose of french bean caused by *Colletotrichum lindamuthianum* appeared in June when temperature and humidity ranged between 11.4 to 21.5°c and and 69 per cent to 90 per cent respectively in Hilly region of U.P. Suhag *et al.* (1985) also studied the epidermiology of pod and leaf blight of radish crop in relation to environmental factors. Bhatt and Chauhan (1985) have also observed that the temperature (29°c ± 1°c) and relative humidity (above 89 per cent) have considerably effect on the development of neck blast of rice.

The review of earlier reports elucidate clearly that environmental temperature ranging between 25°C to 36°C and humidity of 90 per cent followed with heavy rainfall made the fruits prone to fungal invasion. A maximal fungal population was recorded during months of July and August when temperature and relative humidity were optimum. The disease of pointed gourd fruits appeared during May and reached at peak in the month of July-August and then declined. Particularly the species Fusarium was found to be dominant and virulent over other pathogens. However, disease was not recorded in the month of April due to high temperature (21.7°C to 37.7°C) low humidity (30 per cent to 70 per cent) and scanty rainfall (0.5 mm) even in the presence of such virulent pathogen like Fusarium sp. It is obvious form above result that inspite of the pathogen being virulent and host being susceptible, severity of disease will not be epidemic unless and until all the environmental conditions become favourable.

Effect of Physical Factors on Disease Development

Relative humidity and temperature are the important physical factors which greatly influence the infection and development of fruit rots. Besides favouring fungal growth, these factors have also predisposing effects on the hosts. In recent past several workers including Kanwar *et al.* (1973), Jamaluddin (1973), Prasad (1977), Prasad and Roy (1979), Poddar (1980), Ojha (1983), Choubey (1984) have studied the effect of humidity and temperature on rotting of fruits.

The pointed gourd fruits were artificially inoculated with *F. equiseti* respectively and *C. lunata*. The healthy and inoculated fruits were kept at different temperature *i.e.* 0°C, 10°C, 15°C, 20°C, 25°C and 30°C and relative humidities to evaluate the effect of these factors on severity of fruit rots.

Relative Humidity

The severity of infection was noticed to increase with the rising of relative humidities. The Dandalia and Hilly green variety of *T. dioica* infected with *C. lunata* exhibited 0.59 per cent and 1.90 per cent rotting while *F. equiseti* resulted loss in 0.43 per cent and 1.20 per cent respectively at 33 per cent relative humidity after an incubation period of 7 days. However, percentage rotting of TDD and TDH increased upto 32.30 per cent and 37.00 per cent respectively due to *C. lunata* and 36.10 per cent and 35.90 per cent due to *F. equiseti* respectively at 96 per cent RH within the same incubation period.

It has also been observed that the incidence of disease was maximum during July to August when RH is near to 96 per cent. Chand (1968) reported that *Gleosporium fructigenum* produced infection in apple fruits between 80 per cent RH to 100 per cent RH whereas 84.9 per cent RH was critical level below which chilli fruit rot does not develop (Choudhary, 1957). Batra (1978), Gupta and Nema (1979) and Agarwal (1979) have also reported maximum disease development in various fruits between 90 to 100 per cent relative humidity.

Temperatures

Both the varieties of Parwal fruits showed no infection at 15°c temperature however, at 20°c symptoms developed within 7 days but the loss recorded was meagre. Highest percentage loss was observed at 30°C where it was 36.2 per cent & 31.5 per cent due to *C. lunata* and 38.0 per cent and 35.1 per cent due to *F. equiseti* in case of TDD and TDH respectively.

The results obviously reflect that temperature plays a decisive role in incidence of fruit rot. In recent past several workers have studied the effect of temperature on rotting of fruits due to fungal pathogens (Srivastava and Tandon, 1967; Jamaluddin, 1973). The findings of Singh (1977) and Prasad (1977) are also in support of the present results where rotting rate was almost nil at 0°C-10°C. Thus, the rot of the fruits may be prevented to a great extent by storing them at 0°C-10°C.

Integrated Disease Management

In the recent past disease control management of edible horticulture produced have been directed to devise ecofriendly, sustainable and economical methods. Minimizing use of chemicals as pesticides due to their residual hazardous effect the concept of integrated pest management was developed by Patel (1991), Patil (1992), Singh (1992), Roy (1995) Khare (1999) and Roy & Singh (2002). Thus, consensus has now built up with an aim to restrain diseases by manipulating environmental, ecological and genetical factors with minimum uses of chemical pesticides. In order to achieve the target one has to develop knowledge with nature and life cycle of pathogens, interaction amongst host pathogen and environment, geographic location, cultural practices, irrigation and use of organics or fertilizers.

Pointed gourds are short lived therefore are grouped under perishable horticulture produce. The green colour (intact chlorophyll) of fruits determines their market value. The loss in chlorophyll contents of fruits followed by change in colour either pre or post harvest condition is indicative of their deterioration and subsequently reduction in market value. Therefore, it is pre-requisite that methods used to manage the diseases should maintain the green colour of fruits. Keeping the facts in view under the IPM programme the botanicals (eco-friendly) and non-hazardous chemicals (degradable) are preferred over the use of bio-agents.

Use of Botanicals (Phytopesticides)

The use of plant extracts to cure plant disease was first attempted by Dedicators in 470 B.C. (Sharville, 1960), however this practices could not receive proper attention till the beginning of the 20th century. During the last few decades systematic efforts have been made to screen chlorophyllous plants for their antifungal, antibacterial and antiviral activities.

Bio intensive integrated management strategies have been developed by using eco-frieindly botanicals like *Phaseolus vulgaris* (Afifi, 1975), Peanut embryos (Lindsey and Tunrer, 1975), *Curcuma aromatica* (Tirumala Rao, 1976), *Rosa indica* (Dixit, *et al.*, 1976) *Achyranthes aspera* (Chakravarty and Pariya, 1977) etc. to control fungal diseases of several fruits and vegetables Singh (1986) tested extracts of various plants including *Azadiractta indica, Thuja orientalis, Lantana camara, Datura fastuosa* and *Reccinus communis*, against diseases of Singapore banana. Roy (1993) also controls the diseases of banana by different plants extract. Roy (1995) and Sing *et al.* (1993) have evaluated anti-microbial properties of a large number of angiosperms.

On fungal growth and spore germination: Due to known antifungal properties of plants extracts, the present investigation has been conducted to determine the effectiveness of the leaf extracts of some medicinal plants on growth and spore germination of two test fungi *viz. Curvularia lunata* Wakker and *Fusarium semitectum* Berk & Rev., so that positive findings may be practiced to control rot of Parwal fruits. Leaf extracts showing inhibition of spore germination more than 70 per cent as selected for the control of pointed gourd rot.

Leaf extracts of ten medicinal plants *viz., Adhatoda vasica* Nees, *Andrographis peniculata*, Ness., *Azadirecta indica* L., *Catharanthesw roseus* L. (Don), *Cinnamomum camphora* (L.) T. Nees and Ebern, *Ocimum sanctum* Linn., *Plumabago zeylanica* Linn. and *Strychnos nux-vomica* Linn. were screened and evaluated against the spore germination and radial growth of test fungi under laboratory condition, results are depicted in Table 26.1.

A comparative perusal of finding clearly indicate that among ten medicinal plants screened, *A. peniculata,A. indica* and *P. zeylancia* were proved to be effective fungistatic for the fungi under study. The extracts of these plants may be suggested to the growers for controlling the post harvest diseases of fruits, which are easily available and less expensive. The effective leaf extracts were applied on the pre and post infected pointed-gourd fruits.

Table 26.1: Screening of Leaf Extracts on Spore Germination and Growth Behaviour of *F. equiseti* and *C. lunata*

Leaf Extracts	*F. equiseti*				*C. lunata*			
	Growth Behaviour		Spore Germination		Growth Behaviour		Spore Germination	
	Colony Diam. (cms)	Mycelial Dry Wt. (mg)	% Germination	% Inhibition Over Control	Colony Diam. (cms)	Mycelial Dry Wt. (mg)	% Germination	% Inhibition Over Control
Control	4.1	400	72.40		3.5	410	60.29	
Adhatoda vasica Nees	3.0	365	40.22	44.44	2.6	350	30.43	49.52
Andrographis peniculata Nees	1.2	250	11.20	84.53	1.1	290	10.71	82.23
Azadirachta indica Adr. Juss	0.3	190	5.55	92.33	0.1	120	2.0	96.68
Catharanthes roseus L.	0.9	390	8.22	88.64	1.1	385	38.02	36.93
Cinnamamum camphora (L) T.Nees & Ebern	2.2	390	39.09	46.00	2.0	390	37.56	37.70
Lunatana camara L.	1.1	210	10.60	85.36	1.0	300	6.6	88.45
Ocimum sanctum Linn.	1.2	200	15.22	78.98	1.4	220	11.11	81.75
Plumbago zeylanica Linn.	0.8	230	10.01	86.17	1.0	200	9.09	84.92
Strychnos nux-vomica Linn.	2.4	330	19.55	73.30	2.0	350	11.53	80.88
Vitex negundo Linn.	2.1	395	40.21	44.46	2.2	375	38.46	36.20

Pre and post infection treatment: The results (Table 26.2) make it clear that the post infection treatment with the extracts of *A. indica, A. peniculata, L. camara, O. sanctum, P. zeylanica* and *S. nux-vomica* was more promising than the pre infection treatment. In case of TDD variety all the six leaf extracts showed 100 per cent control of fruit caused by *Fusarium equiseti* and *Curvularia lunata*. Other than *A. peniculata*, leaf extracts of all the plants understudy showed more than 80 per cent control in pre-treated fruits (*F. equiseti*). However, *A. peniculata, O. sanctum* and *P. zeylancia* showed 100 per cent control in pre-treated fruits. In case of TDH variety also post infection treatment was more effective.

Table 26.2: Effectiveness of Leaf Extracts in Control of Pointed Gourd Fruit

Leaf Extracts	Percentage Control							
	T. dioica Dandalia				*T. dioica* Hilly green			
	F. equiseti Rot		*C. Lunata* Rot		*F. equiseti* Rot		*C. Lunata* Rot	
	PR	PO	PR	PO	PR	PO	PR	PO
A. indica Adr. Juss	82.43	100.00	76.97	100.00	0.00	82.78	91.84	94.21
A. peniculata Ness	0.00	100.00	100.00	100.00	0.00	93.18	100.00	0.00
L. camara L.	100.00	100.00	0.00	100.00	0.00	0.00	0.00	100.00
O. sanctum Linn.	100.00	100.00	100.00	100.00	0.00	100.00	100.00	100.00
P. zeylanica Linn.	88.88	100.00	100.00	100.00	0.00	93.79	99.00	100.00
S. nux-vomica Linn.	100.00	100.00	0.00	100.00	88.35	0.00	92.84	89.49
Control (inoculated not treated)	0.00	0.00	0.00	0.00	0.00	0.00	0.00	0.00

Post infection treatment with leaf extracts had pronounced inhibitory effect on rotting of pointed gourd fruits (*i.e.*, 100 per cent control) than the pre-infection treatment. In TDD fruit-rot caused by *F. equisati* and *C. lunata* was checked by 100 per cent with the application of extracts of *A. indica, A. peniculata, L. camara, O. sanctum, P. zeylanica* and *S. nux-vomica*, while in TDH *L. camara* and *S. nux-vomica* failed to check *F. equiseti* rot when treated after infection. Similarly *A. peniculata* did not inhibit *C. lunata* fruit-rot during post infection changes.

Inhibitory effect of leaf extracts against both the test fungi understudy might be attributed to the presence of active principle like sequiteerpene, steroids, certain alkaloids (Lantanandene and Lancamarene contained by the leaves of *L. camara*), steroid like substances (Azadirachtin, present in the leaves of *A. indica*), gum resins and phenol. Mahadeva (1982) has reported that plants contain prohibitions which are toxic to microorganisms.

Effect of Non-hazardous Chemicals

On fungal growth and spore germination: The non-hazardous chemicals *viz.* Boric acid, Calcium hyrdoxide, Copper sulphate, Ferrous sulphate, Formalin, Mercuric chloride, Potassium iodide, Potassium permanganate, Sodium hydroxides and Sodium hypochloride were selected to evaluate their efficacy against spore germination of two test fungi *viz.*, *F. equiseti* and *C. lunata.* Some of them were found to be fungistatic whereas other had fungicidal effect on the test fungi. Different concentrations of non hazardous chemicals had variable effects on the test fungi Table 26.3.

A comparative study of results indicate that Mercuric chloride (0.1 per cent), Formalin (0.5 per cent), $CuSO_4$ (2.0 per cent), Lime (2.0 per cent), NaOH (1.5 per cent) were proved to be effective chemicals which showed 100 per cent inhibition of spore germination of the test fungi.

Pre and post infection treatment: The non-hazardous chemicals were also used both in pre and post infection treatment for controlling diseases of pointed gourd. The results (Table 26.4) indicate that chemicals used as post infection treatment was more effective than the pre infection treatment. 2 per cent aqueous solution of potassium iodide, sodium hydroxide, copper sulphate, potassium permaganate, 1.5 per cent of sodium hypochloride, 0.1 per cent of mercuric chloride and 0.5 per cent formalin showed significant reduction (96 per cent to 100 per cent) in *F. equiseti* fruit rot (Dandalia). Other chemicals *viz.* Ferrous sulphate, Borric acid and Calcium hydroxide showed poor inhibitory effect *i.e.* 39.63 per cent, 25.321 per cent and 40.428 per cent respectively. The chemicals, used for pre infection treatment has less response in controlling the fruit rot.

Rotting of Hilly green variety caused by *F. equiseti* was inhibited nearly upto the level of 100 per cent by 2 per cent aqueous solution of potassium iodide, sodium hydroxide, copper sulphate; 1.5 per cent solution of sodium hypochloride, 6.1 per cent solution of mercuric chloride and 0.5 per cent solution of formation when treated after infection.

Similar inhibition due to these chemicals has also been recorded by several earlier workers. Thakur and Chenula (1974) controlled soft rot of potato tubers caused by *Rhozopus arrhizus* treating with Borric acid, Formalin and Mercuric chloride. They have also noticed that pre infection treatment with formalin protected potato tubers from soft rot for a period of 75 days.

General Discussion

Like other green cucurbitaceous vegetables, pointed gourd fruits are also very prone to fungal attack due to high water content (70 per cent) and carelessly post harvest handling. The quantitative loss in case of *T. dioica* fruits either due to fungal attack or during storage or marketing was recorded about 40 per cent or more. Agarwal and Agarwal (1978) have isolated 43 fungi associated with *T. dioica* fruits from the field and markets adjacent to Muzaffarpur (North Bihar). Poddar (1980) and Ojha

(1983) have recorded 10-15 per cent post harvest loss while working on different cucurbitaceous vegetables of this area. Rao and Bhide (1984) have estimated the loss to the tune of 25 per cent in fruits and vegetables before these reach to the consumers *i.e.* a loss of nearly Rs. 5 Crores annually.

Table 26.3: Screening of Non-hazardous Chemicals on Spore Germination and Growth Behaviour of *Curvularia lunata* and *Fusarium equiseti*

Leaf extracts	Conc. in %	C. lunata			F. equiseti		
		% of Spore Germination	% Inhibition Over Control	Colony Diameter	% of Spore Germination	% Inhibition Over Control	Colony Diameter
Control	—	58	—	5.6	62.9	—	6.00
NaOH (Sodium Hydroxide)	0.5	26.66	54.03	—	28.21	55.15	—
	1.0	12.5	78.45	—	20.5	48.78	—
	1.5	9.5	83.62	—	9.11	85.52	—
	2.0	0.0	100.00	—	2.50	96.02	—
Potassium iodide	1.0	12.5	78.45	—	15.90	74.72	—
	1.5	7.21	87.57	2.2	10.80	82.83	2.5
	2.0	0.00	100.00	—	5.1	91.19	—
Sodium hypochloride	0.5	27.27	52.98	—	20.21	67.87	—
	1.0	0.99	98.29	2.0	0.80	98.58	—
	1.5	0.00	100.00	—	0.00	100.00	—
Ferrous sulphate	0.5	27.77	52.12	—	30.22	51.95	—
	1.0	16.66	71.27	—	12.51	80.11	—
	1.5	11.50	80.17	—	5.20	91.73	—
	2.0	9.09	84.33	—	0.00	0.00	—
Potassium permanganate	0.5	40.89	29.5	—	50.23	20.14	—
($KMnO_4$)	1.0	36.66	36.79	0.5	40.44	35.71	4.1
	1.5	12.77	77.98	—	21.85	65.26	—
	2.0	0.00	100.00	—	12.5	80.13	—
Calcium hydroxide	0.5	12.75	78.01	—	12.5	80.13	—
$Ca(OH)_2$	1.0	8.0	86.20	2.0	9.1	85.53	1.0
	1.5	5.0	91.38	—	0.0	100.00	—
	2.0	0.0	100.00	—	0.0	100.00	—
Copper sulphate	0.5	21.73	62.53	—	15.9	74.72	—
($CuSO_4$)	1.0	2.8	95.17	1.2	4.8	92.37	—
	1.5	0.5	94.14	—	0.0	100.00	—
	2.0	0.0	100.00	—	0.0	100.00	—
Formalin	0.1	16.0	72.41	—	12.75	74.73	—
	0.15	0.0	100.00	—	0.5	99.20	—
	0.5	0.0	100.00	—	0.0	100.00	—
Boric acid	0.5	49.1	15.34	—	60.2	4.29	—
	1.0	28.5	50.86	4.6	30.3	51.83	5.00
	1.5	24.0	58.62	—	21.1	66.45	—
	2.0	22.5	61.21	—	20.6	67.2	—

Table 26.4: Effectiveness of Non-hazardous Chemicals in Control of *T. dioica* Fruit Rot

Chemical	*Concentration*	*Percentage Control*							
		T. dioica Dandalia				*T. dioica Hilly Green*			
		F. equiseti Rot		*C. lunata Rot*		*F. equiseti Rot*		*C. lunata Rot*	
Potassium	2%	0.00	96.405	35.427	96.406	0.00	100.00	44.858	97.332
Sodium hydroxide	2%	0.00	100.00	42.231	98.00	0.00	100.00	0.00	87.23
Copper sulphate	2%	56.385	100.00	43.194	45.299	0.00	100.00	100.00	85.985
Potassium permanganate	2%	0.00	100.00	92.428	45.057	0.00	43.075	65.340	68.945
Ferrous sulphate	2%	0.00	39.697	52.024	100.00	0.00	0.00	40.396	19.693
Boric acid	2%	0.00	25.321	30.462	15.230	0.00	43.072	0.00	100.00
Calcium hydroxide	2%	0.00	40.420	35.268	47.591	0.00	0.00	0.00	75.23
Sodium hypochloride	1.5%	40.194	100.00	35.429	100.00	83.00	90.925	0.00	100.00
Formalin	0.5%	0.00	100.00	81.359	94.87	0.00	91.098	79.155	84.769
Control (inoculated not treated)	0	0.00	0.00	0.00	0.00	0.00	0.00	0.00	0.00

Extensive survey of markets for post harvest disease of pointed gourd fruits revealed that in all five hyphomycetous fingi, *viz., Fusarium equiseti* (Corda) Sacc., *F. semitectum* Berk and Rev., *F. moniliforme* Sheldon, *Curvularia lunata* (Wakker) Boedijn and *Alternarai alternata* (Fr.) Keissler; one ascomycetous fungus *i.e. Colletotrichum gloeosporioides* Fens; one member of Phycomycetes *i.e. Phythium aphanidermatum* (Edson.) Fitz. Patrick and tow saprophytic fungi *viz., Aspergillus falvus* Link ex Fries and *A. niger* Van Tiegh. were found to be associated with the soft root of pointed gourd. The incidence of fungi was observed always higher in the market (market disease) than the field. Hilly green fruits although having hard rind but were found more prone to fungal disease both in field and market.

The appearance of different diseases, especially soft rot of fruits was observed starting from the month of May when temperature ranged from 24.5°C-36.5° followed with an early break of Monsoon. In the month of July there was maximum incidence of fungi when temperature and humidity of the environment were relatively high. The severity and subsequent spread of soft rot of pointed gourds caused by *F. equiseti* and *C. lunata* depend upon the favourable range of humidity (69 per cent-87 per cent) and temperature (28.5°C-30.5°C) from May to September. The environmental factors which supports epidemic of soft rot of pointed gourd are either temperature 25°C-35°C with a long period of heavy rain and lack of bright sunlight during harvest or moderate temperature with long period of heavy rain during storage period. Such significant impact on disease development has also been earlier reported by Saharan and Kaushik (1981), Sindhan and Bose (1981), Bhatt and Chauhan (1985) and Bedi (1986). Epidemiological study of soft-rots of pointed gourds conducted may be of immense importance for disease forecasting.

Under laboratory condition also, temperature and humidity exerted a profound influence on soft rot disease. Maximum rotting of the fruits (Dandalia and Hilly green) was recorded at room temperature (28 ± 2°C) and RH 75 per cent-96 per cent. *F. equiseti* was found more virulent as compared to the *C. lunata*. Mislivec and Tiote (1970), Helberg and Kolk (1972), Christensen (1973), Welling (1974), Sinha (1979) and Mehrotra (1983) have come to similar conclusion with regard to mould growth and activity at high temperature and relative humidity. The pathogens were incapable of growing at low temperature and humidity as the inoculated fruits did not show any symptoms of rotting at 33 per cent RH and 15°C temperature. Tandon (1967) and Choubey (1984) recommended the storage of fruits and green vegetables at lower temperature. Rao and Bhide (1984) stated that lower temperatures check rate of metabolism in living organisms. According to them such temperature minimize losses due to fungal invasion in two ways *(i)* it delays the advent of senescence and consequently retrains the resistance of fruit and *(ii)* it slows down the rate of advancement of the pathogens. Therefore, green fruits of pointed gourd may be stored at lower temperature (15-20°C) and relative humidity (33 per cent) for longer period.

In the present past the use of fungicides is avoided for well known reasons. However, the use of non-hazardous chemicals and botanicals are encouraged by the plant pathologists. Among medicinal plants tested, leaf extracts of *A. indica, a. peniculata, O sanctum, C. roseus, L. camara* and *P. zeylanica* were found effective (90 per cent) in inhibiting the spore germination of *F. equiseti* and *C. lunata*. The leaf extracts of *A. indica, A. peniculata, O. sanctum, C. roseus, P. zeylanica* and *S. nux-vomica* were tried as pre and post infection treatment on both varieties of *T. dioica* fruits. The latter showed significant result whereas the former did not show the promising effect. In case of Dandalia variety all six leaf extracts tested showed 100 per cent control when treated after infection caused by both the test fungi. In case of Hilly green, post infection treatment was also at significant level. The leaf extracts of *L. camara* and *S. nux-vomica* could not inhibit *F. equiseti* rot significantly, treated after infection. Similar response of leaf extract was recorded in controling fruit rot by Pandey (1982), Ojha (1983) and Gourinath and Manoharachary (1984). Such antimycotic property of leaf extracts of plants have been reported due to the presence of active compounds like phenol, steroids, essential oil, alkaloid etc.

The application of aqueous solution of non hazardous chemicals on spore germination and incidence of fruit-rot disease has been found to have significant results. Sodium hydrochloride, Formalin and Mercuric chloride exhibit 100 per cent control of spore germination. Ferrous sulphate and Boric acid may reduce spore germination but not up to significant level. The effective chemicals used in the present investigation as post infection treatment shows 96 per cent - 100 per cent control in both the varieties of pointed gourds. Similar results have also been recorded by Kaushik *et. al.* (1972) Thakur and Chenulu (1970, 1974) and Sumbali and Mehrotra (1983). Pre infection application of chemicals is noticed to have less response in controlling fruit-rot diseases. Moreover, on the basis of efficacy to control Parwal fruit rot diseases; non hazardous chemicals have been classified in the following manner.

Table 26.5: Non-hazardous Chemicals (Therapeutants)

Hydroxide (alkaline)	*Sulphate*	*Halogen*	*Miscellaneous*
1. Sodium hydroxide	3. Copper sulphate	5. Sodium hypochlorite	7. Formalin
2. Calcium hydroxide	4. Ferrous sulphate	6. Potassium iodide	8. Boric acid
			9. Potassium permanganate

REFERENCES

Afifi, A.F. (1975). Effect of volatile substances from species of labiatae on rhizospheric and phyllospheric fungi of *Phaseolus vulgaris*. *Phytopath. Z.*, **83**: 296-302.

Agarwal, A. (1979). Ph.D. thesis, University of Jabalpur, Jabalpur

Agarwal, G.P. and P. Agarwal (1979). In "*Physiology of Parasitism*" Today and Tomorrows Printers and Pub. Delhi pp. 356-367.

Agarwal, S.R. (Smt) & S.S. Prasad (1978). Investigation of the diseases of *Trichosanthes dioica* Roxb. *Proc. Nat. Acad. Sci. India.* **48B**: IV.

Bambawale, O.M. and P.S. Bedi (1982). Epidemiology of early blight of potato in the Punjab. *Indian Phytopath.*, **35(4)**: 574.

Batra, S. (1978). Ph.D. Thesis, University of Jabalpur, Jabalpur

Bhatt, J.C. and Chauhan, V.S. (1985). Epidemiological studies on neck blast of rice in U.P. hills. *Indian Phytopath.*, **38(1)**: 126-130.

Chakravorty, D.K. and Pariya, S.N. (1977). Inhibition of germination of phytopathogenic fungi in some medicinal plant extracts. *Zeitschriftfur pflanzenOkrari-khaiten* and *pfl-anzenschutz*, **84**: 221-223.

Chand, J.N., M.R. Kandal and R.K. Agarwal (1968). Epidermilogy and control of bitter rot of apple caused by *Gleosporium fructigenum* Bark. *Indian Phytopath.*, **21**: 257-263.

Chaudhary, B. (1968). Diseases of vegetables during transit, storage and marketing and their control. *Bull. Indian. Phytopath Soc.*, **4:** 52-56.

Choubey, N. (1984). Ph.D. thesis, Bhagalpur University, Bhagalpur.

Choudhary, S. (1957). Studies on development and control of fruit rot of chillies. *Indian Phytopath.*, **10(1)**: 58-62.

Christensen, C.M. (1973). Loss of viability in storage microflora. *Seed Sci. and Technol.*, **1**: 547-562.

Dixit, S.N. Tripathi, S.C. and Upadhyay (1976). The antifungal substance of rose flower (*Rosa indica*). *Eco. Bot.*, **30**: 371-374.

Gourinath A. and C. Manoharachary (1984). Effect of angiospermic leaf extracts on three pathogenic fungi. *Indian Bot. Reptr.*, **3(1)**: 95-97.

Gupta, C. and K.G. Nema (1979). Effect of different temp and RH on development of fruit of papaya caused by B. theobromae and C. papaya. *Indian Phytopath.*, **32(1)**: 160-167.

Harrison, M.D., L.W. Livingston and N. Oshima (1965). Epidemiology of potato early blight. *Am. Potato J.*, **42**: 279-291

Helberg, A. and H. Kolk (1972). Mycoflora on deteriorating grains. *Acta Agric. Scand.*, **22**: 137-149.

Jamaluddin (1973). D. Phil Thesis, University of Allahabad, Allahabad.

Kanwar, J.S., D.P. Thakur and O.P. Kadian (1973). A note on the effect of temperature and RH on the development of soft rot of pomegranate fruits due to *Rhizopus arrhizus* Fischer. *Indian Phytopath*, **26(4)**: 742-443.

Kaushik, C.D., D.P. Thakur and J.N. Chand (1972). Parasitism and control of *Pestalotia psidii* causing cankerous disease of ripe guava fruits. *Indian Phytopath.*, **25**: 61-64.

Khare, M.N. (1999). Seed borne pathogens of pulses, diseases caused by them and their integrated management. National symposium on integrated Pest Management in North East India. Deptt. of Life Sciences, Assam University, Silchar.

Kulkarni Srikant, K. Ramakrishnan and R.K. Hedge (1982). Epidemiology and control of brown leaf spot of rice in Karnataka. *Indian Phytopath.*, **35(1)**: 80-82.

Lindsey, D.L. and R.B. Turner (1975). Inhibition of growth of *Aspergillus flavus* and *Trichoderma viridae* by peanut embryos. *Mycopathol.*, **55**: 149-152

Mahadevan, A. 1982). In: *Biochemical Aspects of Plants Disease Resistance*, Today and Tomorrows Printers and Pub. Delhi pp. 110-164.

Mislivec, P.B. and J. Tuite (1970). Temperature and relative humidity requirements of species of *Penicillium* isolated from yellow dent corn kernels. *Mycologia.*, **42**: 75-83.

Ojha, N.L. (1983). Ph.D. Thesis, Bhagalpur University, Bhagalpur

Pandey, K.N. (1982). Antifungal activity of some medicinal plants on stored seeds of *Eleusine coracana. Indian Phytopath.*, **35(3)**: 499-501.

Pant, D.N. (1950). On the occurrence of a genus of fungi imperfecti *Blastotrichum corda* in India. *Indian Phytopath.*, **3**: 179-181.

Patel, K.D. (1991). Studies on post-harvest fungal diseases of guava fruits. Ph.D. Thesis, Rajasthan Agril, University, Bikaner.

Patil, P.K. (1992). Studies on *Rhizopus* rot and *Botryodiplodia* rot of mango fruits, Ph.D. Thesis, Rajasthan Agril. University, Bikaner.

Poddar, K.D. (1980). Ph.D. Thesis, Bhagalpur University, Bhagalpur.

Prakash, O., Mishra, B. and A.P. Mishra (1973). Anthracnose disease of *C. melo. Indian J. Mycol. & Pl. Pathol.*, **3**: 217-218

Prasad, M.M. (1977). Post infection changes in sugar content of Banana fruits. *Curr. Sci.*, **46(6)**: 197-198.

Prasad, M.M. and Roy, A.K. (1979). Effect of temperature and humidity of fungal deterioration of two vegetables. *Nat. Acad. Sci. Letters.*, **2(12)**.

Roy, A.K. (1993). Post harvest diseases, losses and control in bananas (var. Singapuri) International Conference on post-harvest handling of Tropical fruit, Chiang Mai, Thailand.

Roy, A.K. (1995). Effectiveness of the neem for the management of vegetable diseases. Neem for the management of crop diseases (Ed. V. Mariapapan), Associated Publishing, Co. N. Delhi, India, pp. 63-67.

Roy, A.K. and Singh, H.N.P. (2002). In: *Frontiers in Microbial Biotechnology and Plant Pathology*, Scientific Publishers, Jodhpur (India) pp. 133-144.

Saharan, G.S. and J.C. Kaushik (1981) Occurrence of epidemiology of powdery mildew of *Brassica. Indian Phytopath.*, **34(1)**: 54-57.

Sahu Kritagyan, S.P. and Singh, S.P. (1980). Fruit rot of pointed gourd (T. dioica) in Bihar. *Indian Phytopath.*, **33(2)**: 308-309.

Sharville, *E.G.* (1960). "The Nature and Use of Modern Fungicides." Burgess Publishing Co., Minnesota, U.S.A.

Sindhan, G.S., and S.K. Bose (1980). Epidemiology of angular leaf spot of French bean caused by *Phaeoisariopsis griseola. Indian Phytopath.*, **33(1)**: 64-68.

Sindhan, G.S., and S.K. Bose (1981). Epidemiology of Anthracnose of French bean caused by *Colletotrichum lindemuthianum. Indian Phytopath.*, **34(4)**: 484-487.

Singh, H.N.P. (1992). Studies on diseases of banana: A pathological and biochemical approach, Ph.D. thesis Bhagalpur University, Bhagalpur.

Singh, H.N.P., Prasad, M.M. and Sinha, K.K. (1993). Evaluation of medicinal extracts against banana rot. *Indian Bot. Soc.*, **72**: 163-164.

Singh, R.H. (1977). In "*Physiology of Micro-organism*" (Ed. K.S. Bilgrami). Today and Tomorrow's Printers and Publishers, New Delhi, pp. 285-288.

Sinha, R.K. (1979). Ph.D. Thesis, Bhagalpur University, Bhagalpur.

Srivastava, M.P., Sudhir Chandra and R.N. Tondon (1964). Post harvest diseases of some fruits and vegetables. *Proc. Nat. Acad. Sci. India.* **34B**: 339-342.

Suhag, L.S., Ram Singh and Y.S. Malik (1985). Epidemiology of pod and leaf blight of radish crop caused by *Alternaria alternate Indian Phytopath.*, **38(1)**: 148-149.

Sumbali, Geeta and R.S. Mehrota (1983). Control of post harvest diseases of temperature fruits by chemicals and treated wrappers. *Indian Phytopath.*, **36(2)**: 270-273.

Tandon, R.N. (1967. Final technical report on harvest diseases of tropical and subtropical fruits. University of Allahabad, pp. 158.

Thakur, D.P. and V.V. Chenulu (1970). Chemcal control of soft rot of apples and mango fruits caused by *Rhizopus arrhizus. Indian Phytopath.*, **23**: 58-61.

Thakur, D.P. and V.V. Chenulu (1974). Chemcal control of soft rot of potato tubers caused by *Rhizopus arrhizus. Indian Phytopath.*, **27(3)**: 375-378.

Wager, W.A. (1945). Early blight in potatos. *Fmq. S. Afr.*, **20**: 318-320.

Waraitch, K.S. and K.S. Nandpuri (1975). Fusarium fruit rot of muskmelon (*C. melo* L.). *J. Res. Punjab Univ.*, **12**: 131-134.

Welling, B. (1974). *Aspergillus* and *Penicillium* sps. on stored Danish barley grains. Identification of species of fungi isolated from 40 barley samples in the harvest year 1970. *Tidssk. Planteavl*, **78**: 1-8.

Index

CAPTION OF COLOUR PLATES

Fig. 8.1: Effect of CF on Size and Shape of a Cell (200X)
A: Control cell
B, C, D, E and F: Cells growing in MS liquid medium supplemented with 20%, 40%, 60%, 80% and 100% CF.
Note: Gradual increase in Plasmolysis of cell

Fig. 15.1: Identification of Bacterial Antagonists for *Xanthomonas oryzae* pv. *oryzae* in a Laboratory Dual Plate Assay

Fig. 15.2: A PCR-based Screening Method for the Detection of 2,4-diacetylphloroglucinol (DAPG)–Production by Plant-Associated Bacteria.
Lane M: Molecular weight marker (1 kb ladder)
Lane 1: Amplification of *PhlD* gene in *Pseudomonas fluorescens* strain CHAO (used as reference strain)
Lane 3 & 12: DAPG producing Indian strains of plant-associated bacteria

Fig. 15.3: Biological Activity of Purified Compound from Culture Fluids of a *P. fluorescens* Strain PTB 9 Inhibiting the Growth of the Rice Bacterial Blight Pathogen, *Xanthomonas oryzae* pv. *Oryzae*–Diffusion Technique
A: Check (65% methanol)
B: Purified DAPG (50 l/mL)

Fig. 15.5: *In vitro* Screening of *R. fluorescens* Strain PTB 9 and its Antibiotic Deficient Mutants Against *Xanthomonas oryzae* pv. *oryzae*
A: Wild type strain PTB 9 shows growth inhibition of *Xoo*
B: Phl^- mutants strains do not shows pathogen inhibition

Fig. 6: PCR Analysis for the Insertion of Transposon in Mutants of *P. fluorescens* Strain PTB 9 Defective in DAPG Production
Lane M: Molecular weight marker (1 kb ladder)
Lane 1: Strain PTB 9a
Lane 2: Strain PTB 9b
Lane 3: Strain PTB 9c
Lane 4: Wild type strain *P. fluorescens* PTB 9

Fig. 20.1: Apple Scab Symptoms
(A) Foliage: (1) Lesions on under surface; (2) Lesions on mature leaves, distortion and yellowing
(B) Fruits: (3) Lesions on fruit-lets; (4) Lesion on developing fruit; (5) Scab on fruit pre-harvest; (6) Scab free apple.